TSUNAMI – ANALYSIS OF A HAZARD – FROM PHYSICAL INTERPRETATION TO HUMAN IMPACT

Edited by **Gloria I. Lopez**

Tsunami – Analysis of a Hazard – From Physical Interpretation to Human Impact
http://dx.doi.org/10.5772/2614
Edited by Gloria I. Lopez

Contributors

Aida Alvinius, Erna Danielsson, Gerry Larsson, Cut Husna, Eni Nuraini Agustini, Hiroya Matsuo, R. Omira, M.A. Baptista, S. Mellas, F. Leone, N. Meschinet de Richemond, B. Zourarah, J-P. Cherel, Khaled T. Ramadan, A. Tan, A.K. Chilvery, M. Dokhanian, S.H. Crutcher, Yury Korolev, Po-Fei Chen, Wen-Tzong Liang, Bor-Yaw Lin, Tetsuya Kogure, Yukinori Matsukura, Barbara Keating Helsley, Charles E. Helsley

CBS, Edition **2017**

Published by InTech

Janeza Trdine 9, 51000 Rijeka, Croatia

Publishing Process Manager Dragana Manestar
Technical Editor InTech Prepress, Novi Sad
Cover InTech Design team

Additional hard copies can be obtained from orders@intechopen.com

Tsunami – Analysis of a Hazard – From Physical Interpretation to Human Impact,
Edited by Gloria I. Lopez
p. cm.
ISBN 978-953-51-0865-8

INTECH
open science | open minds

Contents

Preface

This book has been published in order to deepen efforts towards the understanding of tsunami dynamics that seems to be never enough. This multi-disciplinary volume compiles a collection of scientific papers showing the state-of-the-art of tsunami research at different levels. The various contributions cover an array of themes that span from geological evidence to post-trauma human care, encompassing pre-tsunami analyses and modeling to post-tsunami management and preparedness techniques. "Tsunami – Analysis of a Hazard: from physical interpretation to human impact" presents evidence and case studies from different regions of the World: from the isolated Hawaiian Islands and Northern Indian Ocean, to the edges of the Atlantic and Eastern Mediterranean.

Gloria I. Lopez
School of Marine Sciences
University of Haifa,
Israel

Section 1

Preparedness, Emergency Response and Post-Trauma Care – The Human Aspect of a Tsunami

Chapter 1

Organisational and Individual Leadership Challenges During the Rescue Operation Following the 2004 Tsunami

Aida Alvinius, Erna Danielsson and Gerry Larsson

Additional information is available at the end of the chapter

http://dx.doi.org/10.5772/51140

1. Introduction

Disaster prevention and management involves both technical and social dimensions, implying that research inputs from both sides may provide important contributions. This chapter presents an attempt to develop a theoretical understanding of how Swedish response organisations managed the leadership challenges that they encountered during the rescue operation following the 2004 tsunami in South East Asia. Informants from three Swedish authorities, some at strategic level in Sweden and some at field level in Thailand, were interviewed while the rescue operation was still going on. Our text consequently examines their interaction and deals with the soft side of disaster management from a social science perspective.

The tsunami of 26 December 2004 had momentous consequences for large parts of Southeast Asia – from personal tragedies to far-reaching consequences for countries' economies and populations. In addition to Southeast Asia, the tsunami also affected many countries far from its epicentre. A recent analysis by the United Nations lists almost 250,000 dead people (http://www.who.int/hac/crises/international/asia_tsunami/en/). Of the European countries, Sweden suffered the greatest number of casualties in the tsunami. Some 20,000 Swedes were on holiday in Thailand when the tsunami struck. According to a Swedish Governmental Official Report (SOU, 2005:104), it could be confirmed in 2005 that 543 Swedes died and 18 are still missing (Alvinius, Danielsson & Larsson, 2010a; Alvinius, Danielsson & Larsson, 2010b).

2. Our study – method

2.1. Selection of informants

Following the guidelines of grounded theory (Glaser & Strauss, 1967), the selection of participants was guided by a desire to find informants with a wide variety of experiences and occupational roles during the rescue operation. The organisations involved in this study are the Swedish Rescue Services Agency, the Swedish Armed Forces and the National Board of Health and Welfare in Sweden. Data were collected by interviews, following a prepared interview guide. During the tsunami in Southeast Asia in 2004, the task of the Swedish Rescue Services Agency was to lead and coordinate the operational work, i.e. the emergency team which was sent to Southeast Asia, and to manage logistics and home transportation of Swedish tourists who had suffered great losses in terms of both health and belongings. In accordance with Swedish law, the Swedish Armed Forces contributed personnel, equipment and transport to support organisations such as the Swedish Rescue Services Agency, the Swedish National Police Board and the Swedish Civil Aviation Administration. In this supporting role was the Swedish Armed Forces Operative Unit, which supervises missions in Sweden and abroad on a daily basis. The National Board of Health and Welfare was also involved during the rescue operation in Thailand, its task being to work with emergency management within the healthcare system, for example, with the prevention of infectious diseases, social services, provision of information and allocation of government grants to the county councils, home transport of injured Swedish tourists and provision of medical care in the field (Alvinius, Danielsson & Larsson, 2010a).

Organisation	Number of informants	Female/ Male	Their position (during the rescue operation)
1. The Swedish Rescue Services Agency	5	1/4	• Director-general • Head of Operational Management • An emergency service function called Focal Point • Two Crisis Managers
2. Swedish Armed Forces	6	0/6	• Logistics Coordinator • Medical Doctor on Duty • Liaison Officer sent from Operative Unit (OPU) and located at the Swedish Rescue Services Agency • Two managers for the operational section on duty • The head of the logistics department
3. National Board of Health and Welfare	6	2/4	• Director-general • General Manager of Administration • Head of Social Services Department • Head of Crisis Management Department • Two Operative Managers
(N) informants	**17**	**3/14**	

Table 1. Involved organisations, informants and their positions during the 2004 Tsunami

2.2. The formal structure of the Swedish rescue operation

The political level consisted of the Ministry of Defence (heading the Swedish Rescue Services Agency, SRSA, and the Swedish Armed Forces), the Ministry of Health and Welfare (heading the National Board of Health and Welfare), the Ministry of Foreign Affairs and the Prime Minister's Office, directly subordinate to the Prime Minister. The formal structure of the rescue operation is shown in Figure 1.

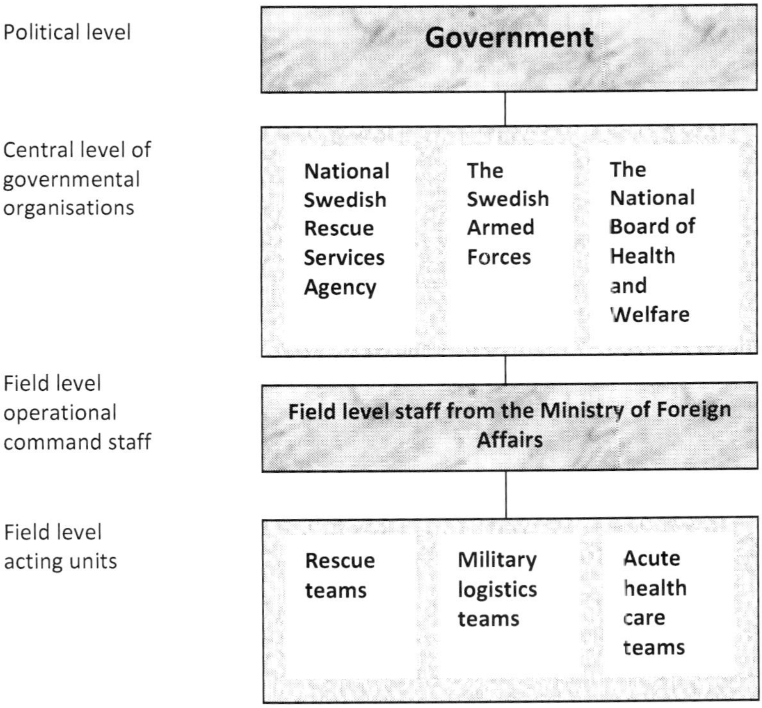

Figure 1. Formal structure of the Swedish rescue operation

This chapter focuses on leadership in the three aforementioned governmental agencies and their respective field teams, rather than on the political level. Each of those three organisations had headquarters in Sweden and the management of a rescue operation in the field in Thailand (see Figure 2). While a formal right of command existed within each organisation, there was no formal right of command for collaboration between them. The local operational command staff set up by the Ministry of Foreign Affairs mainly acted to assist coordination between the broader ranges of involved organisations. Leadership within and between organisations occurred directly (face-to-face) and indirectly (by subordinate leaders/managers).

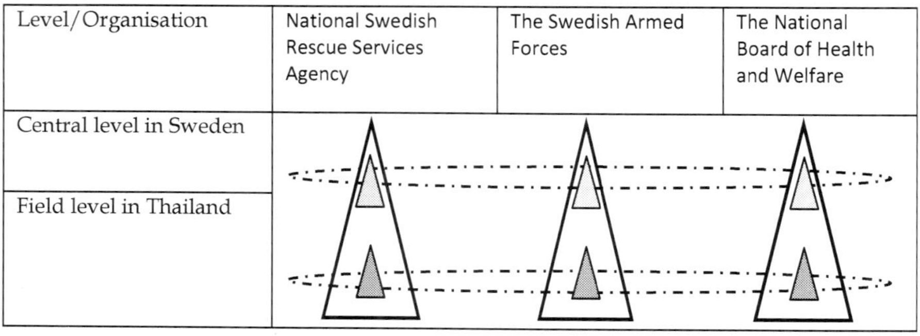

Figure 2. Chain of command within the organisation and leadership through collaboration between organisations at central and field levels

2.3. Outline of the chapter

This chapter deals with the organizational and individual leadership challenges identified during the rescue operation. Firstly, we will describe the difficulties that bureaucratic response organizations need to manage when a disaster such as tsunami occurs. Secondly, we will describe individual aspects of leadership challenges which concern stress caused by disaster and emotion management. A number of illustrations will highlight both aspects.

3. Organisational leadership challenges

3.1. Bureaucratic organizations and disaster management

Swedish political leadership has received much criticism for its handling of Swedish citizens caught up in the tsunami. The main issue raised by an evaluation conducted by a parliamentary commission was that the Government did not have an efficient organisation for the management of a major disaster such as the tsunami. It was concluded that Sweden, in comparison to other countries, was characterised not only by the lack of an effective emergency organisation, but also by the lack of functional leadership at executive level (SOU, 2005:104).

This problem is not unique to the Swedish Government. Emergency response organisations usually face the challenge of having a bureaucratic structure characterised by precision, discipline, reliability and predictability on the one hand (Andrzejewski, 1954), and the demands of meeting extreme situations where predefined directives and routines simply cannot cover all possibly emerging contingencies on the other (Ben Ari, 2005; Kapucu, 2006). The greater the uncertainty, the more difficult it is to routinize activity by preplanning (Morgan, 1986). The very structure of bureaucratic response organisations tends to stifle the creativity and flexibility required to manage large-scale disasters (Kendra & Wachtendorf, 2003). Some of the inadequacies tend to be intra-organisational. Performance may be confounded by a set of built-in conflicts that severely constrain policy-making, internal activities and actual operations. Typical clashes are based on irrational considerations such as internal departmental competition, political rivalries, jealousies, and interests. These built-in conflicts in bureaucratic organisations hinder collaboration, coordination and optimal use of resources (Kirschenbaum, 2004; Drabek & McEntire, 2002).

Rescue operations performed in foreign countries constitute a special case. In domestic disasters, rescue teams typically arrive at the site first and higher management staffs are built up successively. When a disaster takes place abroad, however, a different sequence typically occurs. Firstly, there has to be acceptance on a political level. Secondly, orders are given to the central governmental agencies, and thirdly the field teams can deploy. For this kind of rescue operation, which primarily aims to assist people from one's own country rather than the local population, research in regard to strategic and operational leadership is lacking. When rescue organisations take on an assignment, they constantly deploy and redeploy in partly new forms that need to function each time a disaster occurs. Studying a special case like the Tsunami can therefore add value to the broader picture of disaster management, which is important since large-scale disasters today are often handled by a cooperation of emergency rescue units from all over the world.

3.2. Improvisation and constraints of managing disaster

The complexity of crisis management has been highlighted in previous research. Drabek (1983, 1985) showed that disaster management should be capable of responding to episodic and very temporary change. These kinds of episodic, extreme events create the need for both planned and improvised collaboration and cooperation. Mendonça and Wallace (2004) pointed out that insight into how organisational context may shape improvisation, coordination and collaborative response activities is needed both within and between organisations because response activities can be expected to range from planned to improvised. They emphasise the importance of improvisation during the response phase and suggest a plan that would lead to increased understanding of how improvisation as a response to extreme events may be observed and supported. As Kendra and Wachtendorf (2006, p. 29) state, "Improvisation is required to connect that which is understood retrospectively and that which is emerging." Uhr et al. (2008) used social network as a method of analysing an emergency response system in order to identify people who served as links between two response organisations in the field. They observed the value of this function, which, in organisational theoretical terms, could be labelled 'boundary spanners' (Williams, 2002; Uhr et al., 2008). Other studies call for vertical and horizontal improvements in communication across various levels of government (May, 1985). McEntire (1998) notes that coordination will be increasingly relevant in disaster response operations when there have been some sort of pre-disaster ties between the agencies involved. These ties help generate familiarity with one another and help develop norms that emphasise collaboration. They generate a willingness to work with others to meet the needs of relief recipients and help create a network or means within which information regarding a disaster situation can be easily shared and quickly transmitted to others. Thus, a key to effective collaboration appears to be the interaction and coordination of response organisations with each other before a disaster strikes (Auf der Heide, 1989; Drabek & McEntire, 2002).

Our study was conducted before the Swedish Governmental Official Report was released. We discovered that many high-level leaders in Swedish Government were on Christmas vacation when the tsunami struck South East Asia in 2004, and even the prime minister was away and had not watched the news. However, for some other organizations that are not normally operational during vacations, the holiday was actually advantageous as it meant that day-to-day work did not hinder disaster management activities (Alvinius, Danielsson & Larsson, 2010a).

> But it was both when the actual organisation wasn't working, or rather, it wasn't active because there were a lot of public holidays and I'd say that was perhaps an advantage in this context because the capacity was there for it. There weren't that many other activities requiring immediate attention so we didn't need to put anything to one side because we could just do that. /.../. But if it had been high season, like, with meetings from morning to night, the situation would have been really stressful because you would have been, you would have been forced to organise yourself differently.

Nevertheless, there were many situational constraints hindering leadership performance and causing stress, for example, infrastructural/environmental aspects such as lack of resources, the impediment of routines and day-to-day activities, media as an external stressor, delays, time differences and understaffing.

Asserting individual and organisational boundaries hampered successful collaboration in some cases. Territorial thinking came into play between authorities both during and after the disaster initiative and was accentuated in subsequent discussions once normalcy had gradually crept back into the authorities' day-to-day activities. A shortage of computers and mobile phones was another obstacle to overcome. Staff in Sweden quickly made provision for the necessary equipment to be sent over to the field staff in Thailand, thus establishing communication between command centre and the field, but the geographical distance between the two countries meant there was a six-hour time difference. When field staff in Thailand needed extra resources it could mean staff back in Sweden had to work at 3 am. According to one informant, this time difference was seen to be an obstacle:

> So it has involved an adjustment, a time adjustment – there's six hours difference between us and Thailand, which means they're up and about over there at 3 am.

Understaffing at Bangkok airport made it more difficult to redeem materials sent over from Sweden. The need for more staff at the airport was evident:

> There have been a few problems with other organisations only sending things over to a large airport like Bangkok, without a recipient. That makes it really difficult to find the stuff and you have to be careful about transport documentation and things like that, and communication.

Leadership at the highest organizational level in Sweden geared up too slowly to the extreme situation that prevailed following the tsunami, sending a rescue team to Thailand a couple of days after its occurrence. Nevertheless, when news of the disaster broke, activities began which could be described as expectancy-based preparations. These were partly performed according to general and established plans (structure), and were partly innovative because of the novelty of the situation (allowing freedom of action). Within hours, frustrations began to surface due to rigidity and lack of decision-making at governmental level, thus delaying the rescue operation. However, once the decision to go ahead with the operation had finally been made, government sentiment now swung towards granting a high degree of latitude in its set-up. When the field groups arrived in Thailand, the initially chaotic situation gradually became more ordered, and leadership at central level in Sweden successively became more routine-based (structure). The present case shows that managing contradictory needs for structure and freedom of action becomes easier when link functions and roles (known as boundary spanners) arise in the formal hierarchy during an emergency situation.

However, there were differences in the need for structure and freedom between the two operational levels. For central level staff in Sweden, structure was important because their tasks revolved around obtaining mandates, leaders at this level thus having to wait for the green light from the relevant political authority in order to act. For workers in the field, the opposite was true, as there the task involved reacting to what had happened. In time, however, the needs of the field came closer to that of central level, and after a while both had established structures.

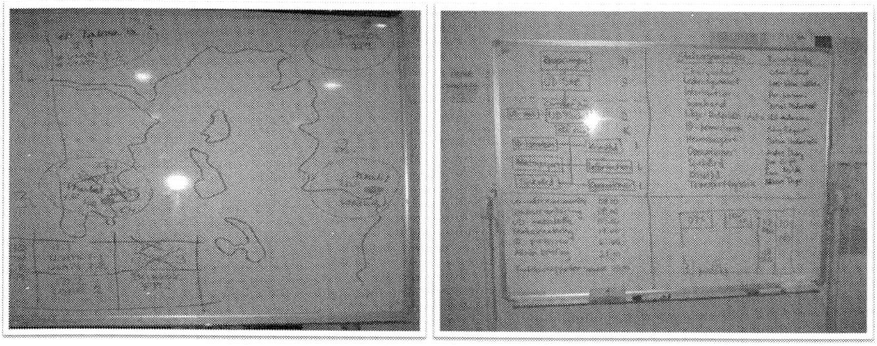

Picture 1. "Field level managing chaos and building gradually more ordered situation".
Photo: Torbjörn Olsson

The ability to act and deal with the chaotic situation on the ground was a necessity for field workers in Thailand as Swedish routines did not work and situations had to be dealt with then and there. We illustrate:

> But with so few people you can't sacrifice the rescue part in favour of establishing structure – that has to come afterwards.

On the one hand, a free rein was needed and a lot of the work was dealt with ad hoc. But on the other hand there was a need for fixed structures.

Routines were an expression of the existence of structure and of its necessity in order for the authorities to act. This manifested itself as a need to follow line organisation and to stay within individual parameters of operation. Decisions had to come from higher authorities – there was a great need to have someone with the courage to 'push the button'. For the work to be carried out there was also a desire to have support staff who could help with everything from making coffee and fetching food, to answering calls from volunteers.

As stated, however, the desire for a free rein was just as apparent as the desire for routines. Making decisions was important, and it was better to make one too many than one too few, as illustrated by utterances such as 'no rules – do it!', 'make a fast decision', informally – 'make it right later', 'no limitations'. According to one of the informants, acting without limitations was characteristic of workers on the ground in Thailand:

> You have to decide quickly, you have to – how should I put it… you can't wait any longer.
>
> So I'd rather make one decision too many than one too few. I think it's really important – especially in situations where you can't wait and have to have the courage to make a decision.

One informant in the field felt that flexibility was a desirable quality in new collaborators, as work structured around a checklist would be perceived as hampering an initiative. In a demanding situation, imagination combined with flexibility was what was necessary out in the field, and to achieve this, mentors were needed to teach less experienced staff to act accordingly. The informant maintained that checklists were not helpful:

> I've had dealings with a lot of young guys who are going to be commanding officers and they want a checklist. You're not getting one, I say; you'll have mentors and you'll be taught this way and from experience etc – that will give you know-how.

3.3. Boundary spanners as facilitators between leaders

When link functions and roles arose in the formal hierarchy, the process of managing contradictory needs for structure and freedom of action became easier for higher levels of leadership. These links were individuals with specific competencies facilitating collaboration and communication within and between organisations. Webb (1991) refers to individuals with these kinds of competencies as boundary spanners. By providing linkages which do not exist on organisational charts, boundary spanners facilitate the sharing and exchange of information and link their organization with the external environment, especially in time of crisis (Aldrich & Herker, 1977; Alvinius, Danielsson & Larsson, 2010b).

As mentioned earlier, there is a need for improvisation in disaster management from an organisational and leadership perspective. Organisational management of environmental demands such as geographical distance, multinational collaboration with different actors such as Non Governmental Organisations (NGO), building a common operating picture and situational awareness on site, requires management of those difficulties, sometimes at the same time (Alvinius, Danielsson & Larsson, 2007). One way to overcome differences between an organisation and its environment is to use boundary spanners. These functions have been shown to be essential to the efficient and effective operation of organisations by improving communication and collaboration between groups (Aldrich & Herker, 1977; Williams, 2002; Alvinius, Danielsson & Larsson, 2010b). We conclude that during the 2004 tsunami, links or boundary spanners contributed greatly to mitigating the difficult circumstances encountered by the Swedish disaster initiative in terms of overcoming great geographical distance and delayed communication between the central and field levels, and in dealing with the scope of disaster and lack of disaster experience among Swedish leaders.

Two kinds of links were identified in our study, labelled planned and spontaneous links respectively. These links connect organisations and individuals both vertically and horizontally; acting vertically between various hierarchical levels within an organisation or between the management of the authorities and the political level, and horizontally between different organisations, or between individuals at the same level within the same organisation/agency (see Table 2).

Link	
Planned Links	**Spontaneous Links**
Predefined position and task a) Liaison officer b) Representative role c) Coordinator d) Negotiator e) Information officer f) Observer g) Translator (organisational differences)	**No predefined position and task** • May belong to another organization • Appears on site/field • Coordination • Collaboration • Support *Purposes* • Getting the picture on site • Sense-making • Estimating the allocation of resources on site • Other competencies, such as translation of another language
Aspects of trust Network of contacts Organisational acceptance, trust is related to a position not only to individuals	**Aspects of trust** Swift trust Willingness to work across organisational boundaries depends on organisational reputation
Negative aspect of Planned Links Too many links mean too many different apprehensions	**Negative aspect of Spontaneous Links** Running the risk of being rejected

Table 2. Illustrates major characteristics of Planned and Spontaneous Links

A planned link is related to an individual's organisational role and is most often approved and accepted by superiors. Planned links refer to individuals with collaboration tasks within their responsibility and mandate, such as liaison officers or negotiators. This linking function may have appeared successful in a previous event, which led to the establishment of this kind of link in the organisation. Planned links may be decision-makers who are able to act outside the framework of the organisation because he or she possesses organisational acceptance and has a wide experience of managing disasters or unexpected events. Individuals with planned link functions in crisis situations belong to the ordinary chain of command during ordinary day-to-day work and may have a managerial position during ordinary working conditions.

Spontaneous links appear to arise when required by the extreme situation. This could happen when areas of responsibility, authority, competences, experiences and resources fail. Spontaneous links often emerge in the field and enjoy the immediate trust of people close by, for example planned links. They are not affiliated with an organisation but they are ordinary people willing to volunteer their special competencies to help during a rescue operation. Spontaneous links have competencies that are different to that of planned links and their existence is temporary. A typical spontaneous link may be a volunteer language translator during a disaster. To make the collaboration process possible, spontaneous links rapidly need to gain trust, but they also run the risk of being rejected if they are not part of an involved organisation. The need for spontaneous links disappears when the crisis is over.

During the tsunami rescue operation a liaison officer, a planned link, was a previously appointed person whose job was to facilitate cooperation between organisations, to liaise between organisational cultures and to make collaboration more efficient, resulting in time savings. The Swedish Armed Forces have a built-in structure for this, which is illustrated in the following quote from a centrally-placed manager of the Swedish Rescue Services Agency:

> Initially, the Swedish Armed Forces had a liaison officer who knew the military language and system placed among the Rescue Services staff. I had discussions with the liaison officer there, and he was familiar with our stuff.

An excerpt from the interview with the liaison officer mentioned in the above quote:

> What I'm most satisfied with is having been able to create the contacts, so the Rescue Services personnel and the Armed Forces personnel had a common entry point, and that was me. So I had a lot of discretion and saved the individual administrators a lot of time.

Spontaneous links happen to be the right individuals in the right place when an unexpected event occurs, which is often experienced by the informants as positive. Strenuous situations and demands can lead to spontaneous links emerging. Actors appear when a need becomes visible, as was the case after the tsunami:

At the end of the day I got in touch with a guy who was in contact with someone down there (Thailand); he suddenly appeared – this Swedish Air Force officer who was there celebrating his 50th birthday. So he turned up and asked if they needed help with anything.

4. Individual leadership challenges

4.1. Stressors and stress reactions

Individual leadership challenges varied according to whether or not the leaders were at home in Sweden or on the ground in Thailand. In the initial phase of the tsunami disaster, uncertainty among the Swedish authorities reigned and there was no understanding of the consequences the Southeast Asian disaster could have for Swedish society. Our informants exhibited stress reactions due to lack of control, for example, they found it difficult to predict what measures needed to be taken. The informants also testified to feeling powerless, for example, in the absence of written directives from higher authorities. Such absence of directives necessitated increased financial responsibility and flexibility, according to several interviewees. Alongside the other causes of stress, the absence of orders led to increased stress among command staff, as illustrated by the following comment:

It can be stressful to make a decision to send over several million kronor's worth of containers, buy air services from an aviation broker and then fly over the containers – being sure that in the end, it won't be me personally paying this 10 000 000 (about one million Euros).

The tsunami catastrophe took its share of Swedish victims, among them parents, leading to a dilemma for the National Board of Health and Welfare in terms of how to manage children who had been orphaned in connection with the disaster. To illustrate:

One of the first questions raised was, what provision was there for taking care of children who had lost their parents – how would it be handled?... and someone got that question from abroad, about what guidelines the National Board of Health and Welfare had.

Stress resulting from media pressure is described in the quote below:

> There was probably a whole bloody shed-load of them ringing, and so it started, the papers and press calling and they push like there's no tomorrow.

Typical reactions ignited by these stress factors included a sense of uncertainty about how to act, uncertainty as to the consequences for one's self and others, and frustration. However, many of those who were involved in the disaster effort but working from their usual workplaces in Sweden described these stress reactions as mild:

> It's not a big problem. You shouldn't attach any importance to it. No! It's not a war situation, there's no stress /.../ Daylight, no-one shooting at me, the lights are on. Sweden's in one piece, I go home, talk to my wife: a few odd conversations but no, not so very stressful.

Picture 2. "Deceased Swedes prior to the first air transport back home. Staff on the ground had the job of collecting the deceased from refrigerated containers and placing them in coffins prior to transporting them home. The coffins were to be draped with the Swedish flag before departure and everything was to be managed with dignity, staff having discussed how to address relatives on the scene at the airport. The first six coffins were then transported to the Uppsala garrison for a memorial service at which the deceased were passed on to their families. Having to put the deceased in coffins and meet their relatives was an emotionally trying task for the staff." Photo: Torbjörn Olsson

Nevertheless, on the ground in Southeast Asia, stress reaction levels were significantly higher. Typical stressors for fieldworkers were their own organisation's lack of understanding at central level as to the tasks they were solving, as well as deficiencies in their own understanding as to the role their organisation was playing and how they should be employed in operations. Reactions such as disappointment were a fact when central level conveyed criticism instead of making concrete decisions:

> They were aware that we were coming – a group like us – but they didn't initially fully comprehend the capacity we had and how we were to be employed and such.

In the field, staff was affected by physical symptoms such as sleep deficit and psychological conditions such as tunnel vision. In some cases, the number of tasks was perceived to be burdensome since they were being carried out under difficult conditions. Intense work shifts and sleeplessness wore out the fieldworkers, even though they were willing and committed to helping.

> I was working with great focus and intensity, most likely had tunnel vision to a degree, worked too much /.../ During those days I was working, calculated it was 96 hours /.../ I was most stressed by the fact that I slept too little so to speak.

The character and extent of the tsunami catastrophe was perceived to have caused high levels of stress and subsequent stress reactions. Many of the fieldworkers testify to having strong feelings:

> So there's like, about 10,000 to 15,000 dead people over there and you see all the destruction /.../ Perhaps you can associate stress with the fact that sometimes, when you've been meeting with a lot of relatives, sometimes for a whole day, when we went out with them and stuff so you got so close to them. You got told so much that you got really affected, you know. We cried several times. We could have parting ceremonies with coffins and flags; we could have four to five a week.

Picture 3. Collection point for the dead in Ban Muang. Staffs were confronted with an unfathomable number of dead people there and it was difficult to take in. At the same time, conditions were static – the people were already dead. The task now was to identify them as carefully and as sensitively as possible. Not much more could be done than to play one's part as well as possible.
Photo: Torbjörn Olsson

As previously mentioned, field workers were committed and willing. They wanted to do a good job and be professional but stretched themselves in the process:

> The commitment was there, and the willingness was there in the whole team to do a professional job and do it well. So I think that was different from the others back in Sweden but on the other hand we really stretched ourselves.

In summary (see Table 3), the factors that triggered stress reactions at central levels of authority in Sweden were a lack of directives, the need to make serious decisions without formal mandates, pressure from the media, and new, unexpected questions and demands from one's own organisation. Factors triggering stress in the field were associated more with central level's lack of understanding for or knowledge of what fieldworkers were doing there and of the competence they possessed, the intense work pace and length of the working day, and the experience of encountering relatives and the deceased at parting ceremonies.

Stressors, stress reactions and emotional management → Organisational level ↓	**Stressors**	**Stress reactions**	**Emotional management**
Central level in Sweden	• Lack of control and formal mandate • Financial responsibility • Orphaned children • Media pressure	• Mild stress reaction • Powerlessness • Frustration • Sense of uncertainty	• Distance from the field prevent strong emotions • Mobilise right resources • Emotional management at the macro–level (Sweden is in sorrow) • Using previous experience and personal networks
Field level in Thailand	• Lack of under-standing at the central level • Rescue actions and number of tasks • Difficult conditions such as climate and dead bodies	• Increasing stress reaction • Sleep deficit • Tunnel vision • Disappointment • Sadness • Pride • Willing and committed to helping	• Sharing strong emotions at the field • Managing own and others emotional reactions. Pointing out if a colleague is too stressed. • Using previous experience and personal networks • Effective leadership and cultural awareness

Table 3. Stress and stress reactions among leaders at the central level in Sweden and the field level in Thailand

4.2. Emotional management

Stressful events like the tsunami impact greatly on both individuals and organisations. Crisis management organisations should be able to adapt rapidly to the pressures emanating from the surrounding environment, i.e. the world outside the parameters of the organisation. Society puts demands on authorities to manage crises that arise, and these demands can trigger emotional processes on an individual, group and organisational basis. Such demands are often about mobilising the right resources in relation to the occurred event, and it is in this context that one often talks about incremental stress and tension (Alvinius, Danielsson & Larsson, 2010a).

Recently, researchers have become increasingly interested in examining the subject of emotions in day to day activities, but also in stressful, life and death situations (Alvinius, Kylin, Starrin & Larsson, 2011). From a societal perspective, there are emotional cultures which must be followed, involving compliance with other people's expectations of emotional expression. To meet these expectations, the individual must subscribe to a certain amount of emotion management. One likely significant aspect of the work carried out by operations staff during the tsunami catastrophe was the management of their own as well as others' emotional reactions. Related research from sociology has elicited the concept of 'emotional labour' (Hochschild, 2003), in which the conclusion is made that service professions (as opposed to manufacturing ones) are subject to the demands of prescribed emotional expression, which more or less requires individuals to modify their own natural feelings. They are expected to outwardly exhibit a certain 'face': a composite of a particular vocabulary, tone, body language, dress etc.

Some examples of emotion management during the tsunami rescue operation are mentioned in the following. Each decision-maker and collaborator serving during the tsunami disaster brought their own experiences and contacts to operations. Previous experience was in their bones, allowing them to act directly. Other informants made mention of their own special experiences, for example, from Africa, where they were used to international aid initiatives, medical care and working under time-critical conditions. This type of experience was an asset to those going on to manage their own emotions during the tsunami disaster rescue operation.

Thus, previous experience of managing international crises was also an asset to individuals and organisations, as testified to below:

> I think a lot of this commitment exists in the Civil Contingencies Agency – has done historically, so to speak /.../ because we've carried out international aid operations.

However, the experiences from the tsunami disaster were, according to our informants, of a completely new kind and described as once-in-a-lifetime occurrence; they had never before encountered this kind of catastrophe in their careers. The fact that so many of the afflicted were Swedes and spoke their native language brought the field workers closer to the victims and more involved in their destinies. Managing the tsunami catastrophe required dealing with a number of emotions such as grief, sadness, anger, joy and pride. Familiarity with individuals and personal networks readily facilitated this emotion management. More generally, it was found that effective leadership was related to effective emotion management. This, in turn, was related to cultural awareness in addition to effective personal emotion-focused coping techniques.

5. Conclusions

The act of maintaining balance between authority and structure on one hand and allowing freedom of initiative on the other was found to be the most significant aspect of leadership during the tsunami rescue operation. Thus it can be concluded that operational managers at faraway disaster sites should be prepared to act independently without higher command. When there is distance and information asymmetry between central strategic leadership and the field, operational commanders need to operate autonomously in order to be effective. The functional aspect of putting the formal chain of command aside and taking personal initiative at lower levels also has theoretical implications. According to most organisational theories (e.g. Cameron et al., 1987), top management tends to take a firmer grip in tough times. Decision-making responsibility is passed upwards, participation decreases and control is emphasised. The present study, just like a military study (Larsson et al., 2007), shows cases where the opposite was functional. We think this is of value in highlighting the importance of flexibility and freedom of decision-making at the initial stage of any disaster, at which point decentralisation should be encouraged.

A second major finding is the identification of planned and spontaneous links, and the different positions that these links can take. They contribute in diverse ways to effective operation by enabling exchange between individuals and groups. When functioning at their best, these links provide the rigid structure of bureaucratically organised emergency response agencies, with the creativity and flexibility required. In short, the two kinds of links contribute to organisational adaptation to environmental conditions. We propose and encourage openness towards spontaneous links and paying increased attention to trust, for instance in selection and organisation development processes. The following specific implication is also suggested: educate organisational gate keepers and boundary spanners acting as planned links in order to avoid an over-emphasis on control-driven compliance. This is a long-term endeavour and points to the importance of organisational climate.

The third major finding is the demonstration of the importance of a so far underestimated operational skill, the ability to manage one's own and others' emotions. Our study shows that leaders utilise various ways of emotional management in order to fulfil the demands partly laid upon them by the disaster environment and partly by their own bureaucratic organisation. The extreme demands of a disaster rescue operation highlight the need to look beyond the purely rationalistic perspective of emergency response organizations. Operational effectiveness is also dependent on the quality of the key actors' emotional labour and emotions-focused coping skills capacity.

Leadership challenges at central level in Sweden	Leadership challenges at field level in Thailand
Need for structure	Need for freedom of action
Planned Links facilitate collaboration	Planned Links facilitate collaboration Spontaneous Links are recognised
Manage one's own emotions	Manage one's own and others' emotions

Table 4. Illustrates main conclusions

The generalizability of the three presented conclusions is, strictly speaking, an empirical question. Thus, we encourage research colleagues to consider and evaluate them in future disaster studies.

Author details

Aida Alvinius, Erna Danielsson and Gerry Larsson
Swedish National Defence College and Mid Sweden University, Sweden

6. References

Aldrich, H. & Herker, D. (1977). Boundary-spanning roles and organization structure, *Academy of Management Review*, Vol. 2, pp. 217-230, ISSN 03637425

Alvinius, A., Danielsson, E., Kylin, C., & Larsson, G. (2007). *Lika lägesbild, olika situationsförståelse,* ILM : Serie I:34, ISBN 1401-5676, National Defence College, Department of Leadership and Management, Karlstad, Sweden

Alvinius, A., Danielsson, E., & Larsson, G. (2010a). Structure versus freedom of action: leadership during the rescue operation following the 2004 tsunami, *International Journal of Emergency Management*, Vol. 7, Nr. 3/4, pp. 304-322

Alvinius, A., Danielsson, E., & Larsson, G. (2010b). The inadequacy of an ordinary organisation: Organisational adaptation to crisis through Planned and Spontaneous links, *International Journal of Organisational Behaviour*, Vol. 15, pp. 87-102, ISSN 1440-5377

Alvinius, A., Kylin, C., Starrin, B., & Larsson, G. (2011). *Rules of Emotional Engagement : samverkan och förtroendeskapande ur ett emotionsteoretiskt perspektiv*, ILM : Serie I:67, ISBN 1653-1523, National Defence College, Department of Leadership and Management, Karlstad, Sweden

Andrzejewski, S. (1954). *Military Organisation and Society,* Routledge & Kegan Paul LTD, ISBN 65012010128, London, UK

Auf der Heide, E. (1989). *Disaster Response: Principles of Preparation and Coordination*, The C.V. Mosby Company, MO, ISBN -13 978-0801603853, St. Luois, USA

Ben-Ari, E. (2005), Epilogue: A "Good" Military Death, *Armed Forces & Society*, Vol. 31, pp. 651-664

Cameron, K.S., Kim, M.U., & Whettten, D.A. (1987). Organizational effects of decline and turbulence, *Administrative Science Quarterly*, Vol. 32, No. 2, pp. 222–240, ISSN 00018392

Drabek, T.E. (1983). Alternative patterns of decision-making in emergent disaster response networks, *International Journal of Mass Emergencies and Disasters*, Vol. 1, No. 1, pp. 277–305, ISSN 00333352

Drabek, T.E. (1985). Managing the emergency response, *Public Administration Review*, Special Issue: Emergency Management: A Challenge for Public Administration, Vol. 45, pp. 85–92, ISSN 1940-2006

Drabek, T.E., & McEntire, D.A. (2002). Emergent phenomena and multiorganisational coordination in disasters: lessons from the research literature, *International Journal of Mass Emergencies and Disasters*, Vol. 20, No. 2, pp. 197–224

Glaser, B., & Strauss, A.L. (1967). *The discovery of grounded theory: Strategies for qualitative research*, Aldine, ISBN -13 978-0202302607, Chicago, USA

Hochschild, A. (2003). *The managed heart: Commercialization of human feeling*, University of California Press, ISBN 0-520-23933-4, Berkeley, California

Kapucu, N. (2006), Interagency Communication Networks During Emergencies: Boundary Spanners in Multiagency Coordination, *American Review of Public Administration*, Vol 36, No. 2, pp. 207-225

Kendra, J.M., & Wachtendorf, T. (2003). *Creativity in Emergency Response after the World Trade Center Attack', in Beyond September 11th: An Account of Post-Disaster Research, Natural Hazards Research and Applications Information Center, Public Entity Risk Institute, and Institute for Civil Infrastructure Systems,* Special Publication No. 39. Boulder, CO: Natural Hazards Research and Applications Information Center, University of Colorado, USA, pp. 121-146

Kendra, J.M., & Wachtendorf, T. (2006). *The Waterborne Evacuation of Lower Manhattan on September 11: A Case of Distributed Sensemaking,* Preliminary paper #355, Disaster Research Center, University of Delaware. USA, Available at http://dspace.udel.edu:8080/dspace/bitstream/handle/19716/3194/Preliminary%20Paper%20%23355.pdf?sequence=1

Kirschenbaum, A. (2004). *Chaos organization and disaster management*, Marcel Dekker, ISBN-13 9780824747152, New York, USA

Larsson, G., Harem, T., Sjöberg, M., Alvinius, A., & Bakken, B. (2007). Indirect Leadership under Severe Stress: A qualitative Study through the Subordinates' Eyes: a qualitative inquiry of the 2004 Kosovo riots, *International Journal of Organisational Analysis*, Vol. 15, No 1, pp. 23-34

May, P.J. (1985). FEMA's role in emergency management: examining recent experience, *Public Administration Review*, Vol. 45, pp. 40–47

McEntire, D.A. (1998). *Towards a Theory of Coordination: Umbrella Organizations and Disaster Relief*, Quick Response Report No. 105, Natural Hazards Research and Information Application Center, University of Colorado, Boulder, Colorado, USA, Available at http://www.colorado.edu/hazards/research/qr/qr105.html

Mendonça, D., & Wallace, W.A. (2004). Studying organisationally-situated improvisation in response to extreme events, *International Journal of Mass Emergencies and Disasters*, Vol. 22, No. 2, pp. 5–29

Morgan, G. (1986). *Images of organization*, Sage, Beverly Hills, California

SOU 2005:104, Swedish Government Official Report (2005). *Sweden and the Tsunami: Examination and Proposals*, Swedish Tsunami Commission, ISBN 91-38-22476-3, Stockholm, Sweden

Uhr, C., Johansson, H., & Fredholm, L. (2008). Analysing emergency response systems', *Journal of Contingencies and Crisis Management,* Vol. 16, No. 2, pp. 80–90

Webb, A. (1991). Co-ordination: a problem in public sector management, *Policy and Politics*, Vol. 19, No. 4, pp. 229–241

Williams, P. (2002). The competent boundary spanner, *Public Administration*, Vol. 80, No. 1, pp. 103–124

http://www.who.int/hac/crises/international/asia_tsunami/en/

Chapter 2

Acute Respiratory Care and Wound Care of Tsunami Patients: What Should the Nurses Do?

Cut Husna

Additional information is available at the end of the chapter

http://dx.doi.org/10.5772/51366

1. Introduction

The earthquake and tsunami on December 26, 2004 killed more than 270,000 people in the 11 countries, injured 500,000 people, and affected approximately 5 million people in the world. The hardest hit was observed in Aceh province, Indonesia that killed approximately 128,000 people, which is about a quarter of the total population and many people were injured or lost their homes [1]. The 2004 Indian Ocean Tsunami also caused extensive damage to five hospitals in Banda Aceh, 122 hospitals and public health centers in few districts. Health providers and clinical networks beyond the hospitals were left in disarray. Communication and transportation system was disrupted and record of most of the nursing staff was lost. The lack of a medical disaster plan, supply of stockpiles and deployment of personnel were painfully evident in the immediate aftermath of the tsunami [2].

The tsunami has several impacts on physical, psychological, psychosocial, and spiritual aspects of the affected patients. The physical impacts of the affected population such as lung injury, head injury, fracture, wounds, water borne diseases, and insect transmitted diseases. In [3] stated that two-thirds of tsunami patients were reported have combined that injuries to the thorax or fractures. Patients who survived from drowning involved in the aspiration of immersion fluids, marine and soil debris into the respiratory tract and all patients displayed signs of pneumonitis and pneumonia.

The tsunami created dramatic events and the ubiquitous threats of mass destruction. These events have heightened the recognition of the role of nurses in disaster management and response. The health care providers including nurses must response to the tsunami affected patients. The nurses should be able to provide disaster nursing response, communicate with disaster command, and evacuate patients as soon as possible. Unfortunately, most of the nurses are unprepared to respond to the disaster emergency. Many cases indicated inadequate nursing care, medical care, poor communication, chaotic management, and meager patient evacuation [4]. For enhancing the capacity to respond to disaster and other public health emergencies, it will require a nursing work force, which is clinically rich in skills and experiences. In addition, nurses should have sufficient skills to respond to tsunami disaster with confidence and authority [5].

The nurses' clinical skills in acute respiratory care involve assessment (airway, breathing, and circulation), intervention (positioning, supplemental oxygenation, fluid management, and antibiotic), and evaluation effectively for airway, breathing, circulation, and disability [6]. Moreover, the nurses' clinical skills in wound care assessment consist of identification characteristic, location, and category of wound. The interventions refer to irrigation, debridement, cleansing, packing, and changing the dressing of wounds [7]. The evaluations are related to evaluate wound condition such as infection, granulation, and signs of wound healing.

In [8], study about wound care for the victims provided by nurses at hospitals of Phang-Nga province, Thailand, found that the nurses had insufficient of knowledge and skills in providing care to the tsunami patients with wounds, particularly while cleaning and suturing the wounds. This results in wounds inflammation and infection due to the penetration of the debris, sand, and mud in the tissues. The nurses also lack of knowledge and technical experiences in dressing the wounds in closed wound with "vacuum assisted closure" (VAC). In reference [9] found that nurses lack of skills in infection prevention and control in mass casualty incident and public education.

Nurses' clinical skills are essential in caring for tsunami patients. Adequate clinical skill helps the nurses in handling the tsunami emergency response effectively. Therefore, the present study assessed the clinical skills for tsunami care and its related factors, and helped in improving clinical skills of nurses in caring for tsunami patients.

Figure 1. Geographic location of tsunami impact of districts/municipalities in Aceh Province

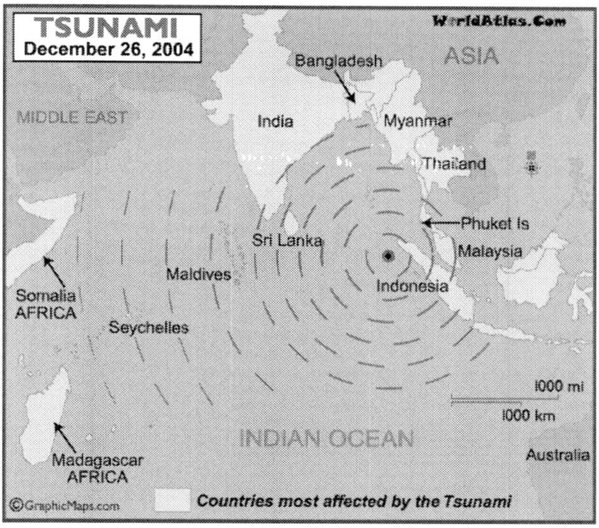

Figure 2. The countries affected by the 2004 Indian Ocean Tsunami

2. Clinical skills of nurses for tsunami care

Clinical skills for tsunami care in this study were adopted from its' related concept and two other studies; reference [10] proposed about nurses' competencies in responding to mass casualty incidents/disaster and [11] stated the main problem patients suffering from tsunami trauma.

2.1. Acute respiratory care

Aspiration is common problems among tsunami patients due to near drowning. Aspiration pneumonia related with near drowning, is an significant cause of death among tsunami patients. Aspiration of seawater, mud, and marine debris into the respiratory tract provides pathogenic inoculums of pulmonary infection, inducing pneumonitis and pneumonia. Tsunami patients not only aspirate water, but also soil and particulate matters [11].

Respiratory distress syndrome, cardiovascular distress, cardiac arrest, salted water aspiration developed complication associated with near drowning. Near drowning condition caused respiratory and cardiogenic shocks which were common problems encountered in tsunami affected patients [12]. Addressed respiratory tract infections and pneumonias were commonly found several days after the tsunami among the patients who survived from near drowning [13]. Tsunami patients involved in the aspiration of immersion fluids as well as marine and soil debris into the respiratory tract, thus producing intra-pulmonary inoculation of bacteria. In accordance, all patients admitted to hospital displayed radiological and clinical signs of pneumonitis and pneumonia [3].

Tsunami lung or tsunami-related aspiration pneumonia is used to describe lung pathology of tsunami patients who have necrotizing, cavitary pneumonia that may create complication with empyema, pneumothorax, and hematogenous spread of infection. Tsunami lung occurs when people being swept by tsunami wave's inhale salt water contaminated with mud and bacteria. The resulting pneumonia like infections normally is treated with antibiotics. However, the 2004-Indian Ocean Tsunami wiped out the medical infrastructure. Therefore, antibiotics to treat the infections in the early stages were not available. Consequently, patients' lung infections festered, entered the bloodstream, and spread to the brain, producing abscesses and neurological problems such as paralysis. The diagnosis of tsunami lung requires chest radiography and computed tomography scan of the brain to confirm abscesses [11]. The diagnosis of tsunami lung requires chest radiography and computed tomography scan of the brain to confirm abscesses [14]. Patients showed high risk of polymicrobial infection of the lungs with various pathogens including Aeromonas spp., Pseudomonas spp., Acinetobacter baumanii, Enterococcus faecium, and Klebsiella pneumonia. The patients were usually affected with a higher frequency of shock, respiratory failure, renal failure, and mortality [11, 3].

Many patients after swallowing tsunami water developed aspiration pneumonia. A study in [6] reported that the tsunami patients have developed aspiration pneumonia and showed febrile and shortness of breath. Intervention focused on intravenous fluids, antibiotics, acetaminophen, and supplemental oxygen. Patients with shortness of breath became unresponsive and stopped breathing, were provided with the advanced cardiac life support protocol (cardiopulmonary resuscitation/CPR), intravenous fluids, and several rounds of epinephrine and atropine.

Aspiration pneumonia, inhalation, and injury may develop acute respiratory distress syndrome (ARDS), the most common respiratory problems for tsunami patients. ARDS is a critical and often fatal complication associated with sepsis, trauma, major surgery, and many serious illnesses. The mortality rate has been reached 90%, but that appears to be decreasing with improvements in treatment techniques. Current treatment for ARDS relies on supportive measures to maintain adequate oxygenation until the lungs recover. Conventional therapy includes endotracheal intubation and volume cycled mechanical ventilation [40].

Aspiration caused salted water pneumonia, asphyxia, and acute respiratory distress syndrome. There were totally 2,311 patients attacked by tsunami, approximately 40% suffered from salted water aspiration, while 2% developed complication associated with near drowning. Mud, sand, and dirt were seen in the mouth and throat of the victims [12]. In [15] found that there are 62 %patients with acute respiratory infections (62%) for epidemic-prone disease surveillance and response after the tsunami in Aceh Province, Indonesia.

The nurses have to assess main signs of the tsunami affected patients which includes respiratory distress profound hypoxemia and diffuse bilateral alveolar infiltrate on chest X-ray (white lung), breathless, tachycardia, and increase effort of breathing, decrease $PaC0_2$, hypoxemia, hypotension, and decrease urine output [17]. The nurses also have to assess for airway, breathing, circulation, and disability in maintaining adequate oxygenation for the patients. Airway is establishing a patent airway by positioning, suctioning, and oxygenation. Breathing related to assess breath sounds and respiratory effort, observe for chest wall trauma or other physical abnormality. Circulation involves monitoring vital signs, especially blood pressure and pulse rate. Disability refers to evaluating the client's level of consciousness (LOC) using ALERT (alert, responsive to voice, responsive to pain, unresponsive) and the GCS, and re-evaluates frequently the client's LOC [17].

In caring for tsunami patients with circulation problems due to acute respiratory distress, the nurses should have sufficient knowledge and skills to assess cardiovascular functions of patients to maintenance circulation, monitoring vital signs and identify signs of shock such as hypovolemic, cardiogenic, and septic shocks. In reference [45] described that shock is the manifestation of the rude unhinging of the machinery of life. Hypovolemic shock associated to cold, clammy skin, reduce peripheral perfusion, tachycardia, oliguria, tachypnea, metabolic acidosis, hypocapnia, and decrease cardiac output. Furthermore, the cardiogenic shock associated to cardiovascular collapse because of the cardiac disease, which is characterized by decrease cardiac output, increase heart rate, tachycardia, vasoconstriction, and increase after load. The septic shock associated with stimulation of the host inflammatory response by an infecting organisms or one of its biological product, the signed with hypovolemic, dissociate, and cardiogenic shock.

The nursing interventions for patients with acute respiratory distress syndrome focused on fluid management involve fluid resuscitation to maintain cardiac output and tissue perfusion. Fluid balance should be considered due to contribution to worsening pulmonary edema. Patients are supported by ventilator support with positive end expiratory pressure (PEEP) to prevent alveolar atelectasis and improve oxygenation, monitor pulmonary artery pressure (PAP) in range 25/10 mmHg and pulmonary artery wedge pressures (PAWP) in range 4-12 mmHg [17]. Anti-infective therapies are provided to patients with acute respiratory problems consisting of ampicillin/sulbactam, carbapenemes, fosfomycin, rifampisin, linezolid and glycopeptides [3].

Moreover, altering the patients' position to prone, improve the oxygenation by reducing the ventilation perfusion mismatch and decrease the shunt. Tissue injury of patient can be prevented by changing patients from a supine to a prone position and thus prevents the use of high inspiratory and expiratory pressures [16]. Anti-infective therapies are provided to patients with acute respiratory problems consisting of ampicillin/sulbactam, carbapenemes, fosfomycin, rifampisin, linezolid and glycopeptides [3].

Nurses' clinical skills in acute respiratory care for tsunami care consist of assessing signs and symptoms of near drowning such as dyspnea, cyanosis, and hypothermia, assessing signs and symptoms of pneumonia, removing foreign body from airway, promoting adequate airway by positioning, suctioning, and oxygenation, providing re-warming intervention by using warm blanket or clothing, administering antibiotic therapy for acute respiratory problems, and evaluating patient's condition such as vital signs, breathing, and circulation.

2.2. Wound care

The patients attacked by tsunami showed severe injury, trauma, and wound. The result of the study in [13] showed that wound caused tsunami has specific characteristic: multiple in number, small to medium size, present on the head, face, and extremities, back, buttock and legs. Most of the wounds belong to lower (49%), and upper extremities (32%). In reference [3] stated that wounds caused tsunami were significantly contaminated with foreign materials such as seawater, mud, sand, soil, vegetation, stick, and corals inside the cavity. In reference [12] mentioned that several traumas to the parts of the body and extremities were sustained when the wave hit and swept the affected patients. The patterns of injury can be ranged from minor abrasion to large lacerated wound. Fracture, dislocation, and tendon injury were also common problems found in tsunami patients. Debridement was needed in nearly 90% of the cases and causes of death were near drowning and blunt trauma.

Patients who survived from the tsunami were severely injured, and the incidence of wound infections was high, being at risk of deadly fungal infections and infections caused by waterborne and highly resistant pathogens. The wound displayed the signs such as pus, foul smell, necrosis, gangrene/subcutaneous emphysema, cellulitis, erythema, swelling or infiltration of adjacent tissue, pain, and delayed healing. Many tsunami patients were injured by debris and polluted seawater, which contributed to wound contamination and further tissue damage. Patients mainly suffered from trauma such as wounds, fractures, head injuries, and many needed surgical treatment [18].

A mixture of sea and fresh water, sewage, sand, soil, foreign material causing polymicrobial infection and immediately contaminated wounds. The most common patients are infected by gram-negative pathogen, namely Escherichia coli, Klebsiella, Proteus species, Staphylococci, and Streptococci and infected by bacteria, such as Pseudomonas, Aeromonas, and Shewanella, were probably washed into wounds. Wounds sustained in tsunami disaster are severely contaminated, therefore adequate debridement and delayed suture are recommended for all patients. Anti-infective therapy such as a potent quinolone combined with clindamycin, carbapenemes and glycopeptides were used frequently to control infection [3, 13].

Tsunami injuries are categorized into two main groups; soft tissue injury and fracture or dislocation. Soft tissue injuries are classified into minor wounds and major wounds. The minor wound is defined as abrasion, scratch, or small-lacerated wound, which has no need for debridement. The major wounds refer to lacerated wounds with or without skin loss, multiple lacerated wounds, or infected wounds excluded necrotizing fasciitis and open fracture. Fracture and dislocation group included closed or open fracture, single or multiple fractures and dislocation [12]).

For tsunami patients with fractures, the nurses have to assess signs and symptoms of fracture such as local pain, local bleeding, local swelling, deformity or dislocation, numbness, and paralysis. Moreover, the nurses should be able to identify the fracture complications such as bleeding, hemorrhage, shock, and death [19]. The nurses should provide adequate treatment of fracture that involves reduction, immobilization, and rehabilitation of the fracture. Reduction aimed to restore alignment and use of internal or external fixation devices to immobilize the fracture. Rehabilitation is particularly important and can prevent complications such as pressure ulcers, deep vein thrombosis, constipation, and urinary stasis. The nursing care for patients with fractures is complex and should include hemodynamic monitoring, pain assessment and management fractures [20].

Many tsunami patients are suffered from trauma and injury in different pain levels. The nurses should have sufficient knowledge and skills to assess tsunami patients for pain levels. In [21] described the numeric pain scale (NPS) which is used to measure pain severity by using whole numbers. The line or bar is marked with whole numbers from 0 to10. Interpretation of pain based on score 0 = no pain, 1 to 3 = mild pain, 4 to 7 = moderate pain, and 8 to 10 = severe pain. In addition, to assess the pain level effectively, nurses must be able to assess pain severity in diverse condition of patients adequately, understand how to monitor physiologic changes associated with pain and its treatment, be prepared to address the psychosocial experiences accompanying pain and know the consequences of inadequate analgesia [22].

Wound care is an important part of the overall treatment for tsunami patients. Wound care consists of changing the dressing, packing wound, irrigating; debridement process of removing dead tissue, cleansing the wound, or providing a protective environment. The local care, an integral part of treatment of wounds, reduce bacterial burden, exudates, protect the wound from iatrogenic damage, and lead to wound closure. The wound healing care by secondary and tertiary intention includes dressing change, packing, irrigation, and debridement. Types of wound care procedures involve dressing change and reapplication of a covering to the wound. Packing is insertion of material into a wound cavity and covering of the wound with a secondary dressing to prevent invasion of organisms. Irrigation refers to applying fluid under pressure to remove adherent materials and by products of wound metabolism from the surface of the wound. Debridement is removal of necrotic material and slough from the wound [7].

Particularly, to management the wounds caused by tsunami should be focus on surgical removal of devitalized tissue and aggressive debridement. During the interim between initial wound surgery and secondary closure, wound were protected using vacuum-assisted closure (VAC) systems. A major benefit of vacuum-assisted closure system is the reduced need for dressing changes. Furthermore, vacuum-assisted closure therapy draws wounds closed by applying controlled negative pressure while smoothly removing infectious material and interstitial fluids, thus allowing tissue decompression. This system promotes cutaneous perfusion and formation of granulation tissue in the wound. Using this system, definitive wound closure could be achieved within the first week of occurrence of wound [3].

The antibiotic and tetanus prophylaxis are pre-requisite to all patients who have sustained wounds. If the wound is tetanus prone, the patient should also receive tetanus immunoglobulin to prevent tolerance to antibiotics. A tetanus-prone wound is one where the conditions for spore multiplication are favorable. These include deep penetrating and puncture wounds where the conditions are likely to be anaerobic, wounds with devitalized tissue, for example, burns, crush injuries, compound fractures, wounds contaminated with dirt or manure, wounds containing foreign bodies, and delayed wound cleaning [23].

Purpose of wound dressing is to provide a condusive environment that will enhance wound healing while offering from trauma. Wound dressings can also help in decreasing or eliminating the pain, reducing the need for dressing changes and providing antibiotic debridement. Choosing the right dressing depend on wound bed characteristics such as wound draining, dry, need moisture, need debridement, and infection [41]. There are some common categories of wound dressings consisting of absorptive dressings, alginate dressings, antimicrobial dressings, biological dressings, collagen dressings, composite dressings, contact layer, hydro gel dressings, impregnated dressings, silicone dressings, transparent film dressings, wound fillers, foam dressings, gauze, non-woven dressings, and hydrocolloid dressings [42].

In caring for tsunami patients with mild and severe injuries, wound dressings may protect the wound from any contaminating agent and microorganism. In reference [42] stated that the gauze may inhibit bacterial penetration and may reduce infection in partial and full thickness of wounds. The gauze, which is easily available and relatively inexpensive, is used widely in wound care. A conservative treatment absorbs minimal exudates and should be placed into the wound in a moist state and removed while still moist. To maintain a moist wound bed, gauze should be moistened with normal saline or a wound gel, but not be so wet that the moisture comes through the cover dressing. The gauze should remain moist while in the wound, so it usually needs to be changed every six hours. When used properly, gauze dressings are conservative and safe on stage II (the topmost layers of skin is severed), III (the fat layer and subcutaneous is severed), and IV (the bone and muscle is severed with death tissue and drainage). The enzymatic debridement is accomplished by coating the wound bed with topical enzymes that digest necrotic tissue.

In reference [43] the nursing management of a client with acute inflammation includes minimizing complications of the edema with inflammation, reducing the inflammatory response, and monitoring systemic response.

2.2.1. Controlling the effect of edema

Rest, ice, compression, and elevation (RICE) system can be used to reduce the effect of edema. Ice is used to control the inflammatory response in extremities, particularly when edema and pain are present. Analgesic may be required for pain control.

2.2.2. Reduce inflammation

Reduce inflammatory agent may be prescribed to stabilize the mast cell and reduce edema in the affected area, for example non-steroid anti-inflammatory drugs (NSAID). Foreign bodies may be removed to reduce the cause of the inflammation.

2.2.3. Monitor system responses

Temperature is monitored and fever is treated with antipyretics. Client's temperature is monitored closely to prevent harm. The client with a fever may also experience malaise, nausea, anorexia, weight loss, tachypnea, and tachycardia. The diet of the client suffering from inflammation should be high in vitamin C, protein, calories, and fluids.

The most of cleansing agents and dressings that were used for patients with trauma on wounds care were normal saline, providone iodine, hydrogen peroxide, and using contact layer gauze, telfa, and enzymatic debridement. Tsunamis injury causes severe acute pain [7]. In caring for patients with acute pain, the nurses should have sufficient knowledge and skills regarding pain management with pharmacological and non-pharmacological approaches. The pharmacological approach focused on analgesia therapy. For curing pain pharmacologic treatments were used such as opioids, sedatives, non steroidal anti inflammatory drugs (NSAIDs) and local treatment included lidocaine injection and topical eutectic mixture of local anesthetic (EMLA) cream [7].

Furthermore, non-pharmacological treatment focused on physical and psychological approaches. In reference [44] stated the physical therapies and other complementary techniques consisted of massage, reflex zone therapy, acupuncture, shiatsu, therapeutic touch and transcutaneous electric nerve stimulation (TENS). Finally, the psychological therapy includes pre-operative information giving, cognitive methods, distraction, music, humor, hypnosis, relaxation training, guided imagery, and biofeedback. In [7] studied pain management for non-pharmacotherapy showed that using humor (18.9%), distraction (15.5%), and deep breathing (12.1%) were effective in curing pain.

Based on the literature review, nurses' clinical skills in wound care for tsunami care consist of assessing signs and symptoms of contaminated tsunami wound with sand, sewage, and mud, assessing characteristics of tsunami injuries (soft tissue injuries and fractures), assessing bone deformities and bleeding for patients with fractures, assessing pain intensity, cleansing tsunami wound by using normal saline, preparing patients for debridement, administering antibiotic and tetanus prophylaxis, and evaluating signs and symptoms of wound healing such as granulation growth.

3. Problem statements

- What is the nurses' clinical skill level of acute respiratory care for tsunami patients among nurses in Banda Aceh, Indonesia?
- What is the nurses' clinical skill level of wound care for tsunami patients among nurses in Banda Aceh, Indonesia?

3.1. Application area

The study was focused on disaster nursing in acute response phase on the 2004 Indian Ocean Tsunami in caring patients in the hospital setting.

3.2. Research design

This study was a descriptive explorative study. The aim of this study were to; (1) identify the nurses' clinical skill level of acute respiratory care for tsunami patients among nurses in Banda Aceh, Indonesia, and (2) identify the nurses' clinical skill level of wound care for tsunami patients among nurses in Banda Aceh, Indonesia.

3.3. Population and setting

The target population of the study was the nurses who worked in acute care, and emergency and critical care setting from a general hospital in Banda Aceh, Indonesia. The acute care setting consisted of medical ward, surgical ward, and neurological ward. The emergency and critical care setting consisted of emergency department (ED), intensive care unit (ICU), and operating room (OR). Three of these settings were directly involved in response phase for tsunami care patients on December 26, 2004. The total number of the nurses was 195. The nurses who met the following inclusion criteria were recruited:

- Had been working as permanent and contracted employees of the hospital.
- Had working experience as a nurse at least a year.

3.4. Sample and sampling

Sample Size

The number of respondents in this study was estimated by using power analysis. The sample size was determined at the level of significance (α) of .05, the power of test (1-β) of .80 and the effect size (ϱ) of .30, which is categorized as medium effect size and is used commonly in nursing research. The sample size of 88 was needed [24]. To avoid low response rate of the sample in the study, the sample size was added by 10%, therefore, the total samples of 97 were recruited.

Sampling Technique

The samples were recruited using systematic random sampling with odd number of each ward. They were randomly selected from each ward as shown in table 2. The reasons to perform this technique because it was identified in the sampling frame, the desired sample size, and the size of the population.

3.5. Instrumentation

The instruments used in this study were developed by the researcher and colleagues. They have three main parts, 1) the Demographic Data Questionnaire [DDQ], 2) the Nurses' Clinical Experience Questionnaire [NCEQ], 3) the Tsunami Care Questionnaire [TCQ].

3.6. Demographic Data Questionnaire (DDQ)

The DDQ was used to gather the following data: age, gender, marital status, religion, educational background, training and education, working experience as a nurse, and attending hospital disaster drill. The questionnaire was developed by the researcher.

Training and Education

The variable training and education was measured by asking respondents to indicate type of trainings and number of times on the particular type of training. Type and number of times were multiplied then summed for all types, producing a training experience index. The higher index score reflects the higher attendance in emergency training and education program.

Attending Hospital Disaster Drill

Hospital disaster drill was measured by using dichotomous choice questionnaire. The higher score reflects the higher attendance in hospital disaster drill.

- Nurses' Clinical Experience Questionnaire (NCEQ)

Clinical experience of nurses was measured by using Nurses' Clinical Experience Questionnaire (NCEQ). The questionnaire consisted of 10 items which was measured by using a 4-point Likert-like scale, and was scored from 0 to 3, which 0 = never, 1 = sometimes, 2 = often, 3 = always. The higher score reflects the higher experience to perform nursing activities in caring for tsunami patients.

- Tsunami Care Questionnaire (TCQ)

The Tsunami Care Questionnaire was used to measure perceived clinical skills of nurses for tsunami care. The instrument is composed of 15 items with two subscales: acute respiratory care (1, 2, 3, 4, 5, 6) and wound care (7, 8, 9, 10, 11, 12, 13, 14, 15).

The respondents were asked to rate each item using a 5-point Likert-like scale, ranging from 1 to 5 where 1 = very poor, 2 = poor, 3 = fair, 4 = good, and 5 = excellent. The total score and subscales score are the summed score of all items and items belong to those subscales. They were then averaged for comparable interpretations and categorized into three levels: low = 1.0 - 2.3, moderate = 2.4 - 3.7, and high = 3.8 - 5.0. The higher score reflects the higher clinical skills for tsunami care perceived by nurses.

3.7. The reliability of the instruments

Nurses' Clinical Experience Questionnaire (NCEQ) and Tsunami Care Questionnaire (TCQ) were tested for internal consistency by using Cronbach's alpha coefficients yielding the values of .93, and .98, respectively. The reliability tests were performed with 20 respondents who had similar criteria to the study respondents. Cronbach's alpha coefficients of the instruments were used with the expected value of at least .70, which were accepted for a new instrument [24].

3.8. Ethical consideration

The respondents who agreed to participate voluntarily in this study were informed about their right to withdraw in this study at any time for any reasons without negative consequences. The researcher explained the purposes of the study, expectations from the respondents' participation and potential harms in this study such as the feelings of the traumatic recall, sadness, depression, and despair in completing the questionnaires. In this study, no respondents experienced emotional or psychological problems during completing the questionnaires. The researcher explained them how to complete the questionnaires. The researcher maintained anonymity of the respondents by using code and all of the information was kept confidential.

3.9. Results

The results of the study will be showed as follow:

- Demographic Characteristics and Experience

  The respondents consisted of 97 nurses who worked in the hospital. Most of the respondents were more than 30 years old (56.7%) with a mean age of 31.9 (SD = 6.6).The majority of the respondents were female (71.1%), around three-fourth of them were married (75.3%), were educated at diploma level (78.4%), and all of them were Islam (100%). With regard to attending emergency training and education of the respondents had six index scores (37.1%) with the mean score of 4.3 (SD = 1.9). Regarding working experience as a nurse of the respondents, more half of them had experience more than 5 years (53.6%) with a mean score of 8.7 (SD = 7.5). Approximately 80.4 percent of the respondents had experience in caring for tsunami patients and 40.2 percent had experience in caring for tsunami patients for 3-4 months with a mean score of 3.0 (SD = 1.9). Obviously, 42.3 percent of the respondents had attended hospital disaster drill with the mean score of 0.4 (SD = 0.5).

- Level of Perceived Clinical Skills for Tsunami Care

  Overall, perceived clinical skills for tsunami care were at a moderate level with the total mean score of 3.52 (SD = 0.86). The highest mean score was on acute respiratory care (M = 3.57, SD = 0.86) and followed by wound care (M = 3.50, SD = 0.96) (Figure 3).

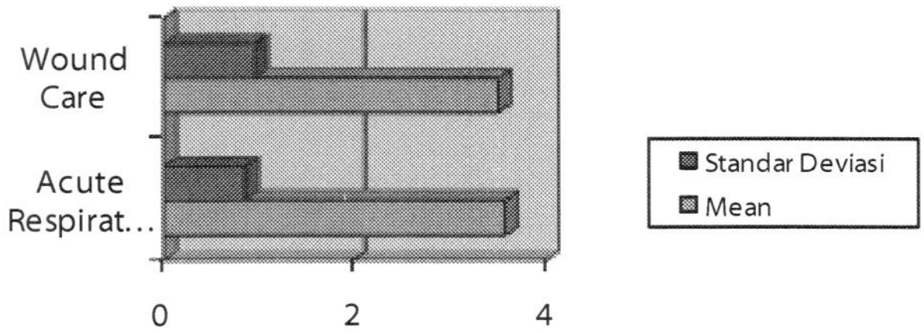

Figure 3. Perceived Clinical Skills

No	Characteristics	Frequency	Percentage
1	Ag Age (years) (M = 31.9, SD = 6.6)		
	(M Min = 23, Max = 50)		
	23-30	42	43.3
	31-40	39	40.2
	>40	16	16.5
2	Gender		
	Female	69	71.1
	Male	28	28.9
3	Ma Marital status		
	Single	24	24.7
	Married	73	75.3
4	Religion		
	Moslem	97	100
5	Ed Educational background		
	Diploma	76	78.4
	Bachelor	21	21.6
6	Wo Working experience as a nurse (years)		
	(M(M = 8.7, SD = 7.5; Min = 2,Max = 27)		
	<5		
	5-10		
	>10	45	46.4
	Experience in caring for tsunami patients	24	24.7
	No	28	28.9
7	Yes		
		19	19.6
		78	80.4
8	Number of months in caring for tsunami patients (M = 3.0, SD = 1.9; Min = 0, Max = 6)		
	None		
	<3		
	3-4	19	19.6
	5-6	23	23.7
	Attending hospital disaster drill	39	40.2
	Yes	16	16.5
9	No		
	Attending emergency training and education (index)	41	42.3
	(M = 4.35, SD = 1.96; Min = 0, Max = 6)	56	57.7
10	None		
	1-4		
	5-6		
		11	11.3
		26	26.8
		60	61.9

Table 1. The Frequency and the Percentage of Demographic Characteristics and Experience (N = 97)

- Perceived Acute Respiratory Care Skills

There was one item of perceived acute respiratory care skills, which had the mean scores at a high level, and the other six items were at a moderate level. The top one items with the highest mean scores was "Administering antibiotic therapy for acute respiratory problems" (M = 3.82, SD = 1.19). The two items with the lowest mean scores were

"Assessing signs and symptoms of near drowning such as dyspnea, cyanosis, and hypothermia" (M = 3.33, SD = 0.74) and "Providing re-warming intervention by using warm blanket or clothing" (M = 3.32, SD = 0.96), which were at a moderate level (Figure 4).

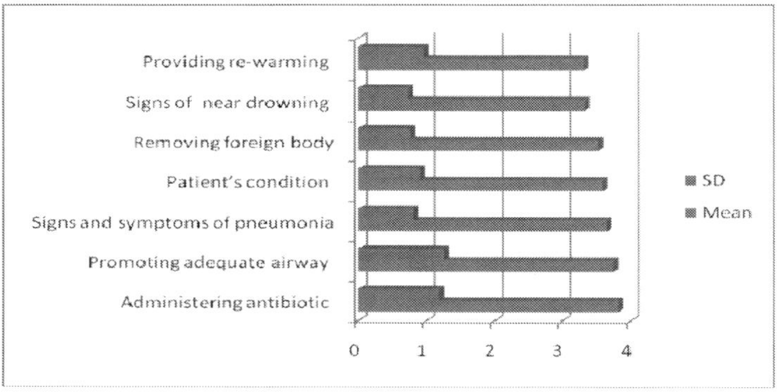

Figure 4. The Levels of Perceived Acute Respiratory Care Skills

- Perceived Wound Care Skills

  All items of perceived wound care skills had the mean scores at a moderate level. The one item with the highest mean score was "Cleansing tsunami wound by using normal saline" (M = 3.67, SD = 1.09). The two items which had an equal mean scores included "Assessing signs and symptoms of contaminated tsunami wound with sand, sewage, and mud" (M = 3.54, SD = 1.25), and "Administering antibiotic and tetanus prophylaxis" (M = 3.54, SD = 1.23). The item with the lowest mean score was "Assessing bone deformities and bleeding for patients with fractures" (M = 3.40, SD = 1.06) (Figure 5).

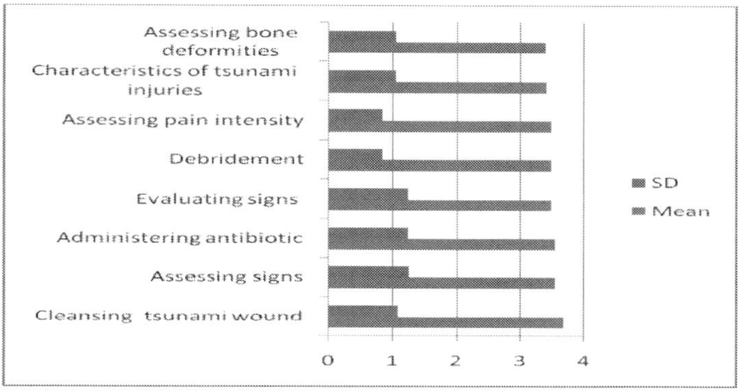

Figure 5. The Levels of Perceived Wound Care Skills

4. Discussion

There were 97 nurses working in dr. Zainoel Abidin general hospital in Banda Aceh, Indonesia who participated in the study. The findings of the study are discussed as follow:

Overall, the findings of this study found that respondents' perceived clinical skills for tsunami care were at a moderate level. There are several reasons might have contributed to the findings of perceived clinical skills for tsunami care in this study.

Educational level of the respondents may have played an important role in the results. Firstly, the majority of the respondents (78.4%) were educated at diploma level only which may have indicated a limited existing knowledge and skill for tsunami response, insufficiency in skill to seeking for appropriate and relevant information, and limited capability to develop further advanced clinical skills. The respondents who only have formal education at the diploma level might not have enough mastery on their knowledge to prepare themselves on the appropriate skills for tsunami care in the response phase.

A recent study found that the nurses whose education was at diploma level showed lower levels of knowledge and skill than baccalaureate, master or doctoral students in clinical management [25]. Moreover, the nurses educated at diploma level showed lower level in theoretical and clinical components for collected the evaluation data, written assignment skill testing, and contribution to conference, self-assessment, and other types of nursing program than baccalaureate and master levels of nurses [26]. Similarly, [27] reported that the nurses who have at the baccalaureate level has strong analytical and creative capacities and better skills in communication, assessment, cultural sensitivity, resourcefulness, the ability to apply knowledge, and scientific reasoning than the diploma level. In addition, the level of education of the nurses might have influenced the nurses' role in responding to an emergency disaster and the willingness to respond [28].

Secondly, the reason that may have influenced the moderate level of perceived clinical skills for tsunami care is age. Almost half of the nurses in this study were aged less than 30 years (42.3%). It could be proposed older nurses have increased clinical skills capability as a result of accumulated experience and knowledge in clinical practice and that younger, less experienced nurses did not have yet enough accumulated clinical skill experiences particularly for tsunami care with complicated problems and serious illnesses. This could imply that the young adults may not have enough capability and competency to perform specific and advanced clinical skills as needed in caring for tsunami patients in acute response phase. However, it is known that young adults have good memory capacity and more opportunity to learn and seek for adequate nursing care in caring for the tsunami patients, so further education may improve the nursing care of tsunami-inflicted wounds. These may have influenced the finding of this study.

Consistently, the previous studies by Marsiske and Willis (as cited in [29]) found that age may have influenced on clinical performance which explicitly emphasized experience and accumulated knowledge structure in maintaining adult's cognition. Young adults have higher memory capacity, speed of processing information and lower inhibition than older adults. In addition, in [25] found that young adults (26-30 years) have lower knowledge and skill levels than middle adults (31-40 years) in clinical management systems.

Thirdly, the reason that may have influenced the moderate level of perceived clinical skills for tsunami care is working experience. The finding of this study showed that nearly half of the respondents had working experience as a nurse for less than 5 years (46.4%). Moreover, 80.4 percent of them had direct experience in caring for tsunami patients in acute response phase for three months in average, because at that time they were recruited and has started working as a nurse in the acute response phase of the tsunami in acute care, emergency and critical care setting.

This indicates that most of the respondents had gained direct clinical skills by transforming experience to performing the nursing skills for tsunami patients. However, they had limitation in the duration of direct clinical experienced in caring for tsunami patients in the acute and emergency response phases. In addition, many experienced nurses with working experience more than 5 years had used advanced clinical skills related to care to the tsunami patients in acute response phase. These advanced skills involve tsunami wound care, and care of patients with tsunami lung or tsunami related aspiration pneumonia due to near drowning.

Similarly, a previous study supports that the experience can help nurses gain sudden insights, acquire new views about the benefits of former learning, absorb information from others, and learn from mistakes and repeated action in similar situations. Experience is a determinant factor in transforming nurses' knowledge to skill [30]. The findings of this study are in accordance with a recent study that showed nurses who had less than 5 years experience showed lower levels of knowledge and skill than those who had experience for more than 10 years nursing experience [25].

Furthermore, the finding of this study was consistent with a previous study by Watson (as cited in [31]) identified three criteria for experience: 1) the passage of time, 2) gaining skills or knowledge, and 3) exposure to an event. Passing of time is commonly used to define experiences in nursing where years of experience are used to categorize the skill of nurses. In addition, in [30] mentioned that the nurses who have worked for half year on resuscitation advanced life support (ALS) training were found to have increased knowledge and skills. In reference [32] also found that most of the nurses who have worked for many years and long periods of practices have high expectations of efficiency and mastery while performing the nursing practice.

Fourthly, the reason that could explain the finding on the moderate level of perceived clinical skills for tsunami care is training and education. Attending emergency training for basic life support (BLS), advanced cardiac life support (ACLS), basic trauma life support (BTLS), and disaster management are important to improve tsunami care skills of the nurses. The finding of this study showed that more than one-third (37.1%) of the respondents had attended only once on the six different topics related to tsunami care of BLS, ACLS, BTLS, disaster management, infection control and prevention, and mental health care for tsunami survivors training. Moreover, 11.3 percent of the respondents were never trained. This could be interpreted that training on emergency response and disaster management for tsunami care were insufficient and lack of continuity).

Consistently, in [33] suggested that emergency training and education should be continued to enable the nurses to develop and apply knowledge and skills to meet demands of their current roles and functions needed for disaster response. Moreover, in [28] mentioned that the nurses need more training and preparation for appropriate response to emergency and disaster. Training and education sessions should address the barriers and perceptions that an event is unlikely to occur and could increase skills of nurses to respond to the disaster. In [34] suggested that emergency training of BLS and ACLS needs to improve by more frequency refresher trainings and allow more times for hand on skills practice. In reference [35] mentioned that the disaster and emergency training provide specific and unique skills set, yet will need to be performed efficiently to respond to emergency or disaster. Lastly, training for emergency medical service (EMS) can result in significant quantifiable team and individual skill gains for a range professional [36].

Lastly, the reason that might have contributed to the moderate level on the finding of the study is attending hospital disaster drill. Approxeimately half of the respondents (42.3%) had attended hospital disaster drill provided by the hospital only once. This indicates that although the respondents have attended only once in their hospital drills, they have gained knowledge and skills to respond to disaster. Attending once on hospital disaster drill might not really influence the clinical skills for tsunami care of the nurses. Frequent and regular attendance on hospital disaster drills at least every one or two years could prepare the nurses with clinically rich knowledge and skills to respond to disaster events. Hospital drill is an essential part of hospital disaster preparedness to provide the sufficient nursing skills in response to disaster. The study in [37] figured out that hospital disaster drill was widely used throughout the world and considered a fundamental tool for evaluation and improvement of disaster response capacity for health care provider including nurses.

Several studies mentioned that disaster drill could provide sufficient knowledge and skills of nurses when responding to mass casualty incidents or disaster. Previous studies supported that disaster drill had equivalent results in prompting critical actions in mass casualty drills and increase the perceived reality to respond to disaster. To respond to mass casualty incidents or disaster, adequate training must be established, implemented, and maintained to ensure their safety and the optimal care of their patients [39]. Moreover, disaster drill also provide the basic tenets of disaster management and triage, decontamination, communication, incident command, and transport patients, and ability to rapidly form multidisciplinary work teams that communicate and function effectively to increase skills in disaster response [38].

Demographic data might have contributed to respondents' perceived clinical skills including diploma level, age, working experience as a nurse, and direct experience in caring for tsunami patients. These factors played an important role in order to improve respondents' perceived clinical skills in caring for tsunami patients.

5. Conclusion

This study employed the descriptive-explorative study to describe the level of perceived clinical skills for tsunami care, among nurses in Banda Aceh, Indonesia. This study was conducted in a general hospital in Banda Aceh, Indonesia. Ninety-seven nurses who worked in hospital of the tsunami-affected area were recruited using the systematic random sampling. The respondents were required to complete a set of questionnaires, which consisted of three parts: 1) the Demographic Data Questionnaire [DDQ], 2) the Nurses' Clinical Experience Questionnaire [NCEQ], 3) the Tsunami Care Questionnaire [TCQ].

A pilot study was conducted and the desired alpha coefficient of .93 for the Nurses' Clinical Experience Questionnaire and .98 for the Tsunami Care Questionnaire were found. The data were analyzed by using descriptive statistics.

6. Summary of the study findings

The sample consisted of 97 nurses with the mean age of 31.9 (SD = 6.6). Majority of them were more than 30 years old (56.7%), and around three fourth of them were married (75.3%). The majority of the respondents were female (71.1%), were educated to diploma level (78.4%), and all of them were Islam (100%). Attending emergency training and education had six index score (37.1%) with the mean score of 4.3 (SD = 1.9). Working experience as a nurse of the respondents was more than 5 years (53.6%) with the mean score of 8.7 (SD = 7.5). The majority of the respondents had experience in caring for tsunami patients (80.4%).

The perceived clinical skills for tsunami care of the respondents showed that the highest mean score was acute respiratory care (M = 3.57, SD = 0.86) with moderate levels and followed by wound care (M = 3.50, SD = 0.96) with moderate levels as well.

7. Limitation of the study

The findings of this study cannot be generalized on the actual clinical skills for tsunami care, because this study was conducted only for perceived clinical skills for tsunami care among nurses who worked in the tsunami-affected area in a general hospital in Banda Aceh, Indonesia.

8. Recommendations

- Nursing Practice

  The finding of this study recommends that the clinical practitioners and hospital policy makers should be aware of the importance on the preparation of clinical skills in response to the tsunami. The nurses' clinical skills for tsunami care needs improvement on knowledge and skills by attending trainings and educational programs and hospital disaster drills regularly. Furthermore, clinical practitioners should encourage nurses to increase their skills in caring for tsunami patients, attending emergency training and education, and attending hospital disaster drill regularly to meet a standard level on preparations in response to the disaster that can happen basically anywhere in the world and particularly in the island countries that highly risk for tsunami disaster.

- Nursing Education

  The findings of this study could serve as evidence or guideline for nurse educators to better prepare nursing students about for nurses' knowledge and skills for tsunami care by providing disaster and emergency trainings. Those training consist of disaster management, basic life support (BLS), advanced cardiac life supports (ACLS), advanced trauma life supports (ATLS), psychological care for tsunami victims, and infection control and prevention in response to tsunami disaster in the hospital setting. Also, the nurse educators can well prepare for the students for exploration the local knowledge and skills of the community that can help for mitigation and response of the disaster in the future by cultural, social, geographical and geological approaches.

- Nursing Research

  These findings present valuable information about perceived clinical skills for tsunami care and its related factors among nurses. Particularly, the emergency training and research center should consider developing and continuing the training program and hospital disaster drill. The hospital and emergency training and research center should also allocate budget for hospital disaster plan to improve nurses' skills in response to disaster. Further study is needed to develop practical guidelines of clinical skills for tsunami care in acute response phase in the hospital setting.

Author details

Cut Husna
Medical & Surgical Nursing Department, Nursing School, Faculty of Medicine, Syiah Kuala University, Darussalam-Banda Aceh, Indonesia

Acknowledgement

Thanks to all of the nurses' who worked in Dr. Zainoel Abidin general hospital Banda Aceh, Indonesia for their participation. Also gratefully thanks to the Director of the Institute for Research and Development on Health and Epidemiology Unit, Faculty of Medicine, Prince of Songkla University, and Graduate School, Prince of Songkla University, Thailand for their assistance for partial granting support to this study.

9. References

[1] World Health Organization (2005). The emergency and its aftermath in moving beyond the tsunami. New Delhi- India: World Health Organization Office for South-East Asia.

[2] Garfield, R., & Hamid, A. Y. (2006). Tsunami response: A year later. American Journal of Nursing, 106(1): 76-79.

[3] Maegele, M., Gregor, S., Yuecel, N., Simanski, C., Paffrath, T., Rixen, D., et al. (2006). One year ago not business as usual: Wound management, infection and psychoemotional control during tertiary medical care following the the 2004 Indian Ocean Tsunami disaster in southeast Asia. Critical Care 1(2): 1-9.

[4] Collander, B., Green, B., Millo, Y., Shamloo, C., Donnellan, J., & DeAtley, C. (2007). Development of an "All-Hazards" hospital disaster preparedness training course utilizing multi-modality teaching. Prehospital and Disaster Medicine, 63-68.

[5] Veenema, T. G. (2006). Expanding educational opportunities in disaster response and emergency preparedness for nurses. Nursing Education Perspectives, 27(2): 93-99.

[6] Chaya, M. (2005). Relieving suffering after the tsunami. Nursing, 35(8): 48-49.

[7] Stotts, N. A., Puntillo, K., Morris, A. B., Stanik-Hutt, J., Thompson, C., White, C., et al. (2004). Wound care pain in hospitalized adult patients. Heart and Lung, 33(5).

[8] Lukthitikul, S., & Hatthakit, U. (2007). Wound care for tsunami victims of nurses at hospitals of Phang-Nga province. Songklanagarind Medical Journal, 25(5): 391- 400.

[9] Rebmann, T., Carrico, R., & English, J. F. (2008). Lessons public health professionals learned from past disasters. Public Health Nursing, 25(4): 344-352.

[10] International Nursing Coalition for Mass Casualty Education (2003). Educational competencies for registered nurses responding to mass casualty incidents. New York: International Nursing Coalition for Mass Casualty Education.

[11] Kaewlai, R., Meennuch, W., Srisuwan, T., Prasitvoranant, W., Yenarkarn, P., & Chuapetcharasopon, C. (2009). Imaging in tsunami trauma. Journal of Medical Ultrasound, 17(1): 1-8.

[12] Prasartritha, T., Tungsiripat, R., & Warachit, P. (2008). The revisit of the 2004 Indian Ocean Tsunami in Thailand: Characteristics of wounds. International Wound Journal 5(1): 8-19.

[13] Wacharong, C., Chukpaiwong, B., & Mahaisavariya, B. (2005). After the tsunamis. The Lancet, 365(9455), 203-203.

[14] Potera, C. (2005). In disaster's wake: Tsunami lung. Environmental Health Perspectives, 113(11): A734-A734.

[15] World Health Organization (2005). Epidemic-prone disease surveillance and response after the tsunami in Aceh Province, Indonesia. Weekly Epidemiological Record, 80(18): 160-164.

[16] Harcombe, C. (2004). Nursing patients with ARDS in the prone position. Nursing Standard, 18(19): 33-39.

[17] Kozier, B., Erb, G., Berman, A., & Snyder, S. (Eds.). (2004). Fundamentals of nursing: Concepts, process, and practice (7 th ed.). New Jersey.

[18] Geertruid, M. H., Marres, M., W., D., de Lange, M., P.H., L., Leenen, M., et al. (2006). Wound infections in repatriated survivors of the tsunami disaster. 18(4): 1-5.

[19] Ignatavius, D. D., & Workman, M. L. (2006). Medical surgical nursing critical thinking for collaborative care (5 th ed.). St. Loius. Missouri: Elsevier Saunders.

[20] Scott, L. (2009). Fracture management. Nursing Standard, 23(27): 59-59

[21] Johnson (2005). Measuring pain Visual Analog Scale versus Numeric Pain Scale: What is the difference? Journal of Chiropractic Medicine, 4(1): 43-44.

[22] Dunwoody, C. J., Krenzischek, D. A., Pasero, C., Rathmell, J. P., & Polomano, R. C. (2008). Assessment, physiological monitoring, and consequences of inadequately treated acute pain. Journal of PeriAnesthesia Nursing, 23(1, Supplement 1), S15-S27.

[23] Sloan, K., & Summers, A. (2006). Tetanus vaccination: The issue of 'just in case' vaccinations in emergency departments. Australasian Emergency Nursing Journal, 9(1): 35-38.

[24] Polit, D. F., & Hungler, B. P. (1999). Nursing research: Principles and methods (6 th ed.). Philadelphia: Lippincott.

[25] Chan, M. F. (2009). Factors affecting knowledge, attitudes, and skills levels for nursing staff toward the clinical management system in Hong Kong. Computers Informatics Nursing 27(1): 57-65.

[26] Oermann, M., Yarbrough, S., Saewert, K., Ard, N., & Charasika, M. (2009). Clinical evaluation and grading practices in schools of nursing: National survey findings part II. Nursing Education Perspectives, 30(6): 352.

[27] Alonzo, A. (2009). Motivational factors in registered nurses completing a baccalaureate completion program. Unpublished Ph.D., University of Kansas, United States, Kansas.

[28] Evers, S., & Puzniak, L. (2005). Bioterrorism knowledge and emergency preparedness among school nurses. The Journal of School Health, 75(6): 232.

[29] Kliegel, M., & Martin, M. (2007). Adult age differences in errant planning: The role task familiarity and cognitive resources. Experimental Aging Research 33: 145-161.

[30] Jensen, M. L., Lippert, F., Hesselfeldt, R., Rasmussen, M. B., Mogensen, S. S., Jensen, M. K., et al. (2008). The significance of clinical experience on learning outcome from resuscitation training: A randomized controlled study. Resuscitation, 80: 238-243.

[31] Considine, J., Botti, M., & Thomas, S. (2007). Do knowledge and experience have specific roles in triage decision-making. Academic emergency Medicine, 14: 722-726.

[32] Bjork, I. T., & Kirkevold, M. (1999). Issues in nurses' practical skill development in the clinical setting. Journal of Nursing Care Quality, 14(1): 72-84.

[33] Gould, D., Berridge, E.-J., & Kelly, D. (2007). The national health service knowledge and skills framework and its implications for continuing professional development in nursing. Nurse Education Today, 27(1): 26-34.

[34] Smith, K. K., Gilcreast, D., & Pierce, K. (2008). Evaluation of staff's retention of ACLS and BLS skills. Resuscitation, 78(1): 59-65.

[35] Kobayashi, L., Shapiro, M. J., Suner, S., & Williams, K. A. (2003). Disaster medicine: The potential role of high fidelity medical simulation for mass casualty incident training. Medicine and Health Rhode Island, 86(7): 196.

[36] Scott, J. A., Miller, G. T., Barry Issenberg, S., Brotons, A. A., Gordon, D. L., Gordon, M. S., et al. (2006). Skill improvement during emergency response to terrorism training. Prehospital Emergency Care, 10: 507-514.

[37] Green, G. B., Modi, S., Lunney, K., & Thomas, T. L. (2003). Generic evaluation methods for disaster drills in developing countries. Annals of Emergency Medicine, 41(5): 689-699.

[38] Kobayashi, L., Shapiro, M. J., Suner, S., & Williams, K. A. (2003). Disaster medicine: The potential role of high fidelity medical simulation for mass casualty incident training. Medicine and Health Rhode Island, 86(7): 196.

39] Gillett, B., Peckler, B., Sinert, R., Onkst, C., Nabors, S., Issley, S., et al. (2008). Simulation in a disaster drill: Comparison of high-fidelity simulators versus trained actors. *Academic Emergency Medicine, 15*: 1144-1151.

[40] Dirkes, S. (2002). Help for ARDS patients. *RN, 65*(8): 52-58.

[41] Baranoski, S. (2008). Choosing a wound dressing, part 1. *Nursing, 38*(1), 60-61.

[42] Mendez-Eastman, S. (2005). Wound dressing categories. *Plastic Surgical Nursing* Retrieved 2, 25, from http://www.dressings.org/Dressings/telfa.htm

[43] Black, J. M. (2005). *Medical surgical nursing clinical management for outcomes* (7th ed. Vol. 1). St. Louis: Elsevier Senders.

[44] Stevensen, C. (1995). Nonpharmacological aspects of acute pain management. *Complementary Therapies in Nursing and Midwifery, 1*(3): 77-84.

[45] Morrison, G. (2007). Understanding shock. *3*(3), 121-125.

Chapter 3

Children and Adolescents-Survivors of the 2004 Indian Ocean Tsunami: Prevalence of Long-Term PTSD and Coping Strategies

Eni Nuraini Agustini and Hiroya Matsuo

Additional information is available at the end of the chapter

http://dx.doi.org/10.5772/51478

1. Introduction

On December the 26th 2004 an submarine earthquake northwest of Sumatra, Indonesia occurred, with a magnitude of 9.0, caused a giant shockwave or tsunami that ruined the shorelines of Indonesia, Sri Lanka, India, Thailand and other countries in Southeast Asia. Most affected country in terms of deaths was Indonesia, around 129.498 died, 37.606 were missing and some 400.000 were internally displaced [1]. Besides the enormous death toll, other far-reaching consequences followed. In any disaster situation children are among the most vulnerable. Exposure to natural disasters has a devastating impact on the psychological and social well-being of children, adolescents. Losing and separating family member, altering routine daily life activity, disrupting community ties, caused children become confuse, scare and distress.

According to Indonesia Social Ministry [2], the number of children orphaned by the tsunami in Aceh was 5,270. In the case of the tsunami disaster, many children and adolescent were suffering from psychological consequences following tsunami aftermath and it was exacerbated by loosing and displacing family, staying in the shelter or tents for months, and lack of social support. Even, the children and adolescent exposed to natural disasters are resilient and recover from early post-trauma symptoms, however the capacity of tsunami to affect mental health is vivid. As a direct consequence of such natural disasters, PTSD is the most common and devastating mental health disorder identified in children and adolescents [3].

According to the Diagnostic and Statistical Manual, Version IV (DSM IV) American Psychiatric Association (APA), PTSD is defined as a clinical syndrome that may develop following extreme traumatic stress. The new definition includes criterion A1 (exposure to a traumatic event) and criterion A2, which is a subjective assessment of the criterion whereby the person report experiencing horror or helplessness at the trauma. PTSD is further characterized as acute when present for less than three months, chronic for more than three months or delayed onset when symptoms develop initially six months or more after the trauma. The first major research of the effects of trauma on children was undertaken by Bloch in 1956 following a tornado in Mississippi. Since the study by Bloch, various authors have undertaken research into the psycho social and physical effect trauma has on children.

Subsequent studies have demonstrated that children and adolescent can develop PTSD following traumatic events. Much of the research has focused on children and adolescent exposed to natural disasters such as hurricanes [4], flooding [5] and earthquakes [6]. They found the symptoms of PTSD among children and adolescences were an exaggerated startle response, repetitive behavior, intrusive thoughts and flash backs about trauma, sleep disturbances, difficult to concentrate, and somatic symptoms. Hence, children who suffer from PTSD demonstrate difficulties in academic achievement, social interaction, and aggressive behaviors [7]. Further this study claims that PTSD may have detrimental effects on their ability to achieve developmental milestones in relation to their peers and on their ability to become fully functioning adults. The self awareness begins to develop during adolescence. This is extremely important for children that have been exposed to chronic trauma, as they develop an understanding of what has and is occurring in their environment. Without the development of self-awareness an adolescent will have difficulty processing and understanding experiences, which may leads to ineffective reasoning skills when interacting with the larger world.

Among these studies, the most common question addressed in past decades is why some people develop PTSD after traumatic events, whereas other do not develop. The presence of many different factors such as gender, severity of trauma, age, social support, and coping strategy may play a role to make somebody more vulnerable to develop PTSD resulted in the individuality of the incidence and especially for coping, coping is individually matter.

Coping is defined as effort to regulate emotion, cognition, behavior, physiology and the environment in response to stressful event or circumstances [8]. Coping conceptualized in the two broad responses: emotion focused coping and problem focused coping [9]. Recently, an empirical evidence has introduced multidimensional of children coping. An investigation proposed four-factor model coping including active coping, distraction, avoidance and support seeking [10]. Active coping involves direct problem solving, cognitive decision making, and restructuring. Distraction refers to replaced action and physical release of emotion. Avoidance involves behavior and cognitive strategies such as withdrawal, denial, and refusing to think about the trauma. Support-seeking consists of emotional sharing and seeking help from others. Children use coping strategies in order to protect and adapt with stressful situation.

The literature clearly explains the consequences of PTSD toward growth and development of children and adolescence. However, there are very few research of the long-term outcome of PTSD in children and adolescence [11]. Furthermore, there is a paucity of study to examine the long-term effect of the tsunami among children and adolescence in Indonesia, especially PTSD and its associated factor. Therefore, we conduct the study of long-term PTSD and its associated factors. Hence, we want to examine the coping strategy of children and adolescence with long-term PSTD.

2. Subjects and methods

2.1. Participants

The participants in this study were 482 students from 4 Junior and Senior High School in Banda Aceh and Aceh Besar district. The data collection was conducted 4 .5 years after tsunami. Permission to conduct the study were acquired by School of Nursing, Faculty of Medicine and Health Sciences Syarif Hidayatullah State Islamic University, School of Nursing, Faculty of Medicine, Syiah Kuala University, the principals of Junior and Senior High School and by the students.

2.2. Ethical consideration

Questionnaires were administered to the participants in their classrooms by the researcher and teacher. Before collecting the data session, students were informed that their participation was voluntary and they had the right to withdraw from the study without any consequences. It was clarified that there were no right or wrong answers to the questions. The participants were encouraged to ask of each item to the researcher if they needed further explanation.

2.3. Self-reported questionnaire

Socio-demographic characteristics. Questionnaire of socio-demographic characteristics consist of age, gender, school grade, losing of parents, somatic response, sources of emotional supports, and history of trauma.

The Child Post Traumatic Stress Reaction Index (CPTSD-RI) consists of 20 items designed to measure children's reactions and distress related to a specific traumatic event [12]. Children are indicating a response on 5 point Likert-type scales from 0 (*none*) to 4 (*most of time*). Total scores range from 0 to 80 with the following severity ranges: 0 to 12 = doubtful, 12 to 24 = mild, 25 to 39 = moderate, 40 to 59 = severe, 60 and above = very severe.

3. Post-traumatic stress disorder symptoms and its associated factors

3.1. Socio-demographic Characteristics (Table 1)

The mean age was 14.4 years (range 11–19). Two hundred and forty-six (51%) were female and 236 (49%) were male. Participants were classified into six groups according to their school grade, junior high school VII–IX and senior high school X–XII.

Background	N	Mean± SD	Range
Age (years)	482	14.4± 1.7	11-19
		Percentage	
Gender:			
Female	246	51%	
Male	236	49%	
School Grade:			
Grade VII	88	18.2%	
Grade VIII	64	13.3%	
Grade IX	62	12.9%	
Grade X	125	25.9%	
Grade XI	80	16.6%	
Grade XII	63	13.1%	
Loosing Parents:			
None	365	75.8%	
Loosing father	57	11.8%	
Loosing mother	18	3.7%	
Losing both of father and mother	42	8.7%	
Supports:			
Support from family	403	84.7%	
Support from teacher	119	24.7%	
Support from relative	80	16.6%	
Support from friend	97	20.1%	
Other	30	6.2%	
History of trauma			
single trauma	31	6.4%	
multiple trauma	451	93.6%	
Somatic response :			
Heartbeat	170	35.3%	
Headache	132	27.4%	
Stomachaches	94	19.5%	
Cold in extremities	94	19.5%	
Sweating palm of hand and feet	79	16.4%	
The best person to share:			
Friend	277	57.5%	
Mother	248	51.5%	
Sibling	234	48.5%	
Father	123	25.5%	
Relative	66	13.7%	
Teacher	37	7.7%	
Others	17	3.5%	

Table 1. Socio-demographic Characteristics

Receiving supports were categorized into five categories: no support, mild (one support), moderate (received two supports), good (received three supports) and very good (received all supports) were reported by 37 (7.7%), 240 (49.8%), 123 (25.5%), 55 (11.4%) and 27 (5.6%), respectively. Support also divided into: familial support, extra familial support and professional support. Familial support that consist of family and relative support were 83.6 % and 16.6% respectively. Extra familial support that consist of teacher, friend and others support were 24.7%, 20.1%, 6.2%, respectively. Professional support was 15.6 % and for social supports level divided into 3 categories: less support level, moderate level, and good level were 57.5%, 36.9%, and 5.6%, respectively.

Thirty-one adolescents (6.4%) suffered a single trauma whereas 451 (93.6%) had various traumatic experiences. The somatic response were reported in 5 categories, such as no somatic response 29.3%, mild somatic response 40.3%, moderate 17.8%, severe 8.9% and very severe 3.7%. The most frequently reported somatic responses were heart beating fast (35.3%) and headache (27.4%). Parental loss was categorized into four categories: lost no parents (75.8%), lost father (11.8%), lost mother (3.7%), lost both parents (8.7%). The best person with whom to share was also evaluated; and participants mostly prefer: friend (57.5%), mother (51.5%) and sibling (48.5%) as the best person to share.

3.2. Post-traumatic stress disorder symptoms (Table 2)

The mean score on CPTSD-RI was 28.9 (range 0–61). Children who reporting none, mild, moderate, severe and very severe symptoms were 54 (11.2%), 124 (25.7%), 196 (40.7%), 103 (21.4%) and 5 (1%), respectively. The most frequently reported symptoms in this study were emotional avoidance, identified events as extreme stressor and regular fea.

PTSD Symptom Severity Level	N	Percentage
None	54	11.2 %
Mild	124	25.7%
Moderate	196	40.7%
Severe	103	21.4%
Very Severe	5	1%

Table 2. PTSD Symptom Severity Level

3.3. PTSD& age

There was no correlation among age and mean score CPTSD-RI (P =0.308). A one-way anova between groups was conducted to explore the impact of school grade on total score on CPTSD-RI. There was no statistically significant difference among groups (P = 0.57).

3.4. PTSD and social support

The severity of symptoms of PTSD in participants who received very good support differed significantly from the severity of symptoms. Among those who received no support;

participants who received very good support had a lower mean CPTSD-RI score than those who received no support (*P* = .004). Independent sample of T test was conducted to predict PTSD symptom severity scores (CPTSD-RI) with familial support, extra familial support, and professional support. Our findings showed that extra-familial support (p: .04) was significant correlate with PTSD symptoms. When comparing the mean score of PTSD symptoms in participants who acquired good support to less support, there was significant difference among supported group, participants who acquired good support were having lower mean score of CPTSD-RI rather than less support group (p: .01).

3.5. PTSD and gender

There was a significant gender difference in the mean score on CPTSD-RI; female students had higher scores than male ($P < 0.05$).

3.6. PTSD and lost of parents

The main score between loosing parents and severity of posttraumatic stress reaction were showed higher mean score CPTSD-RI among participant who lost both of parents (analyses of variance, p: <.05).

3.7. PTSD and number of prior trauma experience

This study showed that the CPTSD-RI score of participants who suffered a single trauma did not differ significantly from those who had various traumatic experiences ($P = 0.704$).

3.8. PTSD and somatic responses

Somatic responses also differed significantly in the mean score on CPTSD-RI; participants with a very severe somatic response had the highest scores on the mean of CPTSD-RI ($P < 0.05$) (Table 3).

4. Long-term PTSD and coping strategy

4.1. Subjects and methods

4.1.1. Participants

As a part of our mix-method study, then we selected participants from the previous study. Eventually we found 10 children who involved in this study from one private junior high school at Aceh Besar district. Along with the teacher, the selection process was held. In this study only children who directly experienced the 2004 Indian Ocean tsunami and lost their parents (one or both parents) were involved. Prior to interview session, since the children were orphaned, the aim of the study and written inform consent were given directly to children under teacher's supervision.

Variables	N	PTSD Mean ± SD	p Value
Gender:	482		
Female	246	31.8 ± 12.5	.000
Male	236	25.8 ± 12.4	
School Grade			
Grade VII	88	26.6± 13.9	
Grade VIII	64	28.5 ± 12.6	
Grade IX	62	29.4 ± 13.3	.57
Grade X	125	29.4 ± 13.3	
Grade XI	80	29.7± 12.1	
Grade XII	63	29.9± 10.8	
Loosing Parents:			
None	365	26.2± 12.1	
Loosing father	57	35.2 ± 12.5	
Loosing mother	18	40.9 ± 10.1	.000
Losing both of father and mother	42	38.6 ± 10.3	
Support level :			
Bad/none of supports	37	31.9 ± 12.1	
Mild	240	30.1 ± 12.9	
Moderate	123	28.6± 12.7	.004
Good	55	23.4± 12.4	
Very Good	27	26.4 ± 12.1	
Somatic response category:			
None	141	22.6± 11.7	
Mild	194	30.3± 12.4	
Moderate	86	32.4± 12.7	.000
Severe	43	32.3±11.3	
Very severe	18	38.7± 10.9	
History of trauma			
single trauma	31	28.0 ± 12.9	
multiple trauma	451	28.9 ± 12.8	.704

Table 3. PTSD and Socio-demographic Characteristics

4.1.2. Design overview

This cross-sectional qualitative study was conducted as the part of our mix method study. We conduct this study on February, 2010. This study were approved by the Kobe University Graduate School of Health Sciences; Faculty of Medicine and Health Sciences, Syarif Hidayatullah State Islamic University; School of Nursing, Faculty of Medicine, Syiah Kuala University; and the principal of junior high school. Data for this study were obtained from clinical structured the PTSD symptoms and coping style post– the 2004 Indian Ocean tsunami.

4.1.3. Interview procedures

The interview was conducted face-to-face on each one child with two reseachers. Some of the children were preferred to answer in Aceh language; therefore we were using both of Indonesian and Aceh language. Children's PTSD symptoms and coping strategies were assessed during an individual meeting that occurred at school. Especially for children whom preferred answer in Aceh language, the local reseacher was interpreted and wrote the child's answer from Aceh language into Indonesian. The reseachers wrote the answers and read back to the child what had been written to make sure that the response was accurately documented. The interviews with children lasted approximately 60 minutes.

4.1.4. Data collection of CAPS-CA

The data collection were consist of 3 items questionnaire; first, basic information like demographic data and the tsunami's experience; two, Clinician-Administered PTSD Scale for Children and Adolescents (CAPS-CA); three, children coping strategy. Prior to data collection, the questions for the interview was piloted with two children at the same school by using both of Indonesia and Aceh language to make sure if the child understood the intendeed meaning of the questions and written the responses. After pilot interview, then we conducted the data collection among 10 children. Data have been collected through observations and clinician structure interview among participants. We were performed clinical diagnoses by using CAPS-CA.

The CAPS is a structured interview that assesses the 17 symptoms of PTSD [13]. We assess both of the frequency and intensity of the 17 diagnostic criteria and two associated features (self blame and trauma spesific fears). Children's coping strategy was asked from one open-ended questions: "what did you do to help you feel better when the intrusive thought occured?".

The duration of each interview was 60-90 minute for 2-3 times of session. All the child response were typed into an Excel database and double checked for accuracy. All children were signed by an indicated number.

4.2. Qualitative analysis of coping strategy

Qualitative study literature states that the step on qualitative study consist of, transcribed data to identify the significant statement of informant; read the transcribed data several times; categorized the statements; determined the potential subthemes from the categories of statements; determined the themes and subthemes; clarrified to verified data from key informant as needed; and revised themes based and verified data as needed.

First of all, we gather coping literatures [8],[9],[10],[15],[16],[17],[18]. Based on coping literatures, then, we analyze the coping strategy among children. The next steps, we divide children's answer into categories. Each category consist of subtypes of coping strategies which represent of children's coping responses, again at this step, we analyze and match children's answer with appropriate coping literatures. Because of the diffrences of children's characteristics, so some of the definitions of the categories and subtypes of coping for children are vary. As a part of cultural consideration, we included a coping subtype to the use of religious or spiritual guidance.

5. Long-term PTSD and coping strategy

5.1. Participants characteristics

Children current age range was 12–16 years, whereas the age at the tsunami was 6-10 years. Five children were females and 5 children were males. Familial loss were separated as: lost father (4 children), lost mother (3 children), lost both of parents (3 children), lost either parents and siblings (4 children).

5.2. CAPS-CA (Table 4)

Children who diagnosed as full PTSD were 5 children, while the remaining children were diagnosed as partial PTSD.

Children	Meet Criteria				Total Score of CAPS-CA	Type of PTSD
	A	B ≥1 (B1-B5)	C ≥3 (C1-C7)	D ≥2 (D1-D5)		
P1	YES	YES (21)	YES (9)	YES (26)	56	Partial
P2	YES	YES (20)	YES (18)	YES (22)	60	Full
P3	YES	YES (20)	YES (12)	YES (14)	46	Partial
P4	YES	YES (25)	YES (26)	YES (21)	72	Full
P5	YES	YES (15)	NO (4)	YES (11)	30	Partial
P6	YES	YES (19)	YES (31)	YES (22)	72	Full
P7	YES	YES (20)	NO (0)	YES (13)	33	Partial
P8	YES	YES (23)	YES (36)	YES (26)	85	Full
P9	YES	YES (18)	YES (15)	YES (15)	48	Full
P10	YES	YES (12)	YES (5)	YES (21)	38	Partial

Table 4. CAPS-CA

5.3. Coping strategy (Table 5)

This qualitative study surfaced the coping strategy among 10 children. We reported 38 coping responses from answer statement of 1 open-ended question about their coping response when the intrusive thoughts occurred. When answering the question, each child answers not only using 1 coping response, but also several coping responses. Of 38 coping responses, we identified the meaning participants coping responses with the meaning of coping strategies based on literatures. Then, we divide into 12 sub types of coping strategy from 38 coping responses and eventually, we determine 4 coping strategies.

Coping strategy (n: 38)
Avoidance coping (n = 20/38)
Paralysis passivity (3/38)
If the intrusive thoughts occurred, sometimes I can't do anything
I prefer to be quite
Thought suppression (2/38)
I had forgot about the tsunami experience
I don't want to remember it anymore
Rumination (3/38)
I can't do anything because it's very hard to eliminate those thoughts
Avoidance other (1/38)
I don't want to hear air plane's sounds
Self isolation (1/38)
I cant's share it with anyone, nobody can understand me
Avoidance action (10/38)
I don't want to go of the sea any more
I don't want to start any conversation related to the tsunami
Active Coping(n = 6/38)
Seeking meaning (5/38)
I use to pray, say Istighfar (ask God forgiveness), and read Quran
Seeking social support (1/38)
Sometimes I try to share with my friend
Adaptive coping(n= 3/38)
Distractive (3/38)
Sometimes I play with my friend
Sometimes I had wrote a poem
Emotion focused(n= 9/38)
Emotional Expressions (5/38)
I use to cry
Self blaming (3/38)
I'm feeling guilty, since I couldn't safe my sister
Wishful thinking (1/38)
I wish I could hold my sister's hand, she would be live now

Table 5. Coping strategy

1. **Avoidance coping.** This is the majority coping among children (20 coping responses). All the children were using avoidance coping like avoid the place, conversation, or any thoughts related with the tsunami experience. For example they state: "I don't want to start any conversation related to the tsunami, except somebody ask me first". or "I can't share it with anyone, nobody can understand me".
2. **Active coping.** Seeking meaning and seeking social support was reported as active coping in this study. Using religious and spiritual sources in order to feel better and

attempts to accept the tsunami event as God will is defined as seeking meaning. Making the religious sources as seeking meaning in this study had showed such as "I use to say Istighfar (ask God forgiveness)", "I will pray" and "I use to read Quran".

3. **Emotion focused coping.**Three children were reported emotional expression or expression feeling, self blame and wishful thinking as their coping. They thought that some part of the tsunami experience was they fault, for example: "I'm feeling guilty, since I couldn't safe my sister" and "I feel regret that I can't safe my parents". Emotional expression or expression feeling can be reported such as "I use to cry" or "I felt sad and cry".
4. **Adaptive coping.** Distractive action as the only adaptive coping have been using among 2 children. Effort to change the attention of their mind into another activity as they stated that "playing with friends' or "writing a poem" can make them feel better when the intrusive thought occurred.

5.4. CAPS-CA and coping strategy (Table 6)

This study showed children with both of partial or full PTSD had several coping strategies. Avoidance coping strategy has been dominated of 10 children as well as the development of PTSD.

Children	Type of PTSD	Coping strategic	Sub category
P1	Partial	Emotion focused	Emotional expression
		Avoidance coping	Paralysis passivity
		Avoidance coping	Rumination
		Emotion focus	Self blaming
P2	Full	Active coping	Seeking meaning
		Adaptive coping	Distraction
		Avoidance coping	Self isolation
P3	Partial	Avoidance coping	Avoidance action
		Active coping	Seeking social support
		Active coping	Seeking meaning
P4	Full	Active coping	Emotional expression
		Avoidance coping	Rumination
		Avoidance coping	Avoidance action
P5	Partial	Emotion focused	Emotional expression
		Avoidance coping	Avoidance other
P6	Full	Avoidance coping	Thought suppression
		Avoidance coping	Avoidance action
		Emotion focus	Self blame
		Emotion focus	Wishful thinking
		Adaptive coping	Distraction
		Active coping	Seeking meaning

Children	Type of PTSD	Coping strategic	Sub category
P7	Partial	Emotion focused	Emotional expression
		Avoidance coping	Paralysis passivity
		Adaptive coping	Distraction
		Emotion focus	Self blame
P8	Full	Active coping	Seeking meaning
		Emotion focused	Emotional expression
		Avoidance coping	Rumination
		Avoidance coping	Avoidance action
P9	Full	Avoidance coping	Paralysis passivity
		Emotion focused	Emotional expression
		Avoidance coping	Avoidance Action
P10	Partial	Active coping	Seeking meaning
		Avoidance coping	Avoidance action
		Avoidance coping	Thought suppression

Table 6. PTSD & Coping strategy

6. Conclusion

This study showed that children and adolescence who live through disaster do experience long-term of PTSD symptoms (63.1%). The persistence of PTSD symptoms in this study appeared to be associated with the memory of tsunami that was ingrained even after 4.5 years; indeed many associated factors contribute with the presence of symptoms [19]. Traditionally, their responses have been misunderstood or trivialized. The idea that "children are resilient and will get over it" is no longer valid. Most children and adolescents will regain normal functioning once basic survival needs were met, safety and security have returned and developmental opportunities were restored, within the social, family and community context [20]. Unfortunately, children and adolescent in these studies did not meet their need immediately. Moreover, they did experience such as displacement, death of family and relative, damage to good and home, and staying in the shelter for months. These circumstances exacerbated their psychosocial consequences.

This study determined adolescents with higher CPTSD-RI score were either lost their parents and less of support. It means that, being lost of family was associated with the less of support for recovery. In particular, results of this study indicate that social support level was also significant correlate with the severity of PTSD. These findings were supported with the previous investigation which showed that social support form family, friends and community are thought to protect against the development of PTSD as family, friend and community can challenge negative beliefs, help to find solutions and encourage behavior [21]. Less social support in this study was determined as a significant risk for psychological consequences such as PTSD. Another study suggested that the severity and persistence of PTSD might be related to the deterioration of social support [22].

However in our study, extra familial supports were considered more significant with the severity of PTSD symptoms compare to familial support and professional support. This could be explained with another previous research that a caregiver's or family member's own trauma could interfere with her or his ability to provide appropriate care and support for their children [23]. Moreover, since many family member of children and adolescent were became tsunami survivors, they do not have more time and energy to pay attention , which in turn children and adolescent were seek support from extra familial supports as their helpfulness support at that moment. Findings in current study confirm previous report that, once extra familial support (e.g., teachers, friends, community leader and religious figures) was increased as perceived helpfulness support then posttraumatic stress reaction in youth would decrease [24]. Children and adolescents in present study were mostly reported friend and mother as the best person to share, this findings seems to be linked with the presence of extra familial support sources. The result enlightened that during adolescence the need for emotional support extend beyond the familial and kin confines to friend and non-kin adults in the community [25]. Which means that adolescent begin to rely on peers than solely on parents for sources idea, value, and behavior.

Our study revealed that children employ similar coping strategy with adult, such as avoidance, emotion focused, active, and adaptive [26]. However, there were some differences between children and adult. Participants in our study tends to be more frequently using avoidance coping and emotion coping. Traumatized children typically employ emotion-focused, avoidant and distractive coping [27]. In contrast, seeking social support was reported less often on children in this study, which means the initiative of children to seek support from other still immature. Nevertheless, our result indicated coping response may be effective in protecting children's mental health in extreme life-endangering condition.

Supporting previous study [28],[29], our study found that gender showed significance difference on CPTSD-RI score, female participants showed high score CPTSD-RI compare to male participants. Reversed with the gender, support, and somatic response, we found in this study, there is no difference between age and severity of PTSD symptoms. Our data indicated that the impacts of disaster may vary depending on the circumstances of characteristics of the adolescents, adolescent's support and their coping.

Findings from this investigation show the need to set up "child-friendly" spaces as soon as possible and activities that normalize the lives of children, give them a sense of safety. Focus and build on interventions that strengthen the population's resiliency and resources. Effective intervention strategies should enhance children and adolescence psychological adaptation despite the tsunami.

Author details

Eni Nuraini Agustini
Psychiatric Nursing Department, School of Nursing, Faculty of Medical and Health Sciencies, Syarif Hidayatullah State Islamic University, Ciputat, Indonesia

Hiroya Matsuo
International Health Department, Kobe University Graduate School of Health Sciences, Kobe, Japan

7. References

[1] Health Minister Department of Indonesia. *Indonesia Health Profile 2006*. Available at: http://www.depkes.go.id/. (Accessed 20 May, 2009)

[2] Dinas Social and Inter-agency joint registered children databaseanalysis, 13 October 2005.

[3] Norris, F. H., Friedman, M. J. and Watson, P. J. (2002) '60,000 Disaster Victims Speak: Part I: An Empirical Review of the Empirical Literature, 1981–2001', *Psychiatry: Interpersonal & Biological Processes* 65: 207–39.

[4] Sallom A. & Lewis M.L. (2010) An explanatory study of African American parent-child coping strategies post-hurricane Katrina. *Traumatology* 16 (1), 31-41

[5] Earls, F., Smith, E., Reich, W. and Jung, K. (1988) 'Investigating Psychopathological Consequences of a Disaster in Children: A Pilot Study Incorporating a Structured Diagnostic Interview', *Journal of the American Academy of Child and Adolescent Psychiatry* 27: 90–95.

[6] Bal, A. (2008). Post-Traumatic Stress Disorder in Turkish child and adolescent survivors three years after the Marmara earthquake. *Child and Adolescent Mental Health*, 13, 134-139.

[7] Anderson T (2005). PTSD on Children and Adolescence. Retrived from: http://www.uic.edu/cuppa/gci/publications/workingpaperseries/pdfs/anderson%2005-04.pdf(Accessed 20 June, 2009)

[8] Compas B. E., Conner-Smith J. K., Saltzman H., Thomsen A. H. & Wadsworth M. E. (2001). Coping with stress during childhood and adolescence: Problems, progress, and potential in theory and research. *Psychological Bulletin 127*, 87-127

[9] Lazarus, R. S., & Folkman, S. (1984) *Stress, appraisal, and coping*. New York: Springer

[10] Ayers T. S., Sandler I. N., West S. G. & Roosa, M.W. (1996) A dispositional and situational assessment of children's coping: Testing alternative models of coping. *Journal of Personalit, 64*, 923-958

[11] Goenjian, A.K., Walling, D., Steinberg, A.M., et al. (2009). Research report Depression and PTSD symptoms among bereaved adolescents 6½ years after the 1988 spitak earthquake. Journal of Affective Disorders 112 81–84

[12] Goenjian, A. K., Pynoos, R. S., Steinberg, A. M., et al. (1995). Psychiatric comorbidity in children after the 1988 earthquake in Armenia. *Journal of the American Academy of Child and Adolescent Psychiatry*, 34(9), 1174-1184

[13] Nader K., Kriegler J.A., Blake D.D., Pynoos R.S., Newman E. & Weathers F.W. (1996) *Clinician Administered PTSD Scale, Child and Adolescent Version*. White River Junction, VT: National Center for PTSD.

[14] Colaizzi, P.F., (1978) Psychological research as the phenomenologist views it.*Existential Phenomenological Alternatives for Psychology* (eds) R.S. Valle and M. King(Oxford University Press, New York), pp48-71

[15] Ebat, A. T. & Moos, R. H. (1991) Coping and adjustment in distressed and healthy adolescents. *Journal of Applied Developmental Psychology* 12, 33-54

[16] Stallard P. & Smith E. (2007) Appraisals and cognitive coping styles associated with chronic post-traumatic symptoms in child road traffic accident survivors. *Journal of Child Psychology and Psychiatry* 48 (2), 194-201

[17] Stallard P. (2003) A retrospective analysis to explore the applicability of the Ehlers and Clark (2000) cognitive model to explain PTSD in children. *Behavioural and Cognitive Psychotherapy* 31, 337–345

[18] Punamaki R. L., Muhammed A. H. & Abdulrahman H. A. (2004) Impact of traumatic events on coping strategies and their effectiveness among Kurdish children. *International Journal of Behaviour Development* 28 (1), 59-70

[19] Agustini, E. N., Asniar, Matsuo, H. (2011) The prevalence of long-term post traumatic stress symptoms among adolescents after the tsunami in Aceh. *Journal of Psychiatric and Mental Health Nursing*. Volume 18, Issue 6, Pages 469–561, e16–e18.

[20] Save the Children (2005). Psychosocial Care And Protection Of Tsunami Affected Children Guiding Principles. Retrived from http://www.savethechildren.org.uk/resources/online-library/search?publication=&page=6&f[0]=field_publication_topic%3A25 (Accessed 1 July, 2009)

[21] Ellis, A. A., Nixon, R. D.V., & Williamson, P. (2009). The effects of social support and negative appraisals on acute stress symptoms and depression in children and adolescents. *British Journal of Clinical Psychology*, 48 (4), 347-361.

[22] Kaniasty, K. Z., & Norris, F. H. (2008). Longitudinal linkages between perceived social support and posttraumatic stress symptoms: Sequential roles of social causation and social selection. *Journal of Traumatic Stress*, 21, 274–281

[23] Scheeringa, M. S., & Zeanah, C. H. (2001). A relational perspective on PTSD in early childhood and adolescents. Journal of Traumatic Stress, 14, 799–815.

[24] Pina, A.A., Villata, I.K., Ortiz, C.D., Gotschall, A.C., *et al* (2008). Social support, discrimination, and coping as predictors of posttraumatic stress reactions in youth survivors of Hurricane Katrina. *Journal of Clinical Child & Adolescents*. 37(3).564-574

[25] Betz, C. L., Hunsberger, M., & Wright, S. C. (1994). Family centered: Nursing care of children (2nd edition). *WB Saunders company*, Philadelphia

[26] Agustini, E. N., Asniar, Hashimoto, T., Matsuo, H. (2010). Time course of long-term PTSD symptoms on youth tsunami survivor. Bulletin of Health Sciences Kobe, 26:81-87

[27] Tolan, P.H., Guerra, N.G., & Montaini-Klovdahl, L.R. (1997). Staying out of harm's way:Coping and development of inner-city children. In S.A. Wolchik& I.N. Sandler (Eds.), Handbook of children's coping: Linking theory and intervention (pp. 453–479). *New York: Plenum Press.*

[28] Davis, L., & Siegel, L.J (2000). Posttraumatic Stress Disorder in children and adolescents: a review and analysis. *Clinical Child and Family Psychology Review*, 3, 135-154

[29] Bokszczanin, A. (2002) 'Long-Term Negative Psychological Effects of a Flood on Adolescents', *Polish Psychological Bulletin* 33: 55–61.

Section 2

Computational Models, Simulations and Early Warning Systems – The Modelling Aspect of a Tsunami

Chapter 4

The November, 1st, 1755 Tsunami in Morocco: Can Numerical Modeling Clarify the Uncertainties of Historical Reports?

R. Omira, M.A. Baptista, S. Mellas, F. Leone, N. Meschinet de Richemond, B. Zourarah and J-P. Cherel

Additional information is available at the end of the chapter

http://dx.doi.org/10.5772/51864

1. Introduction

The Lisbon earthquake occurred on the morning of November, 1st, 1755 and generated a tsunami that hit the southwestern coast of Portugal in less than 20 min after the main shock (Pereira de Sousa, 1919; Baptista et al., 1998). This event remains the largest natural disaster in Europe in the last 500 years, in terms of loss of lives (60 000 to 100 000 deaths) and destruction (Baptista et al., 1998; Chester, 2001).

The strongest earthquake shaking occurred all over the Iberian Peninsula, Morocco, and as far as Hamburg, the Azores and Cape Verde Islands (Martinez-Solares et al., 1979). Recent evaluations of the earthquake magnitude estimate it to 8.5±0.3 (Martinez-Solares and Lopez-Arroyo, 2004).

The tsunami waves caused massive destruction in the southwest Iberian Peninsula and Northwest Morocco (Baptista et al., 1998; El Mrabet, 2005). They travelled as far north to Newfoundland and Cornwall in the UK (Huxham, 1756), and towards south to Antigua and Barbados in the Caribbean (Sylvanus, 1756), and to the coast of Brazil (Kozak et al., 2005; Ruffman, 2006).

The 1755 tsunami remains the largest eye-witnessed historical event in the northeast Atlantic area. Historical documents describe, in detail, the waves along the coasts of Morocco, Portugal and Spain. They mention tsunami run-ups as high as 15 m and wave heights of ~24 m (Soyris, 1755) in some locations. These values raise the question of the reliability of some historical documents reporting the impact of the 1755 Lisbon tsunami.

Various studies have focused on the compilation and revision of the historical documents in order to describe precisely the impact of the 1755 tsunami in Morocco (El-Mrabet, 1991;

2005; Blanc, 2009; Kaabouben et al., 2009). Blanc (2009) published a critical analysis of the historical reports that testify the impact of the tsunami waves in Morocco and concluded that the description of the wave heights was overestimated. Kaabouben et al. (2009), based on the compilation made by El-Mrabet (1991; 2005), presented a careful revision of the historical data of the 1755 event showing that some quoted values were overestimated or inadequately interpreted.

Among the coastal sites of Morocco, the city of Mazagão (Fig. 1), actually El-Jadida, is one where the impact of the 1755 tsunami event is described in some detail by several historical reports (Blanc, 2009). In view of this fact, we consider the El-Jadida site to perform a detailed numerical modeling that aims to examine the reliability of the historical reports and thus to try removing the ambiguity on the historical information describing the impact of the 1755 tsunami waves in Morocco.

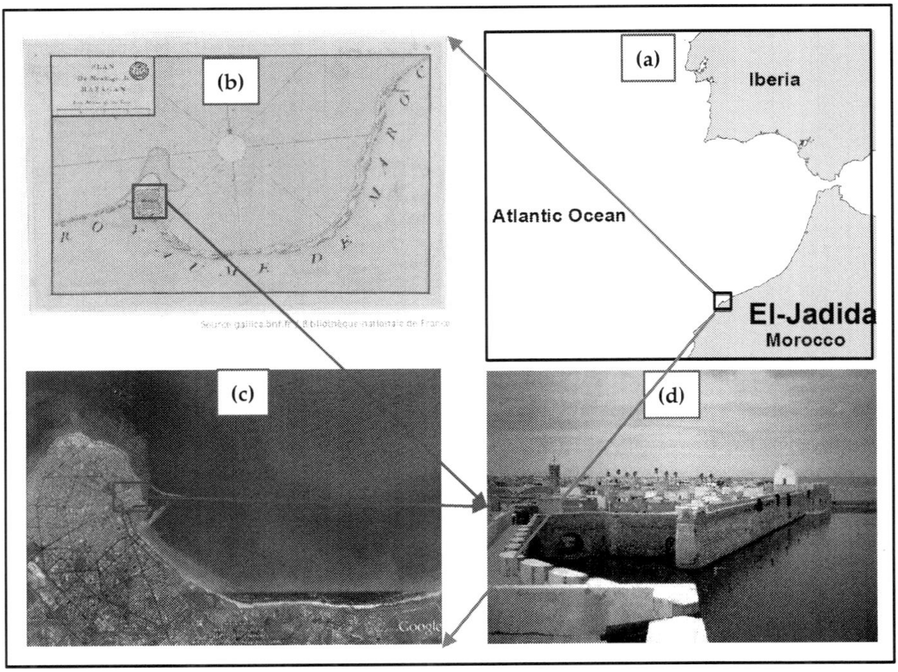

Figure 1. The study area. (a) Regional view of the Gulf of Cadiz zone; (b) The 1750s plan of the city of Mazagão (Source: Bibliotheque Nationale de France); (c) The present-day plan of El-Jadida site (Source: Google Earth 2011); (d) The wall surrounding the old medina (source: http://en.wikipedia.org/wiki/File:Mazagan).

This chapter seeks to clarify the uncertainties of the historical reports using the tsunami hydrodynamic modeling together with detailed paleo-bathymetric/topographic data and a 1755-like earthquake scenario. In order to achieve this, we use the digital terrain model reconstructed from the paleo-bathymetry/topography charts of the studied site before the 1750s, including also man-made constructions existing at that period; and we perform detailed numerical modeling of the tsunami wave characteristics offshore and onshore El-Jadida site. The earthquake scenario corresponding to a ~8.6 magnitude is considered to represent a 1755-like event. Results in terms of maximum wave heights distributions are presented for the candidate scenario at a regional scale, while high resolution inundation map, flood limits and near-shore wave heights are computed for the earthquake scenario at a local scale at El-Jadida. Reliability of historical reports is discussed in light of a comparison of these reports with the tsunami impact obtained from numerical modeling. Finally, we present a new reading of historical documents based on the consideration the historical context period in which the reports have been written. This reading indicates that interpretations of the historical reports may overestimate or underestimate the quoted values if they do not take into account the historical context.

2. Tectonic setting and 1755-like scenario

Morocco, by its peculiar geological context and proximity to the Nubia–Eurasia plate boundary (NEPB), is the western African littoral that is most exposed to earthquake-induced tsunamis (El Alami and Tinti, 1991). The most severe submarine earthquakes felt in Morocco were those generated offshore along the Atlantic coast (El-Mrabet, 1991). Some of these events were tsunamigenic, as was the case of the November, 1st, 1755 event.

The western segment of the NEPB extends from the Azores in the West to the Strait of Gibraltar in the East. Plate Kinematic models suggest a slow convergence velocity (less than 5mm/yr) in the area closer to the Srait of Gibraltar, which characterizes the motion between these two plates (Fernandes et al., 2007). The present-day tectonic regime along the western segment of the NEPB changes from transtension in the West, near the Azores triple junction, to transpression in the East, the Gulf of Cadiz area, with a strike-slip motion in its central segment (Tortella et al., 1997).

It is believed that the western segment of the NEPB, especially its eastern part, is the responsible for the generation of the historical tsunami events known to hit the coastal areas of the Gulf of Cadiz. However, most of these historical earthquakes were not instrumental events, as it is the case of the great Lisbon earthquake and tsunami of November, 1st, 1755.

The main problem encountered when investigating the November, 1st, 1755 tsunami concerns the identification of a single tectonic structure and its mechanism responsible for this event. The earthquake rupture mechanism as well as its location remains not well established in spite of the various attempts to identify this source. In this study, we use the earthquake scenario that produces the worst tsunami at El-Jadida. This scenario is identified from the study of Omira et al. (2009).

Omira et al. (2009) investigated the most credible earthquake scenarios in the region and presented a set of tsunamigenic scenarios based upon the concept of typical faults (Lorito et al., 2008). The tsunami radiation patterns for these scenarios (Fig. 2) clearly show that the

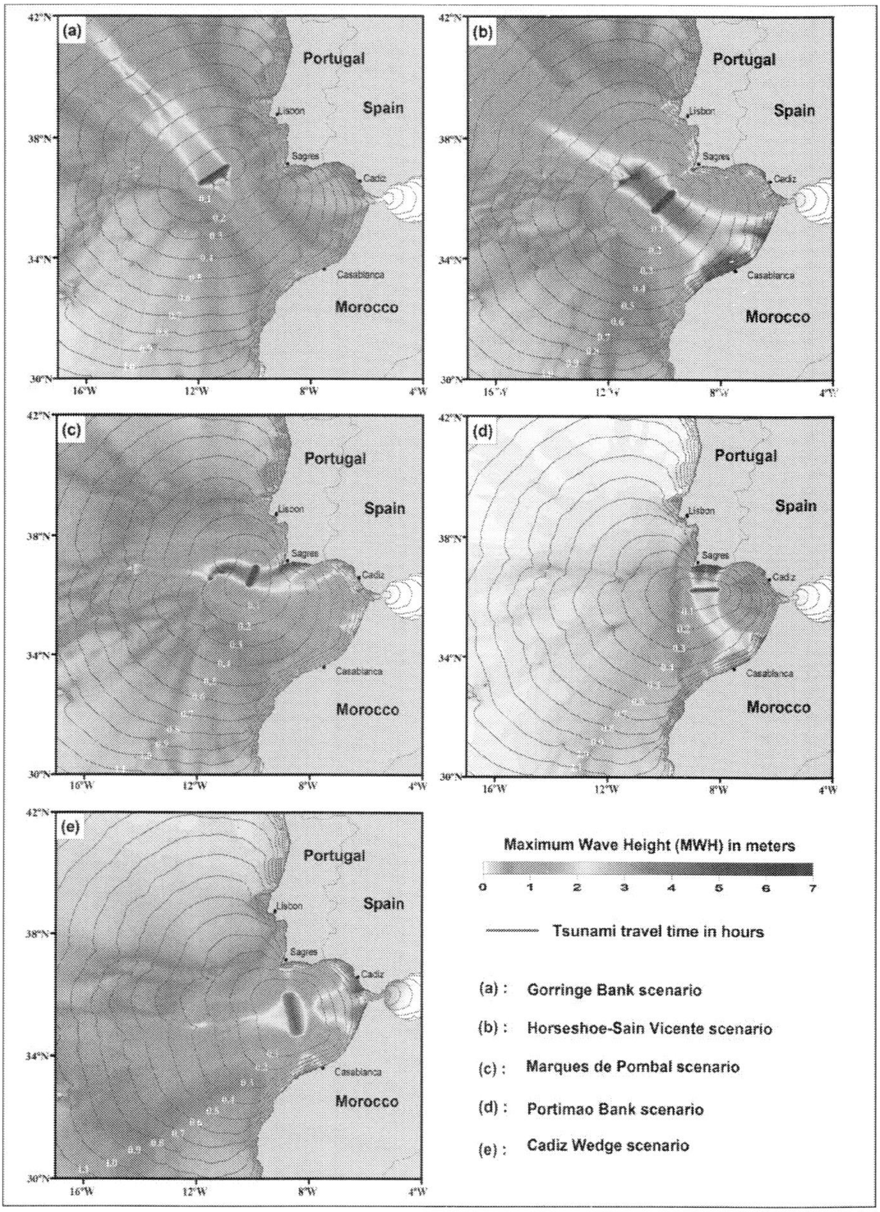

Figure 2. Computed tsunami radiation patterns considering the most credible earthquake scenarios in the Gulf of Cadiz region (in Omira et al., 2009).

Horseshoe fault (HSF) is the most effective in radiating energy towards the Atlantic coast of Morocco. The other scenarios are more effective in radiating energy towards the South of Portugal, as it is the case for the Marques de Pombal fault (MPF) and the Portimão Bank fault (PBF), or towards the Strait of Gilbratar for the Cadiz Wedge fault (CWF). Finally, the Gorringe Bank fault (GBF) remains an exception as most energy radiates towards North America due to the Gorringe Bank feature that prevents the amplification of tsunami energy towards the Gulf of Cadiz. However, the dimensions of the HSF do not account for the magnitude of the 1755 event.

In view of this, we adopt a composite Mw8.6 earthquake scenario of HSF and MPF to represent a 1755-like event. According to Ribeiro et al. (2006), the composite source of the HSF and the MPF is favored as a solution for the 1755 event due to their sub-parallel orientation as well as the almost geometric continuity between both faults that facilitates the strain/displacement transfer between them.

Fig. 3 illustrates the typical faults for the composite source considered in this study. For tsunami modeling requirement, the geometry of the candidate faults is simplified to a rectangle (cf. Fig. 3 for locations) for the computation of the seafloor initial deformation.

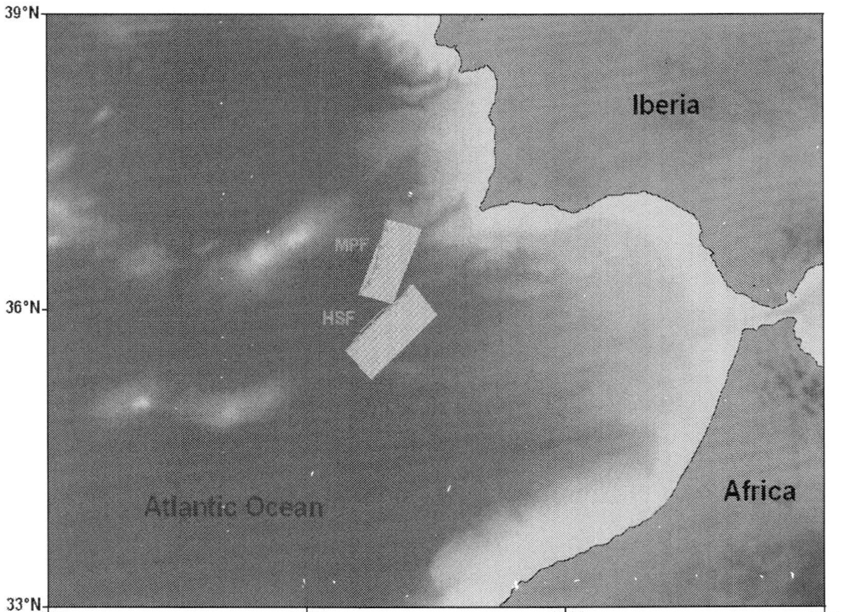

Figure 3. Tsunamigenic scenario (Mw8.6) considered in this study to represent a 1755-like tsunami event. It is a composite source of two thrust faults, namely the Marques de Pombal fault (MPF) and the Horseshoe Fault (HSF).

3. Tsunami modeling

Numerical modeling of tsunami from its generation to the impact is investigated in order to examine the reliability of historical reports against computed waves' amplitudes. To compute the generation of a tsunami triggered by a submarine earthquake we use the half-space elastic theory for the sea bottom deformation combined with the assumption that both deformations of sea bottom and ocean surface are equal. Both linear and non-linear shallow water equations are employed to simulate, respectively, the open-ocean and the near-shore waves' propagation. As the near-shore and inland tsunami propagations are sensitive to the near-shore bathymetry and the coastal topography, we build a set of nested bathymetric/topographic grids of increasing resolutions towards the shoreline to better present the morphology of the study area.

3.1. Preparation of digital terrain model (DTM)

A set of bathymetric/topographic grid layers, which covers the Gulf of Cadiz and the test site area with spatial increasing grid sizes of 320m, 80m, 20m and 5m, is nested for consecutive calculations of tsunami generation, propagation and inland inundation. The parent grid of 320 m resolution, encompassing the sources area offshore the Iberian Peninsula margin, extends from 31ºN to 40ºN and from 5ºW to 15ºW. It was generated from a compilation of multisource height/depth data that includes: i) the GEBCO one minute grid as a starting point, and ii) the SWIM compilation of bathymetric data performed in the region of the Gulf of Cadiz (Zitellini et al., 2009). Two intermediate grids of resolutions 80 m and 20 m are incorporated in the nested grid system for numerical stability requirements. The 80 m sub-grid covers the Atlantic coastal segment of Morocco from Casablanca at the North to Safi at the South, while the 20 m grid focuses on El-Jadida and the surrounding regions. For the site of interest a 5 m resolution grid is generated.

Due to the fact that this study aims to test the reliability of historical reports using numerical modeling, the reconstruction of a paleo-DTM is considered. This paleo-DTM is generated in order to properly represent the most significant coastal features, the shoreline and coastal infrastructures of Mazagão in the 1750s. It was computed from the paleo-bathymetric/topographic carts available before 1755. The finer grid of 5m-resolution incorporates the old medina of El-Jadida (Fig. 4a) where historical descriptions of the 1755 tsunami impact are reported.

The present-day DTM is also built (Fig. 4b) in order to highlight the morphological changes on the El-Jadida coastal area during the 250 prior years. Moreover, this DTM is used to compute inundation in order to show the influence of morphological changes on tsunami impact by comparing results for both paleo- and present-day- DTMs. Details of the generation of the El-Jadida present-day DTM can be found in Omira et al (2012, in press).

Small and large scale bathymetric charts were referenced to the mean sea level, merged on a unique database, and all data was transformed to WGS84/UTM coordinates (fuse 29).

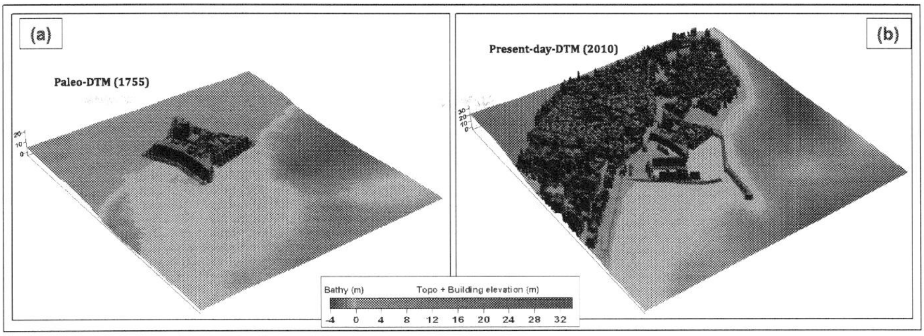

Figure 4. 5-m resolution digital terrain model (DTM) for the study area; (a) The paleo-DTM; (b) The present-day DTM.

3.2. Numerical model

The initial sea surface perturbation is generated for the considered composite submarine earthquake scenario MPF+HSF. The earthquake rupture is supposed to be instantaneous and the generated seabed displacement is computed using the half-space elastic theory (Okada, 1985). The vertical sea bottom displacement is then transferred to the ocean surface with the assumption that both deformations of sea bottom and ocean surface are equal (Kajiura, 1970).

Linear and non-linear approximations of shallow water equations (SWEs) are adopted to simulate the tsunami propagation. In the deep ocean the linear approximation of SWEs, which consists of neglecting both convective inertia force and the bottom friction terms, is valid since the waves travel with amplitudes much smaller than the water depths. While, when tsunami waves approach coastal regions and propagate into shallow water, the nonlinear convective inertia force and bottom friction effects become increasingly important, in such a case the linear approximation is no longer valid and the non-linear SWEs are employed to adequately describe the wave motion near-shore. The adapted version COMCOT-Lx (Omira et al., 2009; 2010; 2011) of the COMCOT code (Liu et al., 1998) is used to solve numerically both linear and non-linear SWEs. This code employs a dynamically coupled system of nested grids and solves SWEs using an explicit leap-frog finite differences numerical scheme.

Computing inundation consists of propagating the incident wave over dry land and evaluating the inland water depth and run-up. Thus, a specific numerical algorithm is needed to update the water depth along the shoreline grid cells at each time step of the computation. In this study, the moving boundary algorithm is adopted (Liu et al., 1995; Wang, 2009). This special treatment is designed to properly track shoreline movements and determine if the total water depth is high enough to flood the neighboring dry cells and hence if the shoreline should moved onshore or not. The inundation computational domain contains the dry "land" cells, the wet "water" cells and the interface between these two types of cells, which defines the shoreline. During the inundation process we suppose that both natural and man-made coastal infrastructure play the role of obstacles and no damage on these structures is modeled.

4. Results and discussions

4.1. Maximum wave heights and tsunami energy distribution

The maximum wave height distribution, which corresponds to the extraction of the maximum sea level perturbation at each grid point from the output tsunami propagation snapshots, is presented in Fig. 5 for the considered tsunami scenarios.

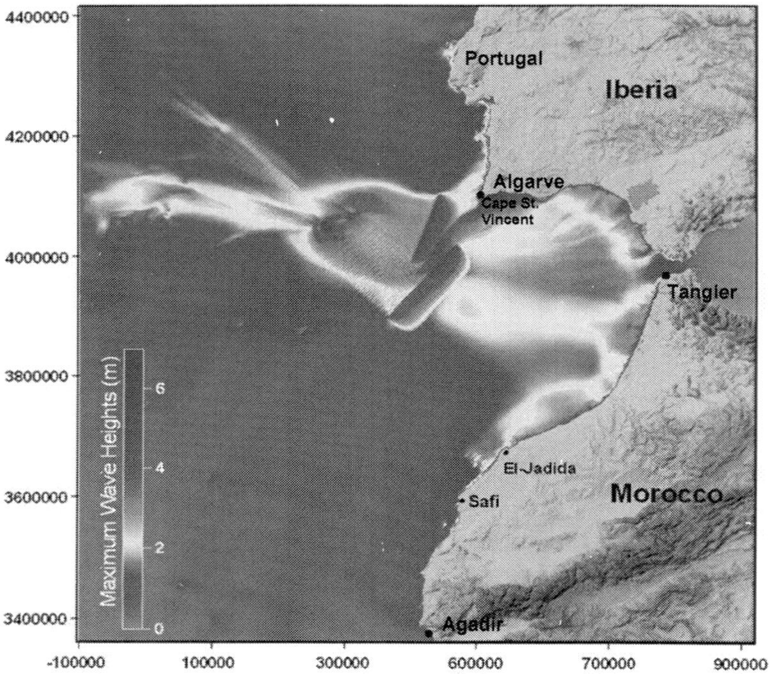

Figure 5. Maximum wave heights distribution in the Gulf of Cadiz region due to the occurrence of HSF+MPF earthquake scenario of Mw8.6.

Examination of this figure indicates that the considered tsunami scenario steered significant tsunami energy towards the Gulf of Cadiz including north-western coast of Morocco and south-western coasts of Iberian Peninsula. The composite HSF+MPF scenario generates significant wave amplitudes that cover almost all the Gulf of Cadiz coastal areas (Fig. 5). Along Moroccan coasts the wave heights range from ~1m to 6m and from 2m to 7m near-shore Algarve region, south of Portugal.

The fault strike and the bathymetry are the principal factors controlling the tsunami energy distribution. The shallower bathymetry SW of Cape St. Vincent acts as wave guide of tsunami energy from the MPF fault to the Algarve coast (Fig. 5). Whereas, the orientation and more southerly location of the HSF fault does not allow this to happen and most tsunami energy from this fault is steered towards the north-western coasts of Morocco (Fig. 5).

Simulation results in Fig. 5 show that the composite earthquake scenario triggers a tsunami that presents some degree of compatibly with both historical tsunami observations and paleo-tsunami studies. This agreement resides in the fact that historical and paleo-tsunami studies indicate, respectively, the coverage of all coasts of the Gulf of Cadiz by significant waves (Baptista et al., 1998) and the presence of tsunami traces and deposits in these coasts (Ruiz et al., 2005; Costa et al., 2011; Medina et al., 2011).

4.2. Tsunami impact

This section consists of evaluating the tsunami impact at a local scale considering the 5m-resolution DTM. Tsunami impact is presented through numerical computations of near-shore wave heights and overland tsunami inundation for both paleo- and present-day DTMs. Simulated wave heights and maximum flow depths are illustrated in Fig. 6 and Fig. 7, respectively.

In Fig. 6, the computed maximum wave amplitudes near-shore El-Jadida range from 2 to 6 m. These values of wave heights, relatively high, are the result of the shoaling effect that is important close to the coast due to the extension of a shallow platform.

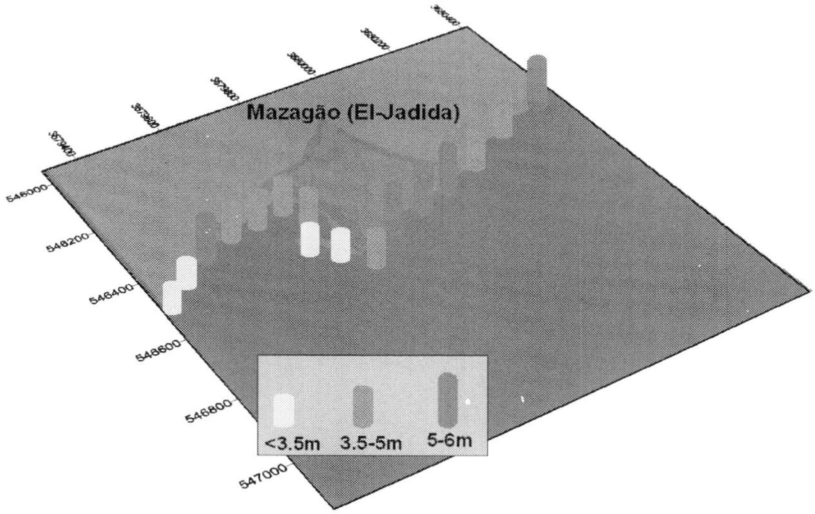

Figure 6. Predicted maximum wave heights along the coast of Mazagão (El-Jadida) for a Mw8.6 scenario using the paleo-digital terrain model.

Fig. 7 illustrates the computed inundation depths considering both the paleo-and the present-day DTMs. The analysis of these results indicates flow depths higher for the paleo-DTM (Fig. 7.a) than for the present-day-DTM (Fig. 7.b) in some areas. This difference in overland flow depths is especially due to the morphological changes that occurred at El-Jadida site from the 1750s to present. Results show also that only the areas surrounding the old medina are flooded as the 11m height wall prevents the entrance of the waves. The computed inland inundation distance reaches 0.7 km.

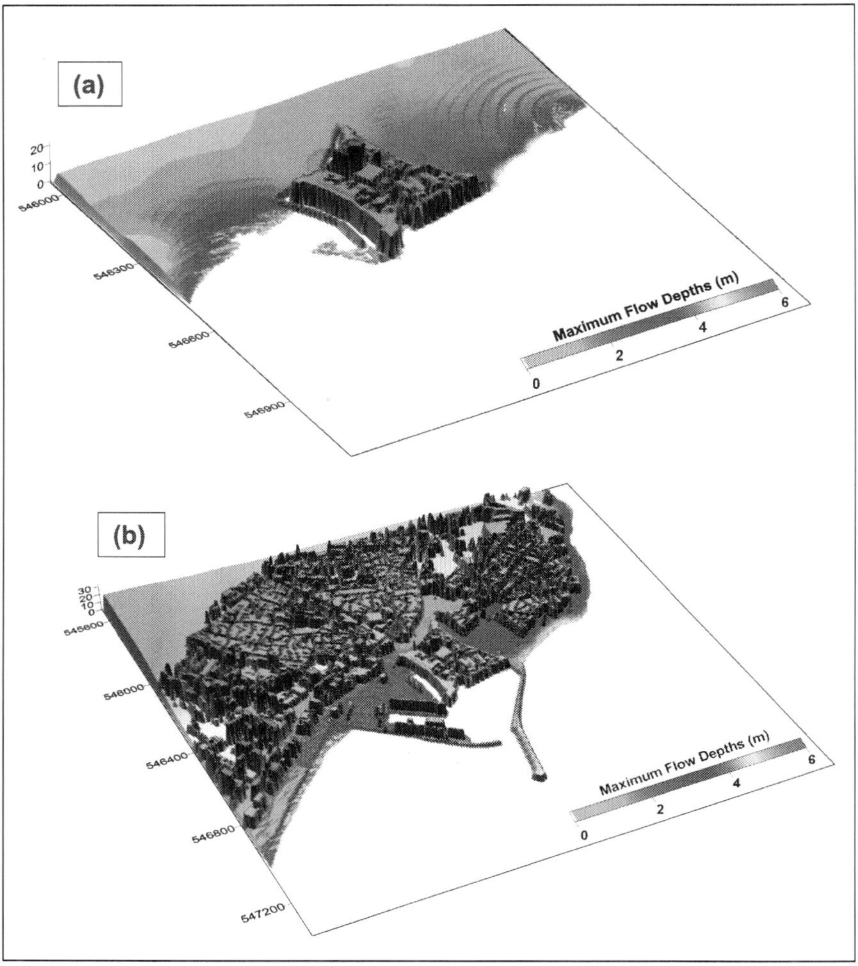

Figure 7. Computed maximum inland flow depths for El-Jadida site considering the HSF+MPF tsunami scenario of magnitude Mw8.6 and (a) Paleo-digital terrain model; (b) Present-day-digital terrain model.

4.3. Predicted values versus historical reports

According to Blanc (2009), reports dealing with the tsunami impact at El-Jadida represent the most detailed among those founded for other sites along the Atlantic coast of Morocco. The letter of Soyris (1755), describing the impact of the November 1st, 1755 tsunami at Mazagão, reports that: "...... the sea increased three times, of seventy five feet, so much that the Portuguese garrison in Mazagão had been compelled to abandon the City....." (Fig. 8). The mentioned wave value of about ~24m (75 feet) is in large disagreement with predicted numerical results indicating maximum wave heights that reach ~6 m in some areas near-shore El-Jadida. Moreover, the computed wave amplitudes at El-Jadida site are well above the estimates of Blanc (2009) established from a specific interpretation of the letter from Mazagão (El Jadida) published in the *Gazeta de Lisboa 1755* that suggests a wave height of ~ 1m above ground at the main gate of the Portuguese fortress (now the old medina). Blanc (2009) also concluded that a 2.5 m wave's amplitude could explain the damage of the 1755 tsunami observed along the Atlantic coasts of Morocco, from Tangier to Agadir (Fig. 5 for location).

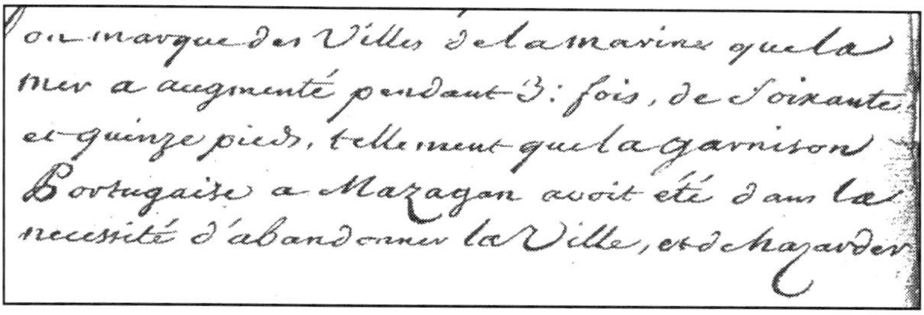
on marque des Villes de la marine que la
mer a augmenté pendant 3: fois, de Soixante
et quinze pieds, tellement que la garnison
Portugaise a Mazagan avoit été dans la
necessité d'abandonner la Ville, et de hazarder

Figure 8. A part of Soyris's letter (1755) that mentions the occurrence of "3 successive waves of 75 feet", about 24 m.

Further south of El-Jadida in the Atlantic coast of Morocco, at Safi town (see Fig. 5), historical documents indicate that the 1755 tsunami waves advanced overland and coming up as far as the great mosque (Gazette of Amsterdam of 26th December, 1756; in Kaabouben *et al.* (2009)). Kaabouben et al. (2009) carefully discussed the tsunami impact at the town of Safi from historical point of view and estimated the inland inundation distance at ~ 1.5 km. However, the analysis of tsunami energy patterns presented in the Fig. 5 shows that less energy is steered towards Safi than towards El-Jadida. This is more or less in agreement with historical documents quoting that the impact of tsunami was greater in the northern than the southern part of the Atlantic coasts of Morocco. With respect to the morphology of each site (El-Jadida and Safi) the inundation limit at El-Jadida reaches ~0.7km inland, which should be less at Safi where the inland inundation may be also overestimated in historical reports.

In order to examine the reliability of the quoted values in that mention wave amplitudes up to 24 m at El-Jadida, we compute maximum tsunami wave heights along the Atlantic coasts of Morocco for an "extreme" case scenario in the region. This scenario is a Mw9.0 corresponding to the composite earthquake source of HSF and MPF that we scaled to reach a 9.0 magnitude (increasing length, width and slip). Result of this simulation is illustrated in Fig. 9 and indicates a maximum wave height of ~ 9m near-shore the coastal areas of El-Jadida.

This analysis shows that, even for an "extreme" earthquake case (Mw9.0), the wave heights at El-jadida do not reach the 24m mentioned in historical documents. These wave heights and inundation distances are in disagreement with numerical simulations.

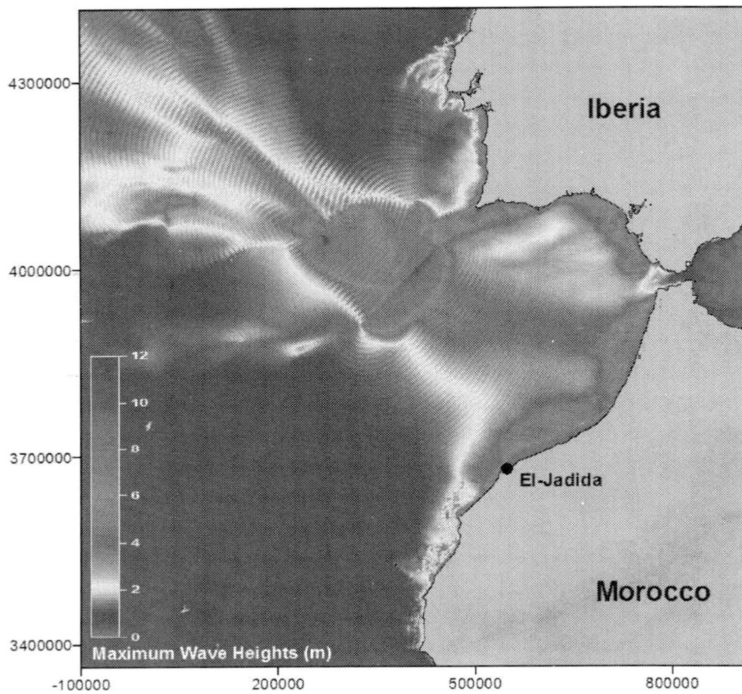

Figure 9. Maximum wave heights computations for the HSF+MPF earthquake scenario of Mw9.0.

4.4. New reading of historical reports: recover the historical data in their original context

Faced to inconsistency in historical information at a first reading, the reliability of the historical sources should be examined. The anachronism in the interpretation of ancient texts can lead to a misinterpretation or inadequate assessment of the value/reliability of some numerical data. Normally, the information presented in a letter is dependent on the system of values (implicit and explicit) shared by the author and reader. This information should not be taken without precautions out of the original context, nor interpreted as such throughout our system of values and/or our contemporary analysis schemes. By whom, for whom and for what purpose the information found in the archives have been collected and formatted? What is the function of this information in the political and religious contexts of the period? Is it possible (and if so how) to reinterpret historical information and translate it into our present-day scientific context? With what margin of error?

The reexamination of historical documents (in English, French, Spanish and Portuguese) for Mazagão site allows highlighting the historical context:

i. Most consulted historical documents, describing the impact in Morocco from the 1755 earthquake and tsunami event, are written by Portuguese Catholic religious to other Catholic religious and these reports were published after "all necessary approvals", ie approvals of the Church and the kingdom.
ii. The reported events concern a contact front between European Christian settlers and "moors" Muslims. Extending the Christian territory is a major issue for the Catholic Church and Extending Christian kingdoms.
iii. We are not in the XXth century in Portugal, Spain or France, where "objectivity" and "science" implicitly mean a break between the studied object and the observer (Meschinet de Richemond and Reghezza, 2010).

The contextual analysis of historical sources allows rejecting the numerical information reported by Soyris (1755) "3 successive waves of over 24m", which are inconsistent with all other qualitative and quantitative testimonies. The analysis of the entire letter of Soyris 1755 (and not only the part referring to tsunami and earthquake impact at Mazagão) can offer a different interpretation of the wave height of ~24m. This overestimation may have a function in the narrative of Soyris and the cultural context of the period. Only a huge wave can justify the fact that the Christian soldiers left their fortress to find shelters in Muslim lands ("...The Portuguese troops at Mazagão had found it necessary to abandon the city and risk their freedom by withdrawing in the mountains. Thanks to God, there is no other harm came to these Christians....") (Soyris 1755). This can also be considered as a sign of the active presence, but always difficult to interpret, of God in the world.

5. Conclusions

This study is addressed to clarify the recently discussed reliability of historical tsunami reports along the coasts of Morocco.

Results of tsunami simulations indicate that the composite tsunami scenario (MPF+HSF) presents some degree of compatibly with the historical tsunami observations along the Gulf of Cadiz coasts concerning the distribution of tsunami energy. This scenario generates large waves along the Atlantic coasts of Morocco, Portugal and Spain. This is in good agreement with tsunami historical reports and also with paleo-tsunami studies indicating, respectively, significant wave heights and the presence of tsunamites in some coastal locations of the region.

On the other hand, the computed maximum wave heights along El-Jadida reach 6m as maximum value for a Mw 8.6 scenario and show a large disagreement with the wave height values quoted in historical documents (~24.6m), when using a mean slip value of 10 meters. Even if we thrust some reports that indicate the overtopping of the walls in the old city this value should be discarded as the height of the wall is circa 11m.

Even though for an extreme earthquake case of Mw9.0 the numerical simulations cannot explain properly the quoted wave heights values at El-Jadida coast.

The contextual analysis is important when dealing with historical documents. It allows us, in this study, to highlight the fact that the quantitative testimony could have various interpretations depending on the context of its reading.

In view of a future tsunami event we believe that the wall around the old medina of El-Jadida will increase the protection against 6 meters tsunami wave (or less than 6m) in this specific part of the city. However, the present-day city extends far away from the old medina area being more exposed to tsunami flows due to the absence of effective sea-defense infrastructures.

Established results are useful for emergency planners and should be taken in consideration to trace tsunami evacuation maps.

Author details

R. Omira
Instituto Português do Mar e da Atmosfera, I. P., Lisbon, Portugal
Instituto Dom Luiz, University of Lisbon, CGUL, IDL, Lisbon, Portugal

M.A. Baptista
Instituto Dom Luiz, University of Lisbon, CGUL, IDL, Lisbon, Portugal
Instituto Superior de Engenharia de Lisboa,Portugal

S. Mellas
LGMSS URAC-45, University Chouaïb Doukkali, El Jadida, Morocco
UMR 220 GRED, Université Paul Valéry-Montpellier III et IRD, France

B. Zourarah
LGMSS URAC-45, University Chouaïb Doukkali, El Jadida, Morocco

F. Leone, N. Meschinet de Richemond and J-P. Cherel
UMR 220 GRED, Université Paul Valéry-Montpellier III et IRD, France

Acknowledgement

This work is funded by TRIDEC (Collaborative, Complex and Critical Decison-Support in Evolving Crises) FP7, EU project and by MAREMOTI (Mareograph and field tsunami observations, modeling and vulnerability studies for Northeast Atlantic and western Mediterranean) French Project. Authors wish to thank REMER (Réseau National des Sciences et Techniques de la Mer) of Morocco that supported this study through a student scholarship. Our gratitude is also addressed to the reviewers for taking time to review this chapter.

6. References

Baptista, M. A.; Heitor, S.; Miranda, J. M.; Miranda, P. & Mendes Victor, L. (1998). The 1755 Lisbon tsunami; evaluation of the tsunami parameters. *J. Geodynamics*, 25, 143-157.

Bergeron, A. & Bonnin, J. (1991). The deep structure of Gorringe Bank (NE Atlantic) and its surrounding area. *Geophys. J. Int.*, 105, 491-502.

Blanc, P.-L. (2009). Earthquakes and tsunami in November 1755 in Morocco: a different reading of contemporaneous documentary sources. *Nat. Hazards Earth Syst. Sci.*, 9, 725–738. http://www.nat-hazards-earth-syst-sci.net/9/725/2009/.

Chester, D. K. (2001). The 1755 Lisbon earthquake. *Prog. Phys. Geogr.* 25, 363–383.

Costa, P. J. M.; Andrade, C. ; Freitas, M. C.; Oliveira, M. A.; daSilva, C. M.; Omira, R.; Taborda, R.; Baptista, M. A. & Dawson A. G. (2011). Boulder deposition during major tsunami events. *Earth Surface Processes and Landforms*. doi:10.1002/esp.2228.

El Alami, S. O. & Tinti, S.(1991). A preliminary evaluation of the tsunami hazards in the Moroccan coasts. *Sc. of Tsunami Hazards*, 31–38.

El Mrabet, T. (1991). La sismicité historique du Maroc (en arabe), Thèse de 3 eme cycle, Faculté des lettres et des sciences humaines, Université Mohammed V. Rabat, 291 pp., (in Arab).

El Mrabet, T. (2005). Les grands séismes dans la région maghrébine, Thèse d'état, Faculté des lettres et des sciences humaines, Université Mohammed V. Rabat, 435 pp., (in Arab).

Fernandes, R.M.S.; Miranda, J.M.; Meijninger, B.M.L.; Bos, M.S.; Noomen, R.; Bastos, L.; Ambrosius, B.A.C. & Riva, R.E.M. (2007). Surface Velocity Field of the Ibero-Maghrebian Segment of the Eurasia-Nubia Plate Boundary. *Geophys. J. Int.*, 169, 1, 315-324.

Gazette Française d'Amsterdam (1776). Suite des Nouvelles d'Amesterdam du 26 Decembre 1755.

Gazeta de Lisboa 1755: Newspaper published in Lisbon, No47, Biblioteca Nacional de Lisboa, Portugal, 1755 (in Portuguese).

Huxham, J. (1756). Philos. Transaction of the Royal Society of London, vol. XLIX, part II, 668-670.

Kaabouben, F.; Baptista, M. A.; Iben Brahim, A.; El Mouraouah, A. & Toto, A. (2009). On the moroccan tsunami catalogue. *Nat. Hazards Earth Syst. Sci.*, 9, 1227-1236. doi:10.5194/nhess-9-1227-2009..

Kajiura, K. (1970). Tsunami source, energy and the directivity of wave radiation. *Bull. Earthquake Research Institute*, 48:835–869.

Kozak, J.T.; Moreira, V.S. & Oldroyd, D.R. (2005). Iconography of the 1755 Lisbon Earthquake. *Academy of Sciences of the Czech Republic, Prague. 82 pp.*

Liu, P L-F., Cho, Y-S., Brrigs, M. J., Kanoglo, U., Synolakis, C E. (1995) Runup of solitary waves on a circular island. *J fluid Mech.* 302: 259-285

Liu, PL-F.; Woo, S-B.; & Cho, Y-S. (1998). Computer programs for tsunami propagation and inundation. *Cornell University*, New York.

Lorito, S.; Tiberti, M. M.; Basili, R.; Piatanesi, A. & Valensise, G. (2008). Earthquake- 18 generated tsunamis in the Mediterranean Sea: Scenarios of potential threats to Southern 19 Italy. *J. Geophys. Res.*, 113, B01301. doi:10.1029/2007JB004943

Martinez-Solares, J.M.; Lopez, A. & Mezcua, J. (1979). Isoseismal map of the 1755 Lisbon earthquake obtained from Spanish data. *Tectonophysics*. 53: 301–313.

Martinez-Solares, J. M. & Lopez-Arroyo, A. (2004). The great historical 1755 earthquake: Effects and damage in Spain. *J. Seismol.*, 8, 275– 294.

Medina, F.; Mhammdi, N.; Chiguer, A.; Akil, M. & Jaaidi, E. B. (2011). The Rabat and Larache boulder fields; new examples of high-energy deposits related to storms and tsunami waves in north-western Morocco. *Natural Hazards,* 59: 725-747. doi: 10.1007/s11069-011-9792-x.

Meschinet de Richemond, N. & Reghezza, M. (2010). La gestion du risque en France: contre ou avec le territoire?. *Annales de Géographie*, 673 : 248-267.

Okada, Y. (1985). Surface deformation due to shear and tensile faults in a half-space. *Bull. Seismol. Soc. Am.*, 75(4): 1135– 1154.

Omira, R.; Baptista, M. A.; Matias, L.; Miranda, J. M.; Catita, C.; Carrilho, F. & Toto, E. (2009). Design of a Sea-level Tsunami Detection Network for the Gulf of Cadiz. *Nat. Hazards Earth Syst. Sci.*, 9, 1327-1338..

Omira, R.; Baptista, M. A.; Miranda, J. M.; Toto, E.; Catita, C.& Catalao, J. (2010). Tsunami vulnerability assessment of Casablanca-Morocco using numerical modelling and GIS tools. *Natural Hazards*, 54,75-95.

Omira, R.; Baptista, M. A. & Miranda, J. M. (2011). Evaluating tsunami impact on the Gulf of Cdaiz coast (Northeast Atlantic). *Pure Appl. Geophys.*, 168: 1033-1043. doi: 10.1007/s00024-010-0217-7.

Omira, R.; Baptista, M.A.; Leone, F.; Mellas, S.; Matias, L.; Miranda, J. M.; Zourarah. B.; Carrilho, F. & Cherel, J-P. (2012, in press). Performance of coastal sea-defense infrastructures in morocco against tsunami threat – Lessons learned from the Japanese March, 11, 2011 tsunami. *Accepted for publication in Natural Hazards.*

Pereira de Sousa, F. L. (1919). O terremoto do 1º de Novembro de 1755 em Portugal, *um estudo demografico, vol. I e II.* Serviços Geologicos de Portugal.

Ruiz, F.; Rodríguez-Ramírez, A.; Cáceres, L. M.; Vidal, J. R.; Carretero, M. I.; Abad, M.; Olías, M. & Pozo, M. (2005). Evidence of high-energy events in the geological record: Mid-holocene evolution of the southwestern Doñana National Park (SW Spain). *Palaeogeography, Palaeoclimatology, Palaeoecology*, 229: 212-229.

Ribeiro, A., Mendes-Victor, L., Cabral, J., Matias, L.,& Terhinha, P.(2006). The 1755 Lisbon earthquake and the beginning of closure of the Atlantic, European Review, 14, no.2,193-205,Cambridge University Press.

Ruffman, A. (2006). From an Ephemerides to 'Observation on The Changes of The Air': Documenting The far-field parameters of the November 1,1755 "Lisbon" Tsunami in the western Atlantic (Abstract). Atlantic Geoscience Society 32nd Colloquium and Annual Meeting, February 3–4, Greenwhich, Nova Scotia. Program with Abstracts, 63–64. *Atlantic Geology*, 42(1), 111.

Soyris, Mr. (1755). Extrait d'une lettre de Maroc en date du 5 novembre 1755: de Soyris a Guys, Archives Nationales de France, Marine B7/403, 1755.

Sylvanus, U. (1756) .The Gentleman's Magazine for December, printed by: D. Henry and R. Cave, St John's gate, 554–564.

Tortella, D.; Torne, M. & Pérez-Estaún, A. (1997). Geodynamic evolution of the eastern segment of the Azores-Gibraltar zone: the Gorringe Bank and the Gulf of Cadiz region. *Marine Geophys. Res.*, 19: 211-230.

Wang, X. (2009). COMCOT user manual-version 1.7. School of Civil and Environmental Engineering, Cornell University Ithaca, NY 14853, USA. http://ceeserver.cee.cornell.edu/pllgroup/doc/COMCOT_User_Manual_v1_7.pdf.

Zitellini, N.; Gràcia, E.; Matias, L.; Terrinha, P.; Abreu, M. A.; DeAlteriis, G.; Henriet, J. P.; Dañobeitia, J. J.; Masson, D. G.; Mulder, T.; Ramella, R.; Somoza, L. & Diez, S. (2009). The quest for the Africa-Eurasia plate boundary west of the Strait of Gibraltar. *Earth and Planetary Science Letters*, 280, 1-4, 15, 13-50.

Chapter 5

Generation and Propagation of Tsunami by a Moving Realistic Curvilinear Slide Shape with Variable Velocitie

Khaled T. Ramadan

Additional information is available at the end of the chapter

http://dx.doi.org/10.5772/50687

1. Introduction

Tsunamis are surface water waves caused by the impulsive perturbation of the sea. Apart from co-seismic sea bottom displacement due to earthquakes, sub-aerial and submarine landslides can also produce localized tsunamis with large and complex wave run-up especially along the coasts of narrow bays and fjords. In recent years, significant advances have been made in developing mathematical models to describe the entire process of generation, propagation and run-up of a tsunami event generated by seismic seafloor deformation [1-3].

The case of particular interest in this chapter is the mechanism of generation of tsunamis by submarine landslides. When a submarine landslide occurs, the ocean-bottom morphology may be significantly altered, in turn displacing the overlying water. Waves are then generated as water gets pulled down to fill the area vacated by the landslide and to a lesser extent, by the force of the sliding mass. Submarine slides can generate large tsunami, and usually result in more localized effects than tsunami caused by earthquakes [4]. Determination of volume, deceleration, velocity and rise time of the slide motion make modeling of tsunamis by submarine slides and slumps more complicated than simulation of seismic-generated tsunami.

Constant velocity implies that the slide starts and stops impulsively, i.e. the deceleration is infinite both initially and finally. Clearly, this is not true for real slides, and a more complex shape of the generated wave is expected [5].

In this chapter, we concern about the tsunami amplitudes predicted in the near-field caused by varying velocity of a two-dimensional realistic curvilinear slide model. The curvilinear tsunami source model we considered based on available geological, seismological, and tsunami elevation. The aim of this chapter is to determine how near-field tsunami amplitudes change according to variable velocities of submarine slide. We discuss the nature and the extent of variations in the peak tsunami waveforms caused by time variations of the frontal velocity and the deceleration for the two-dimensional curvilinear block slide model and compares the results with those for the slide moving with constant velocity. It will show how the changes in the slide velocity as function in time acts to reduce wave focusing. Numerical results are presented for the normalized peak amplitude as a function of the propagation length of the slump and the slide, the water depth, the time variation of moving velocity and the deceleration of the block slide. The problem is solved using linearized shallow-water theory for constant water depth by transform methods (Laplace in time and Fourier in space), with the forward and inverse Laplace transforms computed analytically, and the inverse Fourier transform computed numerically by the inverse Fast Fourier transform (IFFT).

2. Mathematical formulation of the problem

Consider a three dimensional fluid domain D as shown in Figure 1. It is supposed to represent the ocean above the fault area. It is bounded above by the free surface of the ocean $z = \eta(x,y,t)$ and below by the rigid ocean floor $z = -H(x,y) + \zeta(x,y,t)$, where $\eta(x,y,t)$ is the free surface elevation, $H(x,y)$ is the water depth and $\zeta(x,y,t)$ is the sea floor displacement function. The domain D is unbounded in the horizontal directions x and y, and can be written as$D = R^2 \times [(-H(x,y) + \zeta(x,y,t)), \eta(x,y,t)]$. For simplicity, $H(x,y)$ is assumed to be a constant. Before the earthquake, the fluid is assumed to be at rest, thus the free surface and the solid boundary are defined by $z = 0$ and $z = -H$, respectively. Mathematically, these conditions can be written in the form of initial conditions: $\eta(x,y,0) = \zeta(x,y,0) = 0$. At time $t > 0$ the bottom boundary moves in a prescribed manner which is given by $z = -H + \zeta(x,y,t)$. The resulting deformation of the free surface $z = \eta(x,y,t)$ is to be found as part of the solution. It is assumed that the fluid is incompressible and the flow is irrotational. The former implies the existence of a velocity potential $\phi(x,y,z,t)$ which fully describes the flow and the physical process. By definition of ϕ, the fluid velocity vector can be expressed as $\vec{u} = \nabla\phi$. Thus, the potential flow $\phi(x,y,z,t)$ must satisfy the Laplace's equation

$$\nabla^2\phi(x,y,z,t) = 0 \text{ where } (x,y,z) \in R^2 \times [-H,0], \tag{1}$$

subjected to the following linearized kinematic and dynamic boundary conditions on the free surface and the solid boundary, respectively

$$\phi_z = \eta_t \text{ on } z = 0, \tag{2}$$

$$\phi_z = \zeta_t \text{ on } z = -H, \tag{3}$$

$$\phi_t + g\eta = 0 \text{ on } z = 0, \tag{4}$$

with the initial conditions given by

$$\phi(x,y,z,0) = \eta(x,y,0) = \zeta(x,y,0) = 0. \quad (5)$$

So, the linearized shallow water solution can be obtained by the Fourier-Laplace transforms.

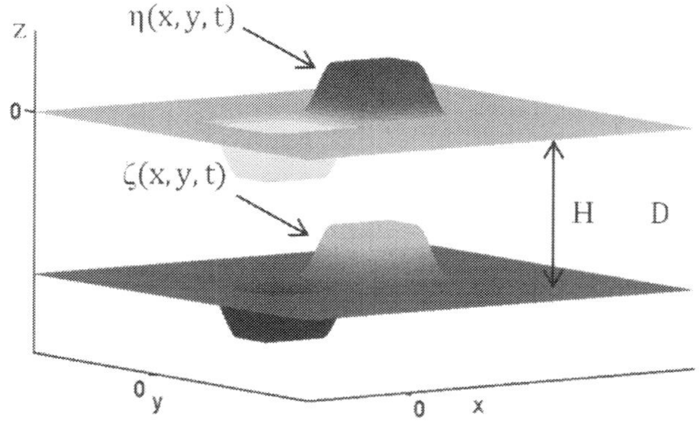

Figure 1. Fluid domain and coordinate system for a very rapid movement of the assumed source model.

2.1. Solution of the problem

Our interest is focused on the resulting uplift of the free surface elevation $\eta(x,y,t)$. An analytical analyses is to examine and illustrate the generation and propagation of a tsunami for a given bed profile $\zeta(x,y,t)$. Mathematical modeling of waves generated by vertical and lateral displacements of ocean bottom using the combined Fourier–Laplace transform of the Laplace equation analytically is the simplest way of studying tsunami development. Equations (1)–(4) can be solved by using the method of integral transforms. We apply the Fourier transform in (x, y)

$$\mathfrak{F}[f] = F(k_1, k_2) = \int_{R^2} f(x,y) e^{-i(xk_1 + yk_2)}\, dx\, dy,$$

with its inverse transform

$$\mathfrak{F}^{-1}[F] = f(x,y) = \frac{1}{(2\pi)^2} \int_{R^2} F(k_1, k_2)\, e^{i(xk_1 + yk_2)}\, dk_1\, dk_2,$$

and the Laplace transform in time t,

$$\pounds\,[\,g\,] = G(s) = \int_0^\infty g(t) e^{-st}\, dt$$

For the combined Fourier-Laplace transforms, the following notation is introduced:

$$\mathfrak{F}(\,\pounds\,(\,f(x,y,t)\,) = \bar{F}\,(k_1, k_2, s) = \int_{R^2} e^{-i(xk_1 + yk_2)} \left[\int_0^\infty f(x,y,t)\, e^{-st}\, dt\right] dx\, dy$$

Combining (2) and (4) yields the single free-surface condition

$$\phi_{tt}(x, y, 0, t) + g\phi_z(x, y, 0, t) = 0 \tag{6}$$

and the bottom condition (3) will be

$$\phi_z(x, y, -H, t) = \zeta_t(x, y, t) \tag{7}$$

The solution of the Laplace equation (1) which satisfies the boundary conditions (6) & (7) can be obtained by using the Fourier-Laplace transforms method.

First, by applying the transforms method to the Laplace equation (1), gives

$$\mathfrak{F}\left\{£\left(\frac{\partial^2\phi}{\partial x^2}\right)\right\} + \mathfrak{F}\left\{£\left(\frac{\partial^2\phi}{\partial y^2}\right)\right\} + \mathfrak{F}\left\{£\left(\frac{\partial^2\phi}{\partial z^2}\right)\right\} = 0 \tag{8}$$

By using the property $\left\{\frac{d^n f}{\partial x^n}\right\} = (ik)^n\,\bar{F}(k)$, Equation (8) will be

$$\bar{\phi}_{zz}(k_1, k_2, z, s) - \left(k_1^{\,2} + k_2^{\,2}\right)\bar{\phi}(k_1, k_2, z, s) = 0 \tag{9}$$

Second, by applying the transforms method to the boundary conditions (6) & (7) and the initial conditions (5), yields

$$s^2\bar{\phi}(k_1, k_2, 0, s) + g\,\bar{\phi}_z(k_1, k_2, 0, s) = 0, \tag{10}$$

and

$$\bar{\phi}_z(k_1, k_2, -h, s) = s\bar{\zeta}(k_1, k_2, s) \tag{11}$$

The transformed free-surface elevation can be obtained from (4) as

$$\bar{\eta}(k_1, k_2, s) = -\frac{s}{g}\,\bar{\phi}(k_1, k_2, 0, s) \tag{12}$$

The general solution of (9) will be

$$\bar{\phi}(k_1, k_2, z, s) = A(k_1, k_2, s)\cosh(kz) + B(k_1, k_2, s)\sinh(kz), \tag{13}$$

where $k = \sqrt{k_1^{\,2} + k_2^{\,2}}$ .The functions $A(k_1, k_2, s)$ and $B(k_1, k_2, s)$ can be found from the boundary conditions (10) & (11) as follows

For the bottom condition (at $z = -h$):

$$\frac{\partial\bar{\phi}(k_1, k_2, -h, s)}{\partial z} = -Ak\sinh(kh) + Bk\cosh(kh) \tag{14}$$

Substituting from (14) into (11), yields

$$-Ak\sinh(kh) + Bk\cosh(kh) = s\bar{\zeta}(k_1, k_2, s) \tag{15}$$

For the free surface condition (at $z = 0$):

$$\frac{\partial \bar{\phi}(k_1,k_2,0,s)}{\partial z} = Bk \text{ and } \bar{\phi}(k_1,k_2,0,s) = A \tag{16}$$

Substituting from (16) into (10), gives

$$A = \frac{-gk}{s^2} B \tag{17}$$

Using (17), Equation (15) can be written as

$$Bk\cosh(kh)\left[1 + \frac{gk}{s^2}\tanh(kh)\right] = s\bar{\zeta}(k_1,k_2,s). \tag{18}$$

From which,

$$A(k_1,k_2,s) = -\frac{g\,s\,\bar{\zeta}(k_1,k_2,s)}{\cosh(kh)[s^2+g\,k\tanh(kh)]},\ B(k_1,k_2,s) = \frac{s^3\,\bar{\zeta}(k_1,k_2,s)}{k\cosh(kh)[s^2+g\,k\tanh(kh)]}.$$

Substituting the expressions for the functions $A(k_1,k_2,s)$ and $B(k_1,k_2,s)$ in (13) yields,

$$\bar{\phi}(k_1,k_2,z,s) = -\frac{g\,s\,\bar{\zeta}(k_1,k_2,s)}{\cosh(kh)(s^2+\omega^2)}\left(\cosh(kz) - \frac{s^2}{g\,k}\sinh(kz)\right), \tag{19}$$

where $\omega = \sqrt{g\,k\tanh(kh)}$ is the circular frequency of the wave motion.

The free surface elevation $\bar{\eta}(k_1,k_2,s)$ can be obtained from (12) as

$$\bar{\eta}(k_1,k_2,s) = \frac{s^2\,\bar{\zeta}(k_1,k_2,s)}{\cosh(kh)(s^2+\omega^2)} \tag{20}$$

A solution for $\eta(x,y,t)$ can be evaluated for specified by computing approximately its transform $\bar{\zeta}(k_1,k_2,s)$ then substituting it into (20) and inverting $\bar{\eta}(k_1,k_2,s)$ to obtain $\eta(x,y,t)$. We concern to evaluate $\eta(x,y,t)$ by transforming analytically the assumed source model then inverting the Laplace transform of $\bar{\eta}(k_1,k_2,s)$ to obtain $\bar{\eta}(k_1,k_2,t)$ which is further converted to $\eta(x,y,t)$ by using double inverse Fourier Transform.

The circular frequency ω describes the dispersion relation of tsunamis and implies phase velocity $c_p = \frac{\omega}{k}$ and group velocity $c_g = \frac{d\omega}{dk}$. Hence, $c_p = \sqrt{\frac{g\tanh(kH)}{k}}$, and

$c_g = \frac{1}{2}c_p\left(1 + \frac{2kH}{\sinh(2kH)}\right)$. Since, $k = \frac{2\pi}{\lambda}$, hence as $kH \to 0$, both $c_p \to \sqrt{gH}$ and $c_g \to \sqrt{gH}$, which implies that the tsunami velocity $v_t = \sqrt{gH}$ for wavelengths λ long compared to the water depth H. The above linearized solution is known as the shallow water solution.

We considered three stages for the mechanism of the tsunami generation caused by submarine gravity mass flows, initiated by a rapid curvilinear down and uplift faulting with rise time $0 \le t \le t_1$, then propagating unilaterally in the positive $x-$ direction with time $t_1 \le t \le t^*$, to a length L both with finite velocity v to produce a depletion and an accumulation zones. The last stage represented by the time variation in the velocity of the accumulation slide (block slide) moving in the $x-$ direction with time $t^* \le t \le t_{max}$ and deceleration α, where t_{max} is the maximum time that the slide takes to stop with minimum deceleration α_{min}. In the $y-$direction, the models propagate instantaneously. The set of physical parameters used in the problem are given in Table 1.

PARAMETERS	FIRST STAGE	SECOND STAGE
Source width, W , km	100	100
Whole width in 1st Stage and Propagation length in 2nd Stage, km	$W' = 100$	$L = 150$
Water depth (uniform), H , km	2	2
Acceleration due to gravity, g, km/sec^2	0.0098	0.0098
Tsunami velocity, $v_t = \sqrt{gH}$, km/sec	0.14	0.14
Moving velocity, v, km/sec , to obtain maximum surface amplitude	0.14	0.14
Duration of the source process, t , min	$t_1 = \frac{50}{v} = 5.95$	$t^* = \frac{200}{v} = 23.8$

Table 1. Parameters used in the analytical solution of the problem.

The first and second stages of the bed motion are shown in Figure 2 and Figure 3, respectively, and given by:

a. First stage: Curvilinear down and uplift faulting for $0 \le t \le t_1$

where for $x \in [-50, 50]$

$$\zeta_{down}(x,y,t) = \begin{cases} -\zeta_0 \frac{vt}{2S}\left(1+\cos\frac{\pi}{50}x\right)\left[1-\cos\frac{\pi}{100}(y+150)\right], & -150 \le y \le -50, \\ -\zeta_0 \frac{vt}{S}\left(1+\cos\frac{\pi}{50}x\right) & , -50 \le y \le 50, \\ -\zeta_0 \frac{vt}{2S}\left(1+\cos\frac{\pi}{50}x\right)\left[1+\cos\frac{\pi}{100}(y-50)\right], & 50 \le y \le 150. \end{cases} \quad (21)$$

and for $x \in [200, 300]$

$$\zeta_{up}(x,y,t) = \begin{cases} \zeta_0 \frac{vt}{2S}\left(1-\cos\frac{\pi}{50}(x-200)\right)\left[1-\cos\frac{\pi}{100}(y+150)\right], & -150 \le y \le -50, \\ \zeta_0 \frac{vt}{S}\left(1-\cos\frac{\pi}{50}(x-200)\right) & , -50 \le y \le 50, \\ \zeta_0 \frac{vt}{2S}\left(1-\cos\frac{\pi}{50}(x-200)\right)\left[1+\cos\frac{\pi}{100}(y-50)\right], & 50 \le y \le 150. \end{cases} \quad (22)$$

For these displacements, the bed rises during $0 \le t \le t_1$ to a maximum displacement ζ_0 such that the volume of soil in the uplift increases linear with time and vise verse in the down faulting.

For $t \ge t_1$ the soil further propagates unilaterally in the positive x- direction with velocity v till it reaches the characteristic length L = 150 km at $t = t^* = 200/v$.

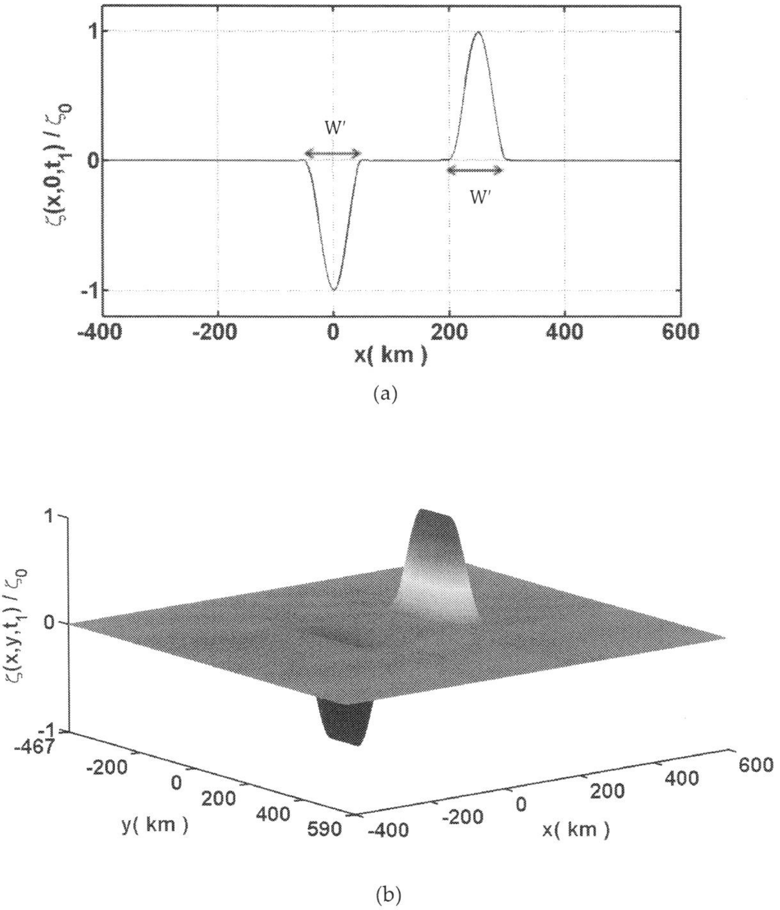

(a) Side view along the axis of symmetry at $y = 0$ (b) Three- dimensional view.

Figure 2. Normalized bed deformation represented by a rapid curvilinear down and uplift faulting at the end of stage one ($t_1 = \frac{50}{v}$)

b. Second stage: Curvilinear down and uphill slip-fault (slump and the slide) for $t_1 \le t \le t^*$

$$\zeta_{down}(x, y, t) = \zeta_{1down}(x, y, t) + \zeta_{2down}(x, y, t) + \zeta_{3down}(x, y, t), \tag{23}$$

where for $y \in [-150, -50]$

$$\zeta_{1down}(x,y,t) = \begin{cases} -\frac{\zeta_0}{4}\left(1+\cos\frac{\pi}{50}x\right)\left[1-\cos\frac{\pi}{100}(y+150)\right], & -50 \le x \le 0, \\ -\frac{\zeta_0}{2}\left[1-\cos\frac{\pi}{100}(y+150)\right], & 0 \le x \le (t-t_1)v, \\ -\frac{\zeta_0}{4}\left[1+\cos\frac{\pi}{50}\left(x-(t-t_1)v\right)\right]\left[1-\cos\frac{\pi}{100}(y+150)\right], & (t-t_1)v \le x \le (t-t_1)v+50, \end{cases}$$

and for $y \in [-50, 50]$

$$\zeta_{2down}(x,y,t) = \begin{cases} -\frac{\zeta_0}{2}\left(1+\cos\frac{\pi}{50}x\right), & -50 \le x \le 0, \\ -\zeta_0, \ 0 \le x \le (t-t_1)v, \\ -\frac{\zeta_0}{2}\left[1+\cos\frac{\pi}{50}\left(x-(t-t_1)v\right)\right], & (t-t_1)v \le x \le (t-t_1)v+50, \end{cases}$$

and for $y \in [50, 150]$

$$\zeta_{3down}(x,y,t) = \begin{cases} -\frac{\zeta_0}{4}\left(1+\cos\frac{\pi}{50}x\right)\left[1+\cos\frac{\pi}{100}(y-50)\right], & -50 \le x \le 0, \\ -\frac{\zeta_0}{2}\left[1+\cos\frac{\pi}{100}(y-50)\right], & 0 \le x \le (t-t_1)v, \\ -\frac{\zeta_0}{4}\left[1+\cos\frac{\pi}{50}(x-(t-t_1)v)\right]\left[1+\cos\frac{\pi}{100}(y-50)\right], & (t-t_1)v \le x \le (t-t_1)v+50, \end{cases}$$

$$\zeta_{up}(x,y,t) = \zeta_{1up}(x,y,t) + \zeta_{2up}(x,y,t) + \zeta_{3up}(x,y,t), \tag{24}$$

where for $y \in [-150, -50]$

$$\zeta_{1up}(x,y,t) = \begin{cases} \frac{\zeta_0}{4}\left(1-\cos\frac{\pi}{50}(x-200)\right)\left[1-\cos\frac{\pi}{100}(y+150)\right] \ 200 \le x \le 250, \\ \frac{\zeta_0}{2}\left[1-\cos\frac{\pi}{100}(y+150)\right], \ 250 \le x \le 250+(t-t_1)v, \\ \frac{\zeta_0}{4}\left[1+\cos\frac{\pi}{50}\left(x-\left(250+(t-t_1)v\right)\right)\right]\left[1-\cos\frac{\pi}{100}(y+150)\right] \ 250+(t-t_1)v \le x \le 300+(t-t_1)v, \end{cases}$$

and for $y \in [-50, 50]$

$$\zeta_{2up}(x,y,t) = \begin{cases} \frac{\zeta_0}{2}\left(1-\cos\frac{\pi}{50}(x-200)\right), & 200 \le x \le 250 \\ \zeta_0, \ 250 \le x \le 250+(t-t_1)v, \\ \frac{\zeta_0}{2}\left[1+\cos\frac{\pi}{50}\left(x-\left(250+(t-t_1)v\right)\right)\right], & 250+(t-t_1)v \le x \le 300+(t-t_1)v, \end{cases}$$

and for $y \in [50, 150]$

$\zeta_{3up}(x, y, t) =$

$$\begin{cases} \frac{\zeta_0}{4}\left(1-\cos\frac{\pi}{50}(x-200)\right)\left[1+\cos\frac{\pi}{100}(y-50)\right], & 200 \leq x \leq 250, \\ \frac{\zeta_0}{2}\left[1+\cos\frac{\pi}{100}(y-50)\right], & 250 \leq x \leq 250+(t-t_1)\,v, \\ -\frac{\zeta_0}{4}\left[1+\cos\frac{\pi}{50}(x-(250+(t-t_1)\,v))\right]\left[1+\cos\frac{\pi}{100}(y-50)\right], & 250+(t-t_1)\,v \leq x \leq 300+(t-t_1) \end{cases}$$

The kinematic realistic tsunami source model shown in Figure 3 is initiated by a rapid curvilinear down and uplift faulting (First stage) which then spreads unilaterally with constant velocity v causing a depletion and accumulation zone. The final down lift of the depression zone and final uplift of the accumulation zone are assumed to have the same amplitude ζ_0.We assume the spreading velocity v of the slump and the slide deformation in Figure 3 the same as the tsunami wave velocity $v_t = \sqrt{gH}$ as the largest amplification of the tsunami amplitude occurs when $v = v_t$ due to wave focusing . The slide and the slump are assumed to have constant width W.

The spreading is unilateral in the x-direction as shown in Figure 3. The vertical displacement,ζ_0, is negative (downwards) in zones of depletion, and positive (upwards) in zones of accumulation. All cases are characterized by sliding motion in one direction, without loss of generality coinciding with the x-axis, and tsunami propagating in the x-y plane.

Figure 4 shows vertical cross-sections (through $y = 0$) of the mathematical models of the stationary submarine slump and the moving slide and their schematic representation of the physical process that we considered in this study, as those evolve for time $t \geq t^*$. The block slide starts moving in the positive x-direction at time $t = t^*$ and stops moving at distance $L' = 150\text{ km}$ while the downhill slide becomes stationary. We discuss the tsunami generation for two cases of the movement of the block slide. First, the limiting case in which the block slide moves with constant velocity v and stops after distance L′ with infinite deceleration (sudden stop) at time $t_{min} = t^* + L'/v$.

Second, the general case in which the block slide moves in time $t_2 \geq t^*$ with constant velocity and then with constant deceleration such that it stop softly after traveling the same distance L′ in time t_3 which depends on the deceleration α and the choice of time t_2.

The velocity v(t) in this case can be defined as

$$v(t) = \begin{cases} v\, t^* \leq t \leq t_2 \\ v - (t - t_2)\alpha, t_2 \leq t \leq t_3 \end{cases}, \tag{25}$$

where $v = v_t = 0.14$ km/sec and α is the deceleration of the moving block slide. We need to determine the time t_3 that the slide takes to reach the final distance L′ and the corresponding deceleration α . This can be done by using the following steps:

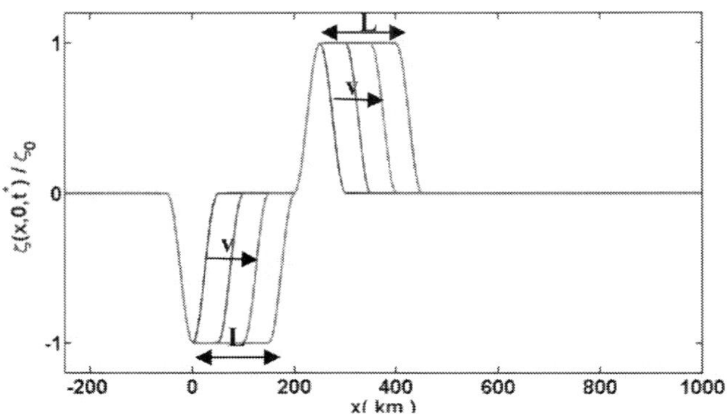

(a)

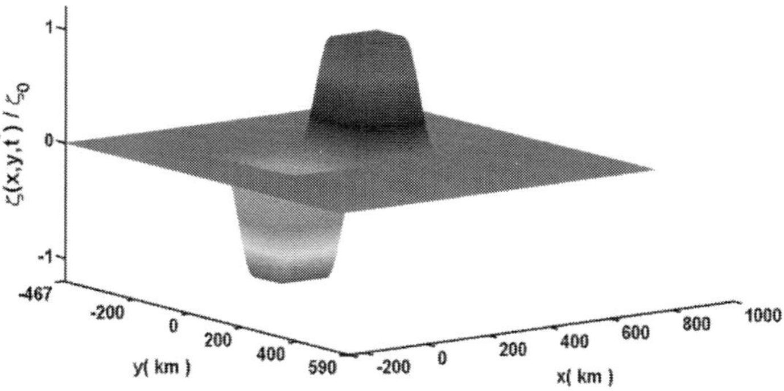

(b)

Figure 3. Normalized Bed deformation model represented by the accumulation and depletion zones at the end of stage two ($t^* = \frac{200}{v}$) (a) Side view along the axis of symmetry at $y = 0$ b) Three-dimensional view.

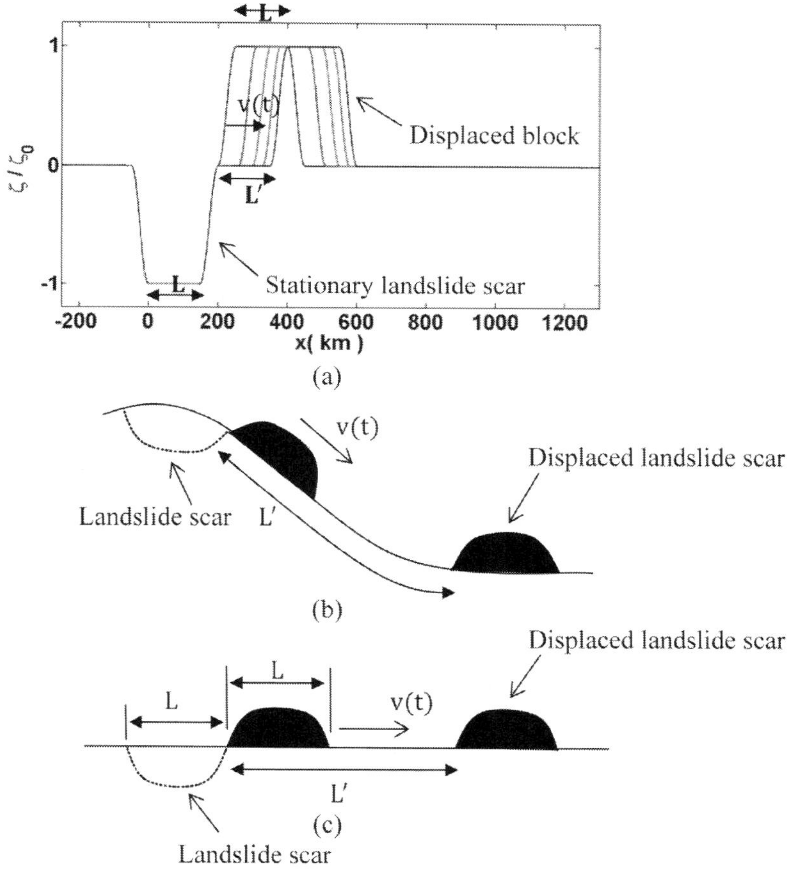

(a) Case 1: Mathematical model of the stationary slump and the moving submarine slide.

(b) Case 2: Physical process of the displaced block moving with a variable slide velocity v(t).

(c) Case 3: Schematic representation of the used model.

Figure 4. A schematic representation of a landslide (bottom) travelling a significant distance L downhill creating a "scar" and a moving uphill displaced block slide stopping at the characteristic length L′.

1. Choosing time t_2 as $t^* \leq t_2 \leq t^* + (L'/v)$ in which the slide moves with constant velocity v where $t^* = t_1 + L/v$ and $t_1 = 50/v$.
2. Getting the corresponding distance $L^* = (t_2 - t^*)v$.
3. Evaluating the remaining distance $L^{**} = L' - (t_2 - t^*)v$.

Substituting L^{**} in the equation

$$L^{**} = (t_3 - t_2)v - \frac{1}{2}\alpha (t_3 - t_2)^2 \tag{26}$$

When the block slide stops moving, then

$$v(t) = v - (t_3 - t_2)\alpha = 0 \tag{27}$$

Eliminating α from equations (7.12) and (7.13), we get relation between t_3 and t_2 which further substituting in equation (7.13), we obtain the deceleration α.

For $t_2 \leq t \leq t_3$, the block slide moves with velocity $v(t) = v - (t - t_2)\alpha$. Table 2 represents different values of t_2 and the corresponding calculated value of t_3 and α.

TIME $\mathbf{t_2}$ (MIN)	TIME $\mathbf{t_3}$ (MIN)	DECELERATION $\boldsymbol{\alpha}$ $\mathbf{(km/sec^2)}$
$t^* = 23.80$	59.51	6.53×10^{-5}
$t^* + 0.1(L'/v) = 25.58$	57.72	7.25×10^{-5}
$t^* + 0.2(L'/v) = 27.37$	55.95	8.16×10^{-5}
$t^* + 0.3(L'/v) = 29.37$	54.15	9.33×10^{-5}
$t^* + 0.4(L'/v) = 30.94$	52.36	1.08×10^{-4}
$t^* + 0.5(L'/v) = 32.72$	50.57	1.30×10^{-4}
$t^* + 0.6(L'/v) = 34.51$	48.79	1.63×10^{-4}
$t^* + 0.7(L'/v) = 36.30$	47.01	2.17×10^{-4}
$t^* + 0.8(L'/v) = 38.08$	45.22	3.26×10^{-4}
$t^* + 0.9(L'/v) = 39.87$	43.44	6.53×10^{-4}
$t^* + (L'/v) = 41.65$	41.65	Infinity

Table 2. Values of $\mathbf{t_2}$ and the corresponding calculated values of $\mathbf{t_3}$ and .

Figure 5 illustrates the position of the slides in the third stage for different choice of deceleration α. In this stage, α_{min} is the minimum deceleration required such that the slide stops after traveling distance L' . In this case $t_2 = t^*$ and $t_3 = t_{max} = t^* + (2L'/v) =$ 59.51 min. For any other $\alpha > \alpha_{min}$, the slide moves with constant velocity with time $t^* \leq t \leq t_2$ and with deceleration α until it stops at time t_3 which is less than t_{max}.

So, the stationary landslide scar for $t^* \leq t$ and the movable block slide with variable velocity $v(t)$ for $t^* \leq t_2 \leq t^* + (L'/v)$ and $t^* + (L'/v) \leq t_3 \leq t^* + (2L'/v)$ can be expressed respectively as

$$\zeta_{stat.landslide}(x, y, t^*) = \zeta_1(x, y, t^*) + \zeta_2(x, y, t^*) + \zeta_3(x, y, t^*), \tag{28}$$

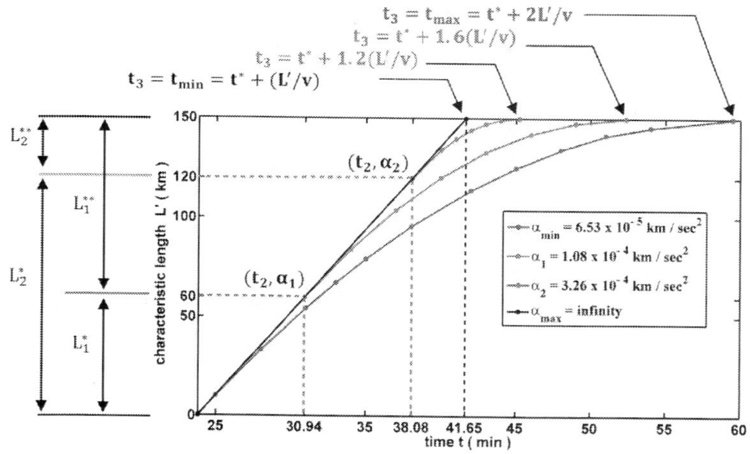

Figure 5. Slide block position against the instants of times $t^* \le t_2 \le t^* + (L'/v)$ and $t_{min} \le t_3 \le t_{max}$.

$$\zeta_{\text{block slide}}(x, y, t) = \zeta_1(x, y, t) + \zeta_2(x, y, t) + \zeta_3(x, y, t), \tag{29}$$

$\zeta_{\text{stat.landslide}}(x, y, t^*)$ is the same as (23) except the time parameter t will be substituted by t^*.

For $\zeta_{\text{block slide}}(x, y, t)$, let $S = \begin{cases} (t - t^*)v \text{ for } t^* \le t \le t_2, \\ (t - t^*)v - \frac{1}{2}\alpha (t - t_2)^2 \text{ for } t_2 \le t \le t_3, \end{cases}$ be

the distance the slide moves during stage three, hence

for $y \in [-150, -50]$

$$\zeta_1(x, y, t) = \begin{cases} \frac{\zeta_0}{4}\left(1 - \cos\frac{\pi}{50}(x-(200+S))\right)\left[1-\cos\frac{\pi}{100}(y+150)\right], & 200+S \le x \le 250+S, \\ \frac{\zeta_0}{2}\left[1-\cos\frac{\pi}{100}(y+150)\right], & 250+S \le x \le 250+S+L, \\ \frac{\zeta_0}{4}\left[1+\cos\frac{\pi}{50}\left(x-(250+S+L)\right)\right]\left[1-\cos\frac{\pi}{100}(y+150)\right], & 250+S+L \le x \le 300+S+L, \end{cases}$$

and for $y \in [-50, 50]$

$$\zeta_2(x, y, t) = \begin{cases} \frac{\zeta_0}{2}\left(1 - \cos\frac{\pi}{50}(x-(200+S))\right), & 200+S \le x \le 250+S, \\ \zeta_0, & 250+S \le x \le 250+S+L, \\ \frac{\zeta_0}{2}\left[1+\cos\frac{\pi}{50}\left(x-(250+S+L)\right)\right], & 250+S+L \le x \le 300+S+L, \end{cases}$$

and for $y \in [50, 150]$

$$\zeta_3(x,y,t_2,t_3) = \begin{cases} \frac{\zeta_0}{4}\left(1-\cos\frac{\pi}{50}(x-(200+S))\right)\left[1+\cos\frac{\pi}{100}\left(y-50\right)\right], & 200+S \le x \le 250+S, \\ \frac{\zeta_0}{2}\left[1+\cos\frac{\pi}{100}\left(y-50\right)\right], & 250+S \le x \le 250+S+L, \\ \frac{\zeta_0}{4}\left[1+\cos\frac{\pi}{50}\left(x-(250+S+L)\right)\right]\left[1+\cos\frac{\pi}{100}\left(y-50\right)\right], & 250+S+L \le x \le 300+S+L, \end{cases}$$

Laplace and Fourier transforms can now applied to the bed motion described by Equations (21)-(24) and Equations (28) & (29). First, beginning with the curvilinear down and uplift faulting (21) and (22) for $0 \le t \le t_1$ where $t_1 = \frac{50}{v}$, and

$$\mathfrak{F}(\pounds(\zeta(x,y,t))=\bar{\zeta}(k_1,k_2,s)=\int_{-\infty}^{\infty} e^{-i(xk_1+yk_2)}\left[\int_0^{\infty}\zeta(x,y,t)\,e^{-st}\,dt\right]dx\,dy. \quad (30)$$

The limits of the above integration are apparent from Equations (21) & (22) and are done as follows:

$$\zeta_{down}(k_1,k_2,s) =$$

$$\int_{-150}^{-50}\frac{1}{2}\left[1-\cos\frac{\pi}{100}(y+150)\right]e^{-ik_2y}dy\left[\int_{-50}^{50}\left(1+\cos\frac{\pi}{50}x\right)e^{-ik_1x}\,dx\int_0^{\infty}-\zeta_0\frac{vt}{2S}e^{-st}\,dt\right]$$
$$+\int_{-50}^{50}e^{-ik_2y}\,dy\left[\int_{-50}^{50}\left(1+\cos\frac{\pi}{50}x\right)e^{-ik_1x}\,dx\int_0^{\infty}-\zeta_0\frac{vt}{S}e^{-st}\,dt\right]$$
$$+\int_{50}^{150}\frac{1}{2}\left[1+\cos\frac{\pi}{100}(y-50)\right]e^{-ik_2y}dy\left[\int_{-50}^{50}\left(1+\cos\frac{\pi}{50}x\right)e^{-ik_1x}\,dx\int_0^{\infty}-\zeta_0\frac{vt}{2S}e^{-st}\,dt\right],$$

$$\zeta_{up}(k_1,k_2,s) =$$

$$\int_{-150}^{-50}\frac{1}{2}\left[1-\cos\frac{\pi}{100}(y+150)\right]e^{-ik_2y}dy\left[\int_{200}^{300}\left(1-\cos\frac{\pi}{50}(x-200)\right)e^{-ik_1x}\,dx\int_0^{\infty}\zeta_0\frac{vt}{2S}e^{-st}\,dt\right]$$
$$+\int_{-50}^{50}e^{-ik_2y}\,dy\left[\int_{200}^{300}\left(1-\cos\frac{\pi}{50}(x-200)\right)e^{-ik_1x}\,dx\int_0^{\infty}\zeta_0\frac{vt}{S}e^{-st}\,dt\right]$$
$$+\int_{50}^{150}\frac{1}{2}\left[1+\cos\frac{\pi}{100}(y-50)\right]e^{-ik_2y}dy\left[\int_{-50}^{50}\left(1-\cos\frac{\pi}{50}(x-200)\right)e^{-ik_1x}\,dx\int_0^{\infty}\zeta_0\frac{vt}{2S}e^{-st}\,dt\right],$$

Substituting the results of the integration for ζ_{down} and ζ_{up} into (20), yields

$$\bar{\eta}(k_1,k_2,s) =$$

$$\frac{1}{\cosh(kH)(s^2+\omega^2)}\zeta_0\frac{v}{L}\frac{1}{2}\left[\begin{array}{l}\left[\frac{(e^{-i200k_1}-e^{-i300k_1})}{ik_1}-\frac{1}{1-\left(\frac{50}{\pi}k_1\right)^2}\left[ik_1\left(\frac{50}{\pi}\right)^2\left(e^{-i300k_1}-e^{i200k_1}\right)\right]\right]- \\ \left[\frac{(e^{i50k_1}-e^{-i50k_1})}{ik_1}+\frac{1}{1-\left(\frac{50}{\pi}k_1\right)^2}\left[ik_1\left(\frac{50}{\pi}\right)^2\left(e^{-i50k_1}-e^{i50k_1}\right)\right]\right]\end{array}\right]\times$$

$$\left[\begin{array}{c}\frac{(e^{i150k_2}-e^{i50k_2})}{ik_2}-\frac{1}{1-\left(\frac{100}{\pi}k_2\right)^2}\left[ik_2\left(\frac{100}{\pi}\right)^2(e^{i50k_2}+e^{i150k_2})\right]+\frac{4\sin(50k_2)}{k_2}+\\ \left[\frac{(e^{-i50k_2}-e^{-i150k_2})}{ik_2}+\frac{1}{1-\left(\frac{100}{\pi}k_2\right)^2}\left[ik_2\left(\frac{100}{\pi}\right)^2(e^{-i150k_2}+e^{-i50k_2})\right]\right]\end{array}\right]. \quad (31)$$

The free surface elevation $\bar{\eta}(k_1,k_2,t)$ can be evaluated by using the inverse Laplace transforms of $\bar{\eta}(k_1,k_2,s)$ given by Equation (31) as follows:

First, recall that $\pounds^{-1}\left\{\frac{1}{s^2+\omega^2}\right\}=\frac{\sin\omega t}{\omega}$ hence $\bar{\eta}(k_1,k_2,t)$ becomes

$\bar{\eta}(k_1,k_2,t)=$

$$\frac{\sin\omega t}{\omega\cosh(kH)}\frac{\zeta_0 v}{2L}\left[\begin{array}{c}\left[\frac{(e^{-i200k_1}-e^{-i300k_1})}{ik_1}-\frac{1}{1-\left(\frac{50}{\pi}k_1\right)^2}\left[ik_1\left(\frac{50}{\pi}\right)^2(e^{-i300k_1}-e^{i200k_1})\right]\right]-\\ \left[\frac{(e^{i50k_1}-e^{-i50k_1})}{ik_1}+\frac{1}{1-\left(\frac{50}{\pi}k_1\right)^2}\left[ik_1\left(\frac{50}{\pi}\right)^2(e^{-i50k_1}-e^{i50k_1})\right]\right]\end{array}\right]$$

$$\times\left[\begin{array}{c}\frac{(e^{i150k_2}-e^{i50k_2})}{ik_2}-\frac{1}{1-\left(\frac{100}{\pi}k_2\right)^2}\left[ik_2\left(\frac{100}{\pi}\right)^2(e^{i50k_2}+e^{i150k_2})\right]+\frac{4\sin(50k_2)}{k_2}+\\ \left[\frac{(e^{-i50k_2}-e^{-i150k_2})}{ik_2}+\frac{1}{1-\left(\frac{100}{\pi}k_2\right)^2}\left[ik_2\left(\frac{100}{\pi}\right)^2(e^{-i150k_2}+e^{-i50k_2})\right]\right]\end{array}\right]. \quad (32)$$

In case for $t\geq t_1$, $\bar{\eta}(k_1,k_2,t)$ will have the same expression except in the convolution step, the integral

become $\int_{t-t_1}^{t}\cos\omega\tau\,d\tau=\frac{\sin\omega t}{\omega}-\frac{\sin\omega(t-t_1)}{\omega}$ instead of $\int_0^t\cos\omega\tau\,d\tau=\frac{\sin\omega t}{\omega}$.

Finally, $\eta(x,y,t)$ is evaluated using the double inverse Fourier transform of $\bar{\eta}(k_1,k_2,t)$

$$\eta(x,y,t)=\frac{1}{(2\pi)^2}\int_{-\infty}^{\infty}e^{ik_2y}\left[\int_{-\infty}^{\infty}e^{ik_1x}\,\bar{\eta}(k_1,k_2,t)\,dk_1\right]dk_2\,. \quad (33)$$

This inversion is computed by using the FFT. The inverse FFT is a fast algorithm for efficient implementation of the Inverse Discrete Fourier Transform (IDFT) given by

$$f(m,n)=\frac{1}{MN}\sum_{p=0}^{M-1}\sum_{q=0}^{N-1}F(p,q)e^{i\left(\frac{2\pi}{M}\right)pm}\,e^{i\left(\frac{2\pi}{N}\right)qn}\qquad p=0,1,\ldots\ldots,M-1,\ q=0,1,\ldots\ldots,N-1,$$

where $f(m,n)$ is the resulted function of the two spatial variables m and n ,corresponding x and y, from the frequency domain function $F(p,q)$ with frequency variables p and q, corresponding k_1 and k_2. This inversion is done efficiently by using the Matlab FFT algorithm.

In order to implement the algorithm efficiently, singularities should be removed by finite limits as follows:

1. As $k \to 0$, implies $k_1 \to 0$, $k_2 \to 0$ and $\omega \to 0$ then $\bar{\eta}(k_1, k_2, t)$ has the limit

$$\lim_{k \to 0} \bar{\eta}(k_1, k_2, t) = 0 \text{ for } t \le t_1, \text{ where } t_1 = \frac{50}{v}.$$

2. As $k_1 \to 0$, then the singular term of $\bar{\eta}(k_1, k_2, t)$ has the following limits

$$\lim_{k_1 \to 0} \left(\frac{e^{-i\,200\,k_1} - e^{-i\,300\,k_1}}{ik_1} \right) = 100 \text{ and } \lim_{k_1 \to 0} \left(\frac{e^{i\,50\,k_1} - e^{-i\,50\,k_1}}{ik_1} \right) = 100.$$

3. As $k_2 \to 0$, then the singular terms of $\bar{\eta}(k_1, k_2, t)$ have the following limits

$$\lim_{k_2 \to 0} \left(\frac{e^{-i\,150\,k_2}}{i\,k_2} - \frac{e^{i\,50\,k_2}}{i\,k_2} \right) = 100, \lim_{k_2 \to 0} \left(\frac{4\sin(50k_2)}{k_2} \right) = 200,$$

$$\text{and } \lim_{k_2 \to 0} \left(\frac{e^{-i\,50\,k_2}}{i\,k_2} - \frac{e^{-i\,150\,k_2}}{i\,k_2} \right) = 100.$$

Using the same steps, $\bar{\eta}(k_1, k_2, t)$ is evaluated by applying the Laplace and Fourier transforms to the bed motion described by (23) and (24), then substituting into (20) and then inverting $\bar{\eta}(k_1, k_2, s)$ using the inverse Laplace transform to obtain $\bar{\eta}(k_1, k_2, t)$. This is verified for $t_1 \le t \le t^*$ where $t^* = \frac{200}{v}$ as follows:

$$\bar{\eta}(k_1, k_2, t) = \bar{\eta}_{down}(k_1, k_2, t) + \bar{\eta}_{up}(k_1, k_2, t), \tag{34}$$

where

$$\bar{\eta}_{down}(k_1, k_2, t) = \bar{\eta}_{1down}(k_1, k_2, t) + \bar{\eta}_{2down}(k_1, k_2, t) + \bar{\eta}_{3down}(k_1, k_2, t),$$

hence,

$$\begin{aligned}
\bar{\eta}_{down}(k_1, k_2, t) = {} & \Bigg[\left[\frac{-\zeta_0}{4\cosh(kH)}\right] \left[\frac{e^{i\,150\,k_2} - e^{i\,50\,k_2}}{i\,k_2} - \frac{1}{1-\left(\frac{100}{\pi}k_2\right)^2} \left[ik_2 \left(\frac{100}{\pi}\right)^2 \left(e^{i50k_2} + e^{i150k_2}\right) \right] \right] + \\
& \left[\frac{-\zeta_0}{\cosh(kh)}\right] \left[\frac{\sin(50k_2)}{k_2}\right] + \\
& \left[\frac{-\zeta_0}{4\cosh(kH)}\right] \left[\frac{e^{-i\,50\,k_2} - e^{-i\,150\,k_2}}{i\,k_2} + \frac{1}{1-\left(\frac{100}{\pi}k_2\right)^2} \left[ik_2 \left(\frac{100}{\pi}\right)^2 \left(e^{-i150k_2} + e^{-i50k_2}\right) \right] \right] \Bigg] \times \\
& \Bigg[\left[\frac{e^{i\,50\,k_1} - 1}{ik_1} + \frac{e^{i\,50\,k_1}}{1-\left(\frac{50}{\pi}k_1\right)^2} \left[ik_1 \left(\frac{50}{\pi}\right)^2 \left(1 - e^{-i50k_1}\right) \right] \right] \cos\omega(t-t_1) + \\
& \frac{2v}{\omega^2 - (k_1 v)^2} \left(\omega\sin\omega(t-t_1) + ik_1 v\cos\omega(t-t_1) - ik_1 v e^{-ik_1(t-t_1)v} \right) + \\
& \left[\frac{e^{-i\,k_1((t-t_1)v)} - e^{-i\,k_1(50+(t-t_1)v)}}{i\,k_1} + \frac{1}{1-\left(\frac{50}{\pi}k_1\right)^2} \left[ik_1 \left(\frac{50}{\pi}\right)^2 \left(e^{-i\,k_1(50+(t-t_1)v)} + e^{-ik_1(t-t_1)v}\right) \right] \right] \cos\omega(t-t_1) \Bigg],
\end{aligned}$$

and $\bar{\eta}_{up}(k_1,k_2,t) = \bar{\eta}_{1up}(k_1,k_2,t) + \bar{\eta}_{2up}(k_1,k_2,t) + \bar{\eta}_{3up}(k_1,k_2,t)$, then

$$\bar{\eta}_{up}(k_1,k_2,t) =$$

$$\left[\begin{array}{c} \left[\frac{\zeta_0}{4\cosh(kH)}\right]\left[\frac{e^{i150k_2}-e^{i50k_2}}{ik_2} - \frac{1}{1-\left(\frac{100}{\pi}k_2\right)^2}\left[ik_2\left(\frac{100}{\pi}\right)^2\left(e^{i50k_2}+e^{i150k_2}\right)\right]\right]+ \\ \left[\frac{\zeta_0}{\cosh(kh)}\right]\left[\frac{\sin(50k_2)}{k_2}\right]+ \\ \left[\frac{\zeta_0}{4\cosh(kH)}\right]\left[\frac{e^{-i50k_2}-e^{-i150k_2}}{ik_2} + \frac{1}{1-\left(\frac{100}{\pi}k_2\right)^2}\left[ik_2\left(\frac{100}{\pi}\right)^2\left(e^{-i150k_2}+e^{-i50k_2}\right)\right]\right] \end{array}\right]\times$$

$$\left[\begin{array}{c} \left[\frac{e^{-i200k_1}-e^{-i250k_1}}{ik_1} + \frac{1}{1-\left(\frac{50}{\pi}k_1\right)^2}\left[-ik_1\left(\frac{50}{\pi}\right)^2\left(e^{-i250k_1}+e^{-i200k_1}\right)\right]\right]\cos\omega(t-t_1)+ \\ \frac{2ve^{-i250k_1}}{\omega^2-(k_1v)^2}\left(\omega\sin\omega(t-t_1)+ik_1v\cos\omega(t-t_1)-ik_1ve^{-ik_1(t-t_1)v}\right)+ \\ \left[\begin{array}{c} \frac{e^{-ik_1(250+(t-t_1)v)}-e^{-ik_1(300+(t-t_1)v)}}{ik_1}+ \\ \frac{1}{1-\left(\frac{50}{\pi}k_1\right)^2}\left[ik_1\left(\frac{50}{\pi}\right)^2\left(e^{-ik_1(300+(t-t_1)v)}+e^{-ik_1(250+(t-t_1)v)}\right)\right] \end{array}\right]\cos\omega(t-t_1) \end{array}\right].$$

Substituting $\bar{\eta}_{down}(k_1,k_2,t)$ and $\bar{\eta}_{up}(k_1,k_2,t)$ into (34) gives $\bar{\eta}(k_1,k_2,t)$ for $t_1 \le t \le t^*$. For the case $t \ge t^*$, $\bar{\eta}(k_1,k_2,t)$ will have the same expression as (34) except the term resulting from the convolution theorem, i.e.

$$\int_{(t-t_1)-t^*}^{t}\cos\omega\tau e^{-ik_1((t-t_1)v-\tau)}\,d\tau =$$
$$\frac{1}{\omega^2-(k_1v)^2}\left[\begin{array}{c}\omega\sin\omega(t-t_1)+ik_1v\cos\omega(t-t_1) \\ -e^{-ik_1vt^*}\left(\omega\sin\omega((t-t_1)-t^*)+ik_1v\cos\omega((t-t_1)-t^*)\right)\end{array}\right],$$

instead of

$$\int_{(t-t_1)}^{t}\cos\omega\tau e^{-ik_1(t-\tau)v}\,d\tau = \frac{1}{\omega^2-(k_1v)^2}\left(\omega\sin\omega(t-t_1)+ik_1v\cos\omega(t-t_1)-ik_1ve^{-ik_1(t-t_1)v}\right).$$

Finally, $\eta(x,y,t)$ is computed using inverse FFT of $\bar{\eta}(k_1,k_2,t)$.

Again, the singular points should be remove to compute $\eta(x,y,t)$ efficiently

1. As $k \to 0$, then $\bar{\eta}(k_1,k_2,t)$ has the following limit

$$\lim_{k\to 0}\bar{\eta}(k_1,k_2,t) = \begin{cases} 0 & (t-t_1)\le t^* \\ 0 & (t-t_1)\ge t^* \end{cases},\ \text{where } t^* = \frac{200}{v}.$$

2. As $k_1 \to 0$, then the singular terms of $\bar{\eta}(k_1,k_2,t)$ have the following limits

$$\lim_{k_1\to 0}\left(\frac{e^{i50k_1}-1}{ik_1}\right) = 50,\ \lim_{k_1\to 0}\left(\frac{e^{-ik_1((t-t_1)v)}-e^{-ik_1(50+(t-t_1)v)}}{ik_1}\right) = 50,$$

$$\lim_{k_1\to 0}\left(\frac{e^{-i200k_1}-e^{-i250k_1}}{ik_1}\right) = 50 \text{ and } \lim_{k_1\to 0}\left(\frac{e^{-ik_1(250+(t-t_1)v)}-e^{-ik_1(300+(t-t_1)v)}}{ik_1}\right) = 50.$$

3. As $k_2 \to 0$, then the singular terms of $\bar{\eta}(k_1, k_2, t)$ have the following limits

$$\lim_{k_2\to 0}\left(\frac{e^{i\,150\,k_2}}{i\,k_2}-\frac{e^{i\,50\,k_2}}{i\,k_2}\right)=100\,,\ \lim_{k_2\to 0}\left(\frac{\sin(50k_2)}{k_2}\right)=50 \text{ and}$$

$$\lim_{k_2\to 0}\left(\frac{e^{-i\,50\,k_2}}{i\,k_2}-\frac{e^{-i\,150\,k_2}}{i\,k_2}\right)=100\,.$$

Finally, $\bar{\eta}_{\text{block slide}}(k_1, k_2, t)$ is evaluated by applying the Laplace and Fourier transforms to the block slide motion described by (29), then substituting into (20) and then inverting $\bar{\eta}_{\text{block slide}}(k_1, k_2, s)$ using the inverse Laplace transform to obtain $\bar{\eta}_{\text{block slide}}(k_1, k_2, t)$.

This is verified for $t^* \le t_2 \le t^* + (L'/v)$ and $t^* + (L'/v) \le t_3 \le t^* + (2L'/v)$.

Then,

$$\bar{\eta}_{\text{block slide}}(k_1, k_2, t) = \bar{\eta}_1(k_1, k_2, t) + \bar{\eta}_2(k_1, k_2, t) + \bar{\eta}_3(k_1, k_2, t). \tag{35}$$

Then,

$$\begin{aligned}
&\bar{\eta}_{\text{block slide}}(k_1, k_2, t) = \\
&\Bigg[\left[\frac{\zeta_0}{4\cosh(kH)}\right]\left[\frac{e^{i\,150\,k_2}-e^{i\,50\,k_2}}{i\,k_2}-\frac{1}{1-\left(\frac{100}{\pi}k_2\right)^2}\left[ik_2\left(\frac{100}{\pi}\right)^2\left(e^{i50k_2}+e^{i150k_2}\right)\right]\right]+ \\
&\qquad\left[\frac{\zeta_0}{\cosh(kh)}\right]\left[\frac{\sin(50k_2)}{k_2}\right]+ \\
&\left[\frac{\zeta_0}{4\cosh(kH)}\right]\left[\frac{e^{-i\,50\,k_2}-e^{-i\,150\,k_2}}{i\,k_2}+\frac{1}{1-\left(\frac{100}{\pi}k_2\right)^2}\left[ik_2\left(\frac{100}{\pi}\right)^2\left(e^{-i150k_2}+e^{-i50k_2}\right)\right]\right]\Bigg] \\
&\times\Bigg[\left[\frac{e^{-i(200+S)k_1}-e^{-i(250+S)k_1}}{ik_1}+\frac{1}{1-\left(\frac{50}{\pi}k_1\right)^2}\left[-ik_1\left(\frac{50}{\pi}\right)^2\left(e^{-i(250+S)k_1}+e^{-i(200+S)k_1}\right)\right]\right]\cos\omega(t_3-t^*) \\
&\quad+\frac{2ve^{-i(250+S)k_1}}{\omega^2-(k_1v)^2}\begin{pmatrix}\omega\sin\omega t_3+ik_1v\cos\omega t_3- \\ e^{-ik_1vt^*}(\omega\sin\omega[t_3-t^*]+ \\ ik_1v\cos\omega[t_3-t^*])\end{pmatrix} \\
&\quad+\left[\frac{e^{-ik_1(250+S+L)}-e^{-ik_1(300+S+L)}}{ik_1}+\frac{1}{1-\left(\frac{50}{\pi}k_1\right)^2}\left[ik_1\left(\frac{50}{\pi}\right)^2\left(e^{-ik_1(300+S+L)}+e^{-ik_1(250+S+L)}\right)\right]\right]\cos\omega(t_3-t^*)\Bigg].
\end{aligned}$$

Finally, $\eta(x, y, t_2, t_3)$ is computed using inverse FFT of $\bar{\eta}(k_1, k_2, t_2, t_3)$.

Again, the singular points should be remove to compute $\eta(x, y, t)$ efficiently

1. As $k \to 0$, then $\bar{\eta}(k_1, k_2, t)$ has the limit

$$\lim_{k\to 0}\bar{\eta}(k_1, k_2, t) = 100\,(100 + 2L)\zeta_0\,,$$

2. As $k_1 \rightarrow 0$, then the singular terms of $\overline{\eta}(k_1,k_2,t)$ have the following limits

$$\lim_{k_1\to 0}\left(\frac{e^{-i(200+S)k_1}-e^{-i(250+S)k_1}}{ik_1}\right)=50 \text{ and } \lim_{k_1\to 0}\left(\frac{e^{-ik_1(250+S+L)}-e^{-ik_1(300+S+L)}}{ik_1}\right)=50.$$

3. As $k_2 \rightarrow 0$, then the singular terms of $\overline{\eta}(k_1,k_2,t)$ have the following limits

$$\lim_{k_2\to 0}\left(\frac{e^{i150k_2}}{ik_2}-\frac{e^{i50k_2}}{ik_2}\right)=100,\ \lim_{k_2\to 0}\left(\frac{\sin(50k_2)}{k_2}\right)=50 \text{ and } \lim_{k_2\to 0}\left(\frac{e^{-i50k_2}}{ik_2}-\frac{e^{-i150k_2}}{ik_2}\right)=100.$$

We investigated mathematically the water wave motion in the near and far-field by considering a kinematic mechanism of the sea floor faulting represented in sequence by a down and uplift motion with time followed by unilateral spreading in x-direction, both with constant velocity v, then a deceleration movement of a block slide in the direction of propagation. Clearly, from the mathematical derivation done above, $\eta(x,y,t)$ depends continuously on the source $\zeta(x,y,t)$. Hence, from the mathematical point of view, this problem is said to be well-posed for modeling the physical processes of the tsunami wave.

3. Results and discussion

We are interested in illustrating the nature of the tsunami build up and propagation during and after the movement process of a variable curvilinear block shape sliding. In this chapter, three cases are studied. We first examine the generation process of tsunami waveform resulting from the unilateral spreading of the down and uplift slip faulting in the direction of propagation with constant velocity v. We assume the spreading velocity of the ocean floor up and down lift equal to the tsunami wave velocity $v_t = \sqrt{gH} = 0.14$ km/sec as the largest wave amplitude occurs when $v = v_t$ due to wave focusing.

3.1. Tsunami generation caused by submarine slump and slide - Evolution in time

We assume the waveform initiated by a rapid movement of the bed deformation of the down and uplift source shown in Figure 2. Figure 6 shows the tsunami generated waveforms during the second stage at time evolution $t = 0.4t^*, 0.6t^*, 0.8t^*, t^*$ at constant water depth $H = 2$ km. It is seen how the amplitude of the wave builds up progressively as t increases where more water is lifted below the leading wave depending on its variation in time and the space in the source area. The wave will be focusing and the amplification may occur above the spreading edge of the slip .This amplification occurs above the source progressively as the source evolves by adding uplifted fluid to the fluid displaced previously by uplifts of preceding source segments. This explains why the amplification is larger for wider area of uplift source than for small source area. It can be seen that the tsunami waveform η/ζ_0 has two large peaks of comparable amplitudes, one in the front of the block due to sliding of the block forward, and the other one behind the block due to spreading of the depletion zone.

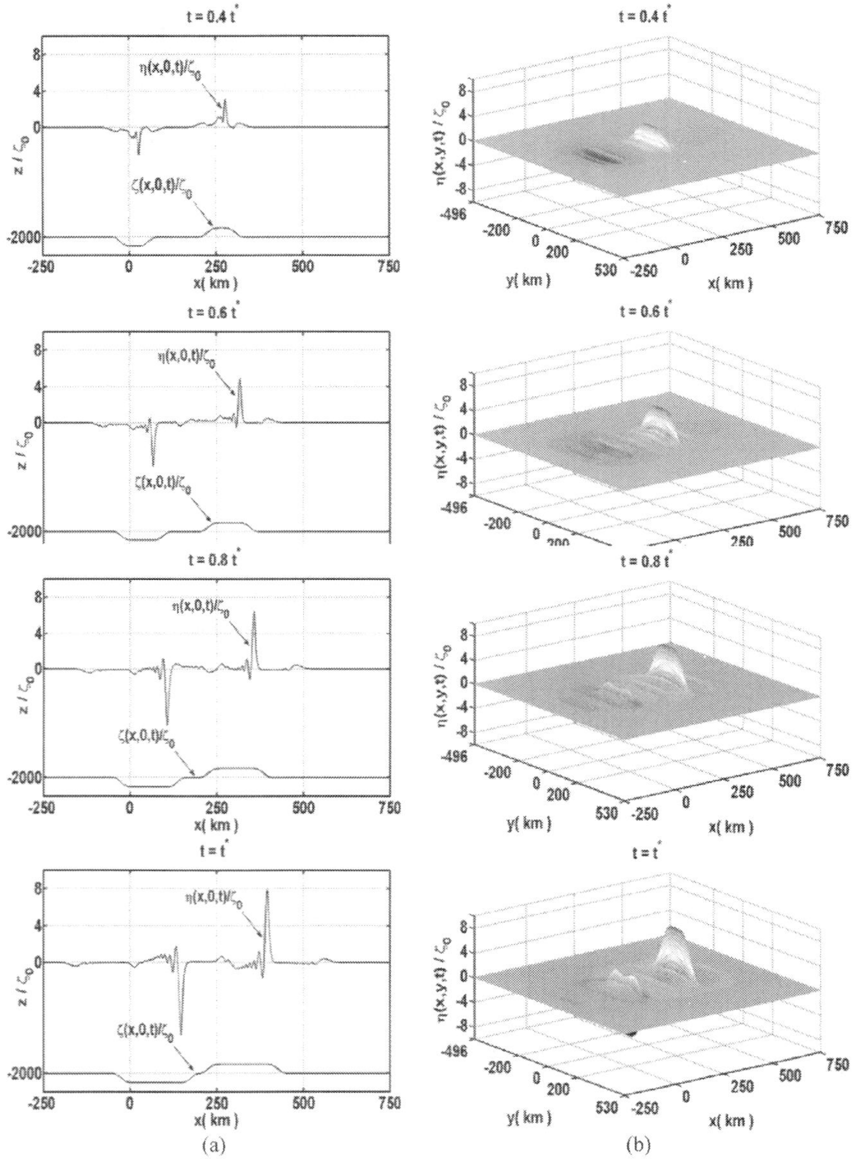

Figure 6. Dimensionless free-surface elevation caused by the propagation of the slump and slide in the x-direction during the second stage with $v = v_t$ at $H = 2$ km , $L = 150$ km, $W = 100$ km, $t^* = 200/v$ sec, (a) Side view along the axis of symmetry at $y = 0$ (b) Three dimensional view.

3.1.1. *Effect of the water depth H*

Figure 7 and Figure 8 illustrate the normalized peak tsunami amplitudes $\eta_{R,max}/\zeta_0$, $\eta_{L,min}/\zeta_0$ respectively in the near-field versus L/H at $t = t^* = t_1 + L/v$, the time when the spreading of the slides stops for H = 0.5, 1, 1.5 and 2 km and for $v = v_t$ and L = 150 km, W = 100 km.

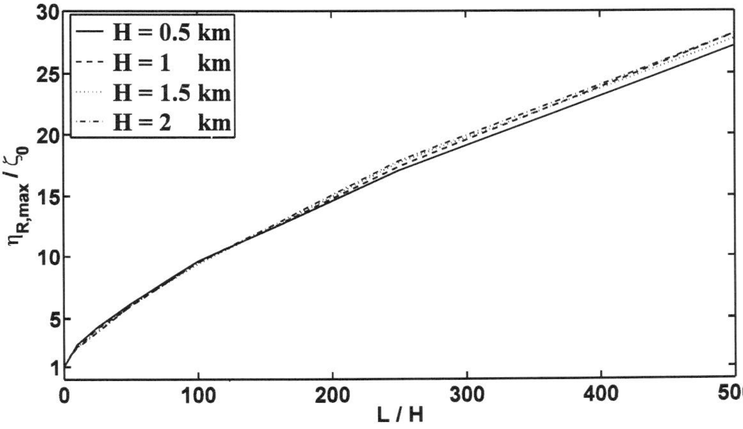

Figure 7. Normalized tsunami peak amplitudes, $\eta_{R,max}/\zeta_0$ at the end of second stage for different water depth H = 0.5, 1 , 1.5 and 2 km at $t^* = t_1 + L/v$ with $v = v_t$ and L = 150 km,W = 100 km.

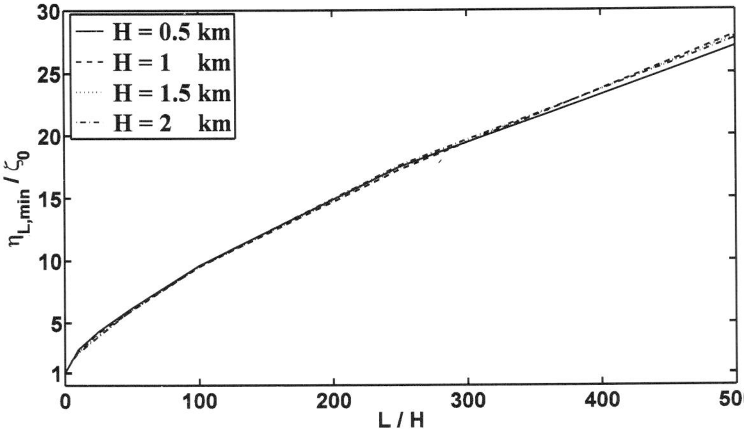

Figure 8. Normalized tsunami peak amplitudes, $\eta_{L,min}/\zeta_0$ at the end of second stage for different water depth H = 0.5, 1 , 1.5 and 2 km at $t^* = t_1 + L/v$ with $v = v_t$ and L = 150 km,W = 100 km.

From Figure 7 and Figure 8, the parameter that governs the amplification of the near-field water waves by focusing, is the ratio L/H. As the spreading length L in the slip-faults increases, the amplitude of the tsunami wave becomes higher.

At L = 0, no propagation occurs and the waveform takes initially the shape and amplitude of the curvilinear uplift fault (i.e. $\eta_{R,max}/\zeta_0 = 1, \eta_{L,min}/\zeta_0 = 1$).

The negative peak wave amplitudes are approximately equal to the positive peak amplitudes ($\eta_{L,min}/\zeta_0 \approx \eta_{R,max}/\zeta_0$) when $v = v_t$ as seen in Figures 7 and 8. The peak tsunami amplitude also depends on the water depth in the sense that even a small area source can generate large amplitude if the water is shallow.

3.1.2. *Effect of the characteristic size*

Figure 9 shows the effect of the water depth h on the amplification factor $\eta_{R,max}/\zeta_0$ for $v = v_t$, with L = W = 10, 50,100 km and L = 150 km, W = 100 km at the end of the second stage (i.e. at $t = t^* = t_1 + L/v$). Normalized maximum tsunami amplitudes for 19 ocean depths are calculated. As seen from Figure 9, the amplification factor $\eta_{R,max}/\zeta_0$ decreases as the water depth H increases. This happens because the speed of the tsunami is related to the water depth ($v = v_t = \sqrt{gH}$) which produces small wavelength as the velocity decreases and hence the height of the wave grows as the change of total energy of the tsunami remains constant. Mathematically, wave energy is proportional to both the length of the wave and the height squared. Therefore, if the energy remains constant and the wavelength decreases, then the height must increases.

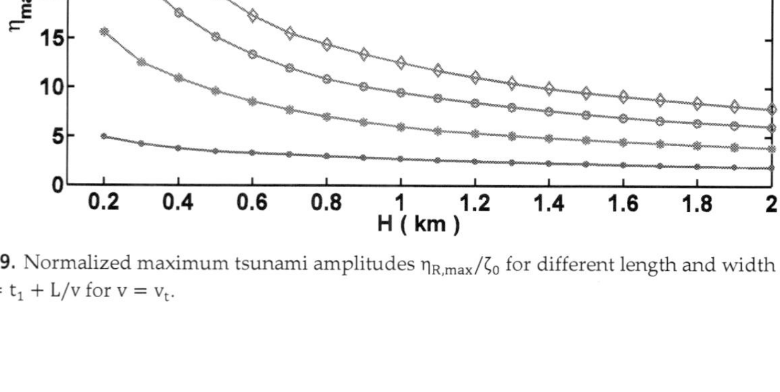

Figure 9. Normalized maximum tsunami amplitudes $\eta_{R,max}/\zeta_0$ for different length and width at $t = t^* = t_1 + L/v$ for $v = v_t$.

3.2. Tsunami generation and propagation-effect of variable velocities of submarine block slide

In this section, we investigated the motion of a submarine block slide, with variable velocities, and its effect on the near-field tsunami amplitudes. We considered the limiting case, in which the slide moves with constant velocity and stops suddenly (infinite deceleration) and the case in which the slide stops softly with constant deceleration for L = 150 km, W = 100 km and $v = v_t$.

3.2.1. *Displaced block sliding with constant velocity v*

Constant velocity implies that the slide starts and stops impulsively, i.e. the acceleration and deceleration are infinite both initially and finally. This means that the slide takes minimum time to reach the characteristic length $L' = 150$ km given by $t_{min} = t^* + (L'/v) = 41.65$ min. We illustrate the impulsive tsunami waves caused by sudden stop of the slide at distance L′ in Figure 10.

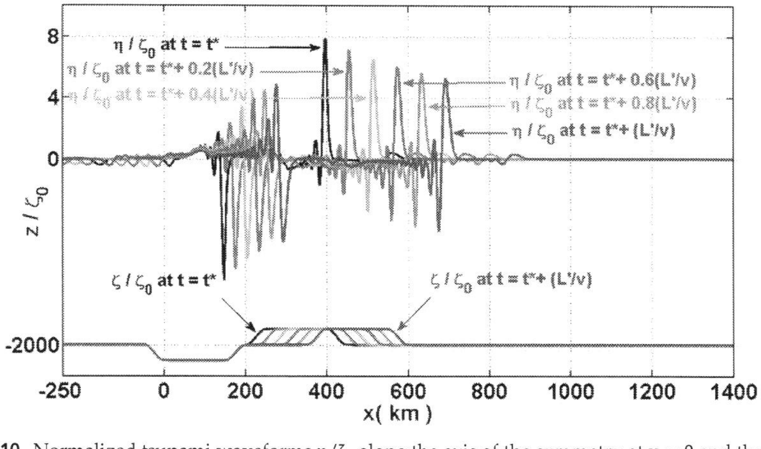

Figure 10. Normalized tsunami waveforms η/ζ_0 along the axis of the symmetry at y = 0 and their corresponding moving slide ζ/ζ_0 with constant velocity v along y = 0, at time $t^* \le t \le t^* + (L'/v)$ for h = 2 km, L′ = 150 km and W = 100 km.

Figure 10 shows the leading tsunami wave propagating in the positive x-direction during time evolution $t = t^*, t^* + 0.2(L'/v), t^* + 0.4(L'/v), t^* + 0.6(L'/v), t^* + 0.8(L'/v), t^* + (L'/v)$ sec at L_E = 0, 30, 60, 90, 120, 150 km respectively, where L_E represents that part of L (see Figure 4c) for $v = v_t = 0.14$ km/sec. It is seen in Figure 10 that the maximum leading wave amplitude decreases with time, due to the geometric spreading and also due to the dispersion. At $t = t_{min} = t^* + (L'/v) = 41.65$, the wave front is at x = 693 km and $\eta_{R,max}/\zeta_0$ decreases from 7.906 at $t = t^*$ to 5.261 at time t_{min}. This happens because the amplification of the waveforms depends only on the volume of the displaced water by the moving source which becomes an important factor in the modeling of the tsunami generation. This was clear from the singular points removed from the block slide model, where the finite limit of the free surface depends on the characteristic volume of the source model

3.2.2. Displaced block moving with linear decreasing velocity with time t

The velocity of the movable slide is uniform and equal v(t) up to time t_{max} as shown in Figure 4a, followed by a decelerating phase in which the velocity is given by

$$v(t) = v - (t - t^{*})\alpha \,, \text{ for } t^{*} \leq t \leq t_{max}.$$

where $v = v_t = 0.14$ km/sec and α is the deceleration of the moving block slide. The block slide moves in the positive x − direction with time $t^{*} \leq t \leq t_{max}$ where $t_{max} = t^{*} + 2L'/v$ is the maximum time that the slide takes to stop after reaching the characteristic length $L' = 150$ km with minimum deceleration α_{min}. Figure 11 shows the leading tsunami wave propagating in the positive x-direction during time evolution $t = t^{*}, t^{*} + 0.2(2L'/v), t^{*} + 0.4(2L'/v), t^{*} + +0.6(2L'/v), t^{*} + 0.8(2L'/v), t^{*} + (2L'/v)$ min in case $t_2 = t^{*}$ (i.e. minimum magnitude of α).

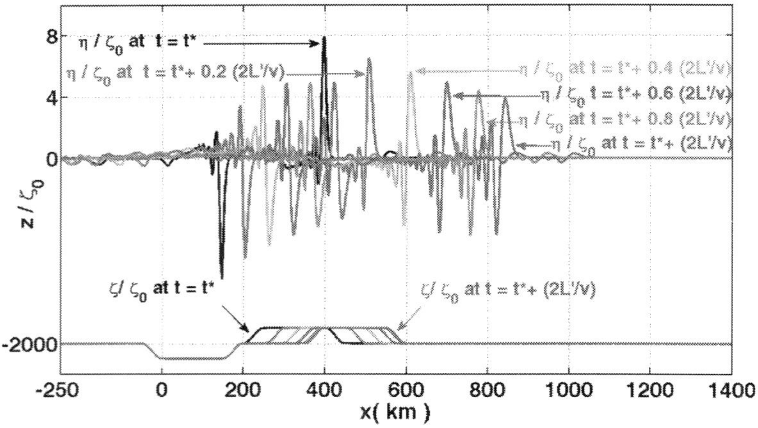

Figure 11. Normalized tsunami waveforms η/ζ_0 and their corresponding moving slide ζ/ζ_0 with variable velocity v(t) along y = 0, at time $t^{*} \leq t \leq t^{*} + 2L'/v$ at H = 2 km, L' = 150 km and W = 100 km.

It was clear from Figure 10 and Figure 11 that at the instant the slide stops, the peak amplitude in case of sudden stop is higher than that of soft stop.

Figure 12 and Figure 13 show the effect of the water depth at $L' = W = 10, 50, 100$ km and $L' = 150$, W = 100 km on the normalized peak tsunami amplitude $\eta_{R,max}/\zeta_0$ when the slide stops moving at length L' instantaneously at $t_{min} = t^{*} + L'/v$ with infinite deceleration and stops moving softly at L' at the time $t_{max} = t^{*} + 2L'/v$ with minimum deceleration for $v = v_t$.

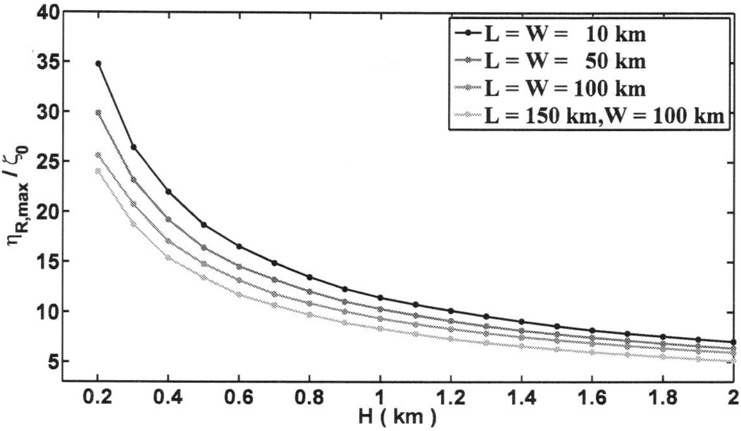

Figure 12. Normalized maximum tsunami amplitudes $\eta_{R,max}/\zeta_0$ when the slide stops suddenly at time $t_{min} = t^* + (L'/v)$ with different slide length L and width W and for $v = v_t$.

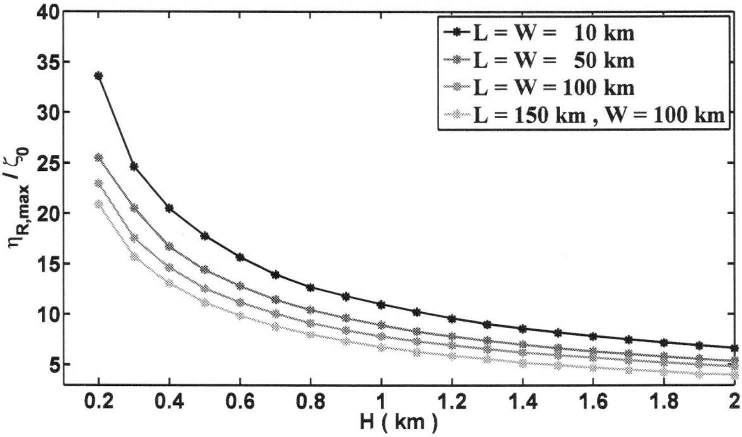

Figure 13. Normalized maximum tsunami amplitudes $\eta_{R,max}/\zeta_0$ when the slide stops softly at time $t_{max} = t^* + (2L'/v)$ with different slide length L and width W and for $v = v_t$.

It is clear from Figure 12 and Figure 13 that the waveforms which caused by sudden stop of the slide motion after they reach the characteristic length L′ at time $t_{min} = t^* + (L'/v)$ have higher amplitude than stopping of the slide with slow motion at time $t_{max} = t^* + (2L'/v)$ .This agrees with the mathematical relation between the wavelength and the wave height where the wave energy is proportional to both the length of the wave and the height squared.

3.2.3. Displaced block moving with constant velocity v followed by variable velocity $\boldsymbol{v(t)}$

In this section, we studied the generation of the tsunami waveforms when the block moves a significant distance with constant velocity $v = v_t$ then continue moving with variable velocity v(t) with constant deceleration until it stops at the characteristic length $L' = 150$ km. Figure 14 shows the tsunami waveforms at the times calculated in Table 2 when the slide reaches the characteristic length $L' = 150$ km.

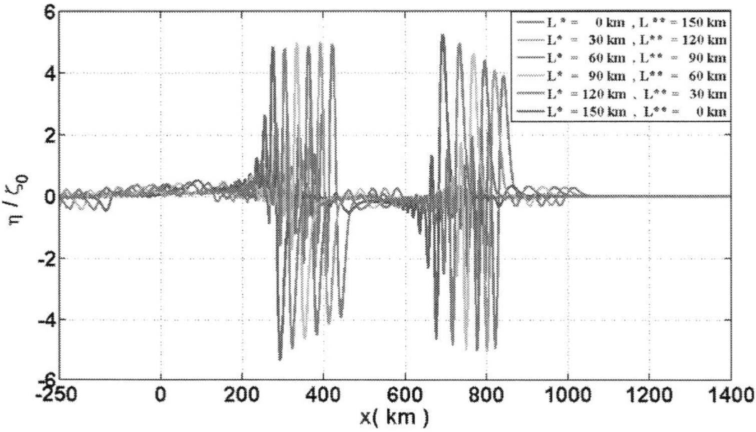

Figure 14. Normalized tsunami waveforms η/ζ_0 along $y = 0$ at $t = t_3$ calculated in Table 7.2 with $L' = 150$ km.

In Figure 14, the first waveform from the left indicates the shape of the wave at the time $t_3 = t_{min} = t^* + (L'/v) = 41.65$ in the limiting case when the slide stops moving suddenly. The last waveform indicates the wave in the other limiting case at the time $t_3 = t_{max} = t^* + (2L'/v)$ when the slide stop moving with minimum deceleration at the distance L′. In between the two limiting cases, the slide begins moving with constant velocity a significant distance followed by a decelerating movement until it stop at the characteristic length $L' = 150$ km at the time $t_{min} \leq t_3 \leq t_{max}$, see Table 2. It is seen how the peak amplitudes of the leading waves decreases gradually from 5.261 to 3.894.

3.3. Tsunami propagation waveforms

In order to compare shape and maximum height of tsunami wave at certain time for different deceleration α, we choose the time $t = t_{max}$. We study the case when the propagating waveforms resulting at the time difference between t_3 and t_{max} (i.e. propagating of tsunami away from the slide) when the slide stops moving at the length L′.

For the limiting case α_{min}, there is no free propagation, while for the other limiting case " the sudden stop", there is free propagation between time t_{min} and t_{max}. For the cases between the two limiting cases, the propagation time is $t_{prop} = t_{max} - t_3$.

Figure 15 shows the shape of the tsunami propagation waveform at $t_{max} = t^* + (2L'/v) =$ 59.51 min (curves in black) for different deceleration α and time t_3 (time at which the slide stops).

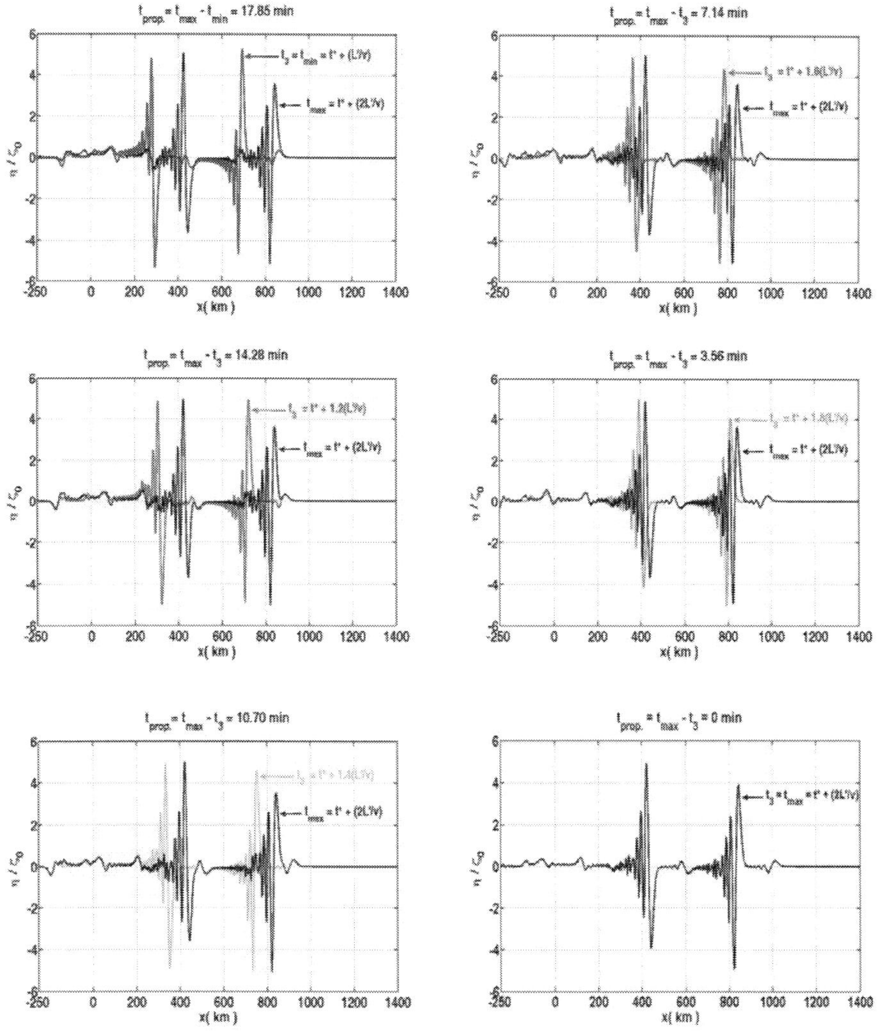

Figure 15. Normalized tsunami propagation waveforms η/ζ_0 along the axis of the symmetry at y = 0 at $t_{prop.} = 0, 3.56, 7.14, 10.70, 14.28, 17.85$ min with L′ = 150 km and W = 100 km.

As the wave propagates, the wave height decreases and the slope of the front of the wave become smaller, causing a train of small wave forms behind the main wave. The maximum wave amplitude decreases with time, due to the geometric spreading and also due to the dispersion.

Figure 16 represents the normalized peak tsunami amplitudes $|\eta|_{min}/\zeta_0$ and η_{max}/ζ_0 of the leading propagating wave in the far-field at time $t = t_{max}$ for the different deceleration α and time t_3 (time at which the slide stops) chosen in Figure 7.15.

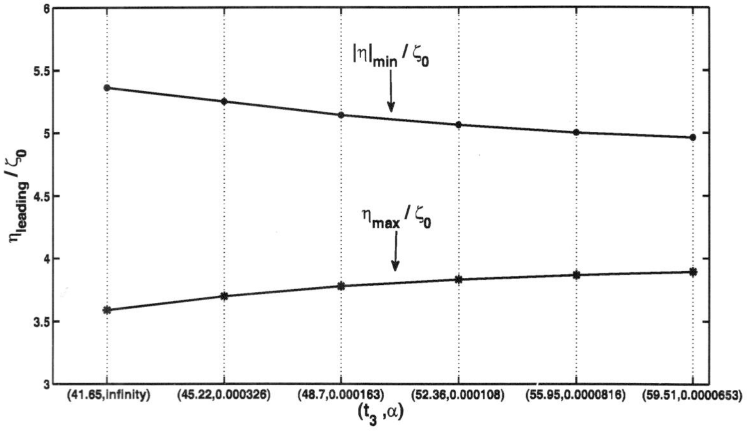

Figure 16. The normalized peak tsunami amplitudes $|\eta|_{min}/\zeta_0$ and η_{max}/ζ_0 at time $t = t_{max}$ for the different deceleration α and time t_3.

It can be seen in Figure 16 that the absolute minimum peak amplitudes of the leading propagated waves at time $t_{max} = t^* + (2L'/v)$, after the block slide stops moving with different deceleration α and time t_3, decreases gradually, while the maximum peak amplitudes increases progressively.

4. Conclusion

In this paper, we presented a review of the main physical characteristics of the tsunami generation caused by realistic curvilinear submarine slumps and slides in the near-field. It is seen that the tsunami waveform has two large peaks of comparable amplitudes, one in the front of the block due to forward sliding of the block, and the other one behind the block due to spreading of the depletion zone. The negative peak wave amplitudes are approximately equal to the positive peak amplitudes. We studied the effect of variable velocities of submarine block slide on the tsunami generation in the limiting cases, in which the slide moves with constant velocity and stops suddenly (with infinite deceleration) and the case in which the slide stops softly at the same place with minimum deceleration. It is seen that the leading tsunami amplitudes are reduced in both cases due to the geometric spreading and also due to the dispersion. We observed that the peak tsunami amplitudes increase with the decrease in the sliding source area and the water depth. We also investigated the more realistic case in which the block slide moves a significant distance with constant velocity v then continue moving with time dependence velocity v(t) and different constant deceleration until it stops at the characteristic length. It is seen how the peak amplitudes of the leading waves decrease gradually with time between the two limiting cases. In this case we demonstrated also the shape of tsunami propagated wave at certain time max t (time at which the slide stops with minimum deceleration). The results show that the wave height decreases due to dispersion and the slope of the front of the wave becomes smaller, causing a train of small wave forms behind the main wave. It can be observed that just a slight variation in the maximum and the minimum tsunami propagated amplitudes after the block slide stops moving with different deceleration α and time t_3, see **Figure 16**. The presented analysis suggests that some abnormally large tsunamis could be explained in part by variable speeds of submarine landslides. Our results should help to enable quantitative tsunami forecasts and warnings based on recoverable seismic data and to increase the possibilities for the use of tsunami data to study earthquakes, particularly historical events for which adequate seismic data do not exist.

Author details

Khaled T. Ramadan
Department of Basic and Applied Science, College of Engineering and Technology,
Arab Academy for Science, Technology and Maritime Transport, Abu Quir Campus, Alexandria, Egypt

5. References

[1] M. S. Abou-Dina and F. M. Hassan, "Generation and Propagation of Nonlinear Tsunamis in Shallow Water by a Moving Topography," *Applied Mathematics and Computation*, Vol. 177, No. 2, 2006, pp. 785-806.

[2] F. M. Hassan, "Boundary Integral Method Applied to the Propagation of Non-linear Gravity Waves Generated by a Moving Bottom," *Applied Mathematical Modelling*, Vol. 33, No. 1, 2009, pp. 451-466.

[3] N. Zahibo, E. Pelinovsky, T. Talipova , A. Kozelkov and A. Kurkin, "Analytical and Numerical Study of Nonlinear Effects at Tsunami Modeling," *Applied Mathematics and Computation*, Vol. 174, No. 2, 2006, pp. 795-809.

[4] Ben-Menahem and M. Rosenman, "Amplitude Patterns of Tsunami Waves from Submarine Earthquakes," *Journal of Geophysical Research*, Vol. 77, No. 17, 1972, pp. 3097-3128.

[5] Kjetil Braathen Haugen, Finn Løvholt and Carl B. Harbitz , " Fundamental mechanisms for tsunami generation by submarine mass flows in idealised geometries", Marine and Petroleum Geology, Vol. 22, Issues 1–2, 2005, pp 209–217.

Chapter 6

Tsunami Propagation Models Based on First Principles

A. Tan, A.K. Chilvery, M. Dokhanian and S.H. Crutcher

Additional information is available at the end of the chapter

http://dx.doi.org/10.5772/50508

1. Introduction

Tsunamis are ocean waves generated by the displacement of a large volume of water due to earthquakes, volcanic eruptions, landslides or other causes above or below the ocean floor (e.g., Karling, 2005; Parker, 2012). The great Indian Ocean tsunami of December 2004 will be remembered for its ferocity, devastation and unprecedented loss of life for a long time (Stewart, 2005; The Indian Ocean Tsunami, 2011). It is also the same tsunami which has galvanized the international community to set up warning systems and undertake preventive measures against the onslaught of future tsunamis in the vulnerable regions around the globe. A surge of scientific studies on all aspects of the tsunami is in evidence in the literature. And a volume entitled *"The Tsunami Threat – Research and Technology"* (Mörner ed., 2011) has been brought out. The current volume entitled *"Tsunami"* (Lopez, ed., 2012) is a sequel to the above in a continued effort to promote understanding and predicting future tsunamis and warning the populace in the potentially vulnerable areas.

There are three distinct stages of a tsunami event: (1) Generation; (2) Propagation; and (3) Inundation/landfall (cf. Cecioni & Belloti, 2011). The generation stage is the most complex and most difficult to analyze, since each tsunami is different and no single mechanism can account for all tsunamis. The inundation stage is also different for different areas affected, and again, no single scenario can describe all affected areas. The propagation stage covers the most extensive area, and is the only one that can be attacked by simple theory and analysis, even though detailed numerical models are found in the literature (see, for example, Imteaz, et al., 2011, and the references therein). These models consist of solving hydrodynamic equations with suitable boundary conditions that necessarily involve tedious numerical integrations. Such models, unfortunately, fall within the realm of the specialists, and are, by and large, outside the reach of the broader audience.

This chapter takes an alternative approach to the study of tsunami propagation in the open ocean. It commences with the theory of water wave propagation in general and applies them to tsunami propagation in particular, using analytical models. It is based upon first principles of physics and avoids numerical analysis, which is thus accessible to the broader scientific community. It only requires the knowledge of general science and basic calculus-based physics. The derivations of the relevant equations are relegated to the appendices for quick reference so that the need to search for them outside this article is kept to a minimum.

2. Theory of water waves and tsunamis

The theory of water waves is well-documented in the literature (e.g., Coulson, 1955; Sharman, 1963; Towne, 1967; Elmore & Heald, 1969). The wave velocity of waves on water of density ρ under the action of gravity and surface tension T is obtained from a linear wave equation by ignoring the non-linear term of $\frac{1}{2}v^2$ in Bernoulli's equation (cf. Coulson, 1955; Elmore & Heald, 1969):

$$v = \sqrt{\left(\frac{g}{k} + \frac{Tk}{\rho}\right)\tanh kh} \tag{1}$$

Here h is the depth of water and k the wave number. In terms of the wavelength $\lambda = 2\pi/k$, we have:

$$v = \sqrt{\left(\frac{g\lambda}{2\pi} + \frac{2\pi T}{\lambda\rho}\right)\tanh\frac{2\pi h}{\lambda}} \tag{2}$$

When the gravity and surface tension terms are equal, λ is called the critical wavelength λ_c:

$$\lambda_c = 2\pi\sqrt{\frac{T}{\rho g}} \tag{3}$$

For water waves, $\lambda_c = 1.73$ cm (cf. Towne, 1967; Elmore & Heald, 1969). The corresponding velocity $v_c = 27$ cm/s (Sharman, 1963).

For waves of $\lambda < \lambda_c < h$, the surface tension term dominates and

$$v = \frac{\omega}{k} = \sqrt{\frac{Tk}{\rho}} \tag{4}$$

Thus

$$\omega = \sqrt{\frac{T}{\rho}}k^{3/2} \tag{5}$$

and the group velocity

$$u = \frac{d\omega}{dk} = \frac{3}{2}\sqrt{\frac{Tk}{\rho}} = \frac{3}{2}u \tag{6}$$

The group velocity is faster than the wave velocity! The individual wave-crests fall behind the group while new crests build up at the forward edge of the group (cf. Towne, 1967). Such waves are called *capillary waves* or *ripples*. They are severely attenuated by viscous effects.

For longer wavelengths ($\lambda > \lambda_c$), the gravity term dominates. We have (vide Appendix A):

$$v = \frac{\omega}{k} = \sqrt{\frac{g}{k}\tanh kh} \tag{7}$$

and

$$\omega = \sqrt{gk\tanh kh} \tag{8}$$

Such waves are called *gravity waves*. Customarily, two kinds of gravity waves are recognized depending upon the wavelength $\lambda = 2\pi/k$: (1) If $\lambda > h$, the waves are called *long waves in shallow water*; (2) If $\lambda < h$, the waves are called *short waves in deep water*.

In the first case, $\tanh kh \approx kh$ and the wave velocity

$$v = \sqrt{gh} \tag{9}$$

The group velocity is, from Eq. (8):

$$u = \frac{d\omega}{dk} = \sqrt{gh} = v \tag{10}$$

Owing to the equality of the wave velocity and the group velocity, such waves are non-dispersive.

In the second case, $\tanh kh \approx 1$. Then

$$v = \sqrt{\frac{g}{k}} \tag{11}$$

and from Eq. (8):

$$u = \frac{d\omega}{dk} = \frac{1}{2}\sqrt{\frac{g}{k}} = \frac{1}{2}v \tag{12}$$

In such waves, the individual waves travel faster than the group and rapidly diminish in amplitude as they cross the group (Towne, 1967).

Tsunami waves in the deep ocean typically have wavelengths of about 200 km and velocity of 700 kilometers per hour. They maintain their form in dispersion-less propagation for long distances with little dissipation of energy. They are thus identified as 'long waves in shallow water' even as they travel above the deepest parts of the ocean, the 'long' and 'shallow' adjectives referring to the relative magnitudes of the wavelength and depth of water. Many of the properties of the tsunamis, such as velocity and dispersion-less propagation follow from this description (e.g., Margaritondo, 2005; Helene & Yamashita, 2006; Tan & Lyatskaya, 2009). However, the linearized theory of water waves is only an approximation to the true situation and a rigorous theory calls for the inclusion of the non-linear term in Bernoulli's equation. For example, the form of these waves having a narrow crest and very wide trough can only be accounted for by a complete non-linear theory (cf. Elmore & Heald, 1969).

Tsunamis also exhibit the characteristics of 'canal waves', first observed by Scott Russell in Scotland in 1834 (Russell, 1844). These disturbances travel like a single wave over long distances and maintain their shapes with little loss of energy. The mathematical theory of such waves, now-a-days called 'solitary waves' or 'solitons', was developed by Boussinesq (1871), Korteweg & de Vries (1895) and Lord Rayleigh (1914). Their analyses show that in a dispersive medium, the non-linear effect can exactly cancel out the dispersive effect to preserve the form of the wave (cf. Stoker, 1957; Lamb, 1993).

3. Tsunami propagation models

One-dimensional propagation models are the easiest to construct and analyze. Even though they may not represent the real situation for tsunami propagation in the open ocean, valuable results can come out of these models. The 'long waves in shallow water' model noted above represents a one-dimensional model (**Model A**) in which the displacement is in the vertical direction and propagation takes place in the horizontal direction (see Appendix A). The Models of Margaritondo (2005) and Helene & Yamashita (2006) are examples of this model. This model correctly predicts the speed of the tsunami by means of Eq. (9). In fact, this equation is widely used in bathymetry. It is also used to calculate the travel times of tsunamis.

In the 'long waves in shallow water' model, the group velocity is equal to the wave velocity according to Eqs. (9) and (10). This helps to explain the dispersion-less propagation of the tsunamis. An alternative derivation of these equations is provided by Margaritondo (2005). By assuming that the vertical displacement of water is proportional to distance from the bottom, Margaritondo (2005) arrives at the same result by applying the energy conservation law. As stated earlier, however, a rigorous derivation would call for the solitary wave solution of non-linear differential equation obtained from Navier-Stokes equation (cf. Stoker, 1957).

Eq. (9) has profound consequences in the inundation phase as the tsunami makes landfall. The wave energy density E within a wavelength λ is

$$E \propto A^2 \lambda \propto A^2 v \propto A^2 \sqrt{h} \tag{13}$$

where A is the amplitude of the wave. Assuming E to remain constant, we have

$$A \propto \frac{1}{\sqrt[4]{h}} \tag{14}$$

A 1 m high wave at a depth of 1000 m would become 5.62 m at a depth of 1 m! That goes to illustrate the devastating effects of tsunamis as they make landfall. Helene & Yamashita (2006) have shown that the Eq. (14) holds true when the depth of the ocean floor varies gradually instead of having abrupt steps, which helps the tsunamis to maintain their characteristics as they approach land.

Helene & Yamashita (2006) have further shown how a tsunami will bend dramatically around an obstacle and strike land in the shadow regions. Since $v \propto \sqrt{h}$ according to Eq. (9), the velocity decreases nearer the coast as the depth decreases. The wave-fronts are able to bend dramatically towards the coast in a diffraction pattern and to strike land in the shadow region. This explains the damage caused in the western coasts of Sri Lanka and India during the Boxing day tsunami of December 2004.

Useful as they are, the one-dimensional models of are not appropriate when the tsunami propagates in the open ocean from a well-defined epicenter since the waves will spread out in concentric circles, which calls for two-dimensional models. In **Model B** (Tan & Lyatskaya, 2009), waves propagate outwards on a flat two-dimensional ocean from the epicenter. When the energy conservation principle is imposed, the energy density of the wave falls off inversely as the distance from the epicenter ρ:

$$I_{ModelB} \propto \frac{1}{\rho} \tag{15}$$

The wave amplitude therefore falls off inversely as the square-root of the radial distance ρ:

$$\psi_{ModelB} \propto \sqrt{I_{ModelB}} \propto \frac{1}{\sqrt{\rho}} \tag{16}$$

The formal derivation of the results (15) and (16) are given in Appendix B where it is mentioned that strictly speaking, they apply at distances away from the epicenter.

For long distance propagation, the curvature of the Earth must be taken into consideration. This is incorporated in **Model C** (Tan & Lyatskaya, 2009) which analyzes tsunami propagation on a spherical oceanic surface. In this model, the energy density of the wave varies as (vide Appendix C):

$$I_{ModelC} \propto \frac{1}{\sin\theta} \tag{17}$$

where θ is the zenith angle measured from the center of the Earth with the epicenter at the north pole. Hence the wave amplitude is a function of the polar angle:

$$\psi_{ModelC} \propto \sqrt{I_{ModelC}} \propto \frac{1}{\sqrt{\sin\theta}} \tag{18}$$

The amplitude variation depends solely on the polar angle. Since the epicenter is considered to be located at the north pole ($\theta = 0$), the wave amplitude gradually falls until it reaches the equator ($\theta = \pi/2$), 10,000 km away, and then starts to rise again as the convergence effect due to the curvature of the Earth comes into play and finally regains its original value at the south pole ($\theta = \pi$). Thus, if the Earth were entirely ocean, barring any losses, the wave amplitude would regain its original value at the anti-podal point.

The comparison of wave amplitudes in Model C (true) and Model B (approximate) is given in Appendix C. We have:

$$\frac{\psi_{ModelC}}{\psi_{ModelB}} = \sqrt{\frac{\theta}{\sin\theta}} \tag{19}$$

Table C.1 shows that the difference in the two solutions is slight for small values of θ. Even at the equator ($\theta = \pi/2$), the enhancement in amplitude due to curvature is only 25%. The enhancement becomes progressively greater until the south pole ($\theta = \pi$) is reached, where it becomes infinite.

When the tsunami is caused by an oceanic plate sliding under a continental plate, the subduction zone can be described by a finite line source instead of a localized point source. In that case, **Model D** is appropriate. In this model, the wave-fronts are ellipses with foci at the end-points of the line source and the energy propagates along con-focal hyperbolas (Appendix D). The wave amplitude near the source will be far different from that of a point source (Model B). If c is the length of the line source, then at distances ρ_b from the center across the line source, one has (vide Appendix D):

$$\frac{\psi_{ModelD}}{\psi_{ModelB}} = \frac{1}{\sqrt[4]{1 + \frac{c^2}{\rho_b^{\,2}}}} \tag{20}$$

The wave amplitude in Model D is finite at the origin as opposed to being infinite in Models B and C. However, away from the source, the solutions for Models B and D rapidly converge, as the wave-fronts become more circular. For $\rho_b > c$, the difference is below 10%, whereas for $\rho_b > 10c$, the two solutions are virtually indistinguishable (Table D.1).

Tsunamis are vast and highly complex geophysical phenomena, each having a character of its own. It is impossible to construct one model for any tsunami even with a high-speed numerical code. Further, each stage of the tsunami – generation, propagation and inundation, has to be modeled and studied separately. Nonetheless, simplified models based on first principles are able to explain individual aspects of this very complex geophysical phenomenon without detailed numerical computations.

4. Model applications

Viewed from space, the Earth is a watery planet with the oceans covering a full 71% of the surface area. The world ocean consists of three inter-connected oceans of the Pacific, Atlantic and Indian oceans, which comprise 51.5%, 25.6% and 22.9% of the water surface, respectively (we disregard 'Southern Ocean' as a separate entity). The Pacific ocean is thus larger than the two other oceans put together. It alone covers 34% or just over a full one-third of the Earth's surface and is comfortably larger than all the landmasses (at 29% of the Earth's surface) put together. The Pacific ocean is thought to be the remnant of 'Panthalassa', the world ocean, when all the landmasses were joined together as 'Pangaea'.

The Pacific ocean provides an ideal venue for tsunami propagation studies for several reasons. First, as stated above, it is the largest body of water, covering a full one-third of the globe. Second, it is bounded by active tectonic plate junctions, studded with volcanoes called the 'Ring of Fire'. Tsunamis produced at these hotspots can traverse the length and breadth of the ocean with relative ease. Third, there are no landmasses or large islands to block or interfere with the propagation of tsunamis formed in the ocean. Fourth, the ocean itself is dotted with small islands which pose little interference with tsunami propagation, but provide valuable platforms for recording tsunami wave amplitudes. Many of these islands are volcanic in origin and are sources of tsunamis themselves. Fifth, the longest stretch of ocean water is found between the Japan archipelago in the north-west and southern Chile on the south-east covering a distance of over 17,000 km or 85% of the distance between the North pole and the South Pole. At both the ends of this diameter lie some of the most active plate tectonic regions and tsunamis from either ends have traversed this favorite racetrack. Last but not least, an astonishing 80% of all tsunamis are recorded in the Pacific ocean.

With this geographical backdrop, we now proceed to study representative tsunami event to illustrate the validity of our propagation models. Model A, even though uni-dimensional, is a valuable tool for all tsunami events, as it correctly furnishes the velocity given the depth of the ocean, or vice-versa. It further predicts the travel times, which are vital for warning purposes. These results are independent of the direction, given the isotropy of space. Model A fails when the amplitude of the wave is to be studied, in which case Model B, C or D is called into consideration. In the following, we provide examples where one of the latter models, in conjuction with Model A, is used to analyze historic tsunami events. The data are taken from the National Oceanic and Atmospheric Administration website at www.ngdc.noaa.gov/hazard/tsu_travel_time_events.shtml.

4.1. Hawaii tsunami of 1975

On 29 November 1975, a magnitude 7.2 earthquake occurred on the southern coast of the island of Hawaii with the epicenter at 19.3°N and 155.0°W at a focal depth of 8 km (cf. Pararas-Carayannis, 1976). The earthquake, the largest local one since 1868, generated a locally damaging submarine landslide tsunami which was recorded at 76 tide gauge stations in Alaska, California, Hawaii, Japan, Galapagos Islands, Peru and Chile. The tsunami caused $1.5 million damage in Hawaii, 2 deaths and 19 injuries (Dudley & Lee, 1988). From the travel times registered, the tsunami reached Guadalupe Island, Mexico, 3864 km away in 5 h 9 m at an average speed of 750 kph, while it took 6 h 8 m to reach Tofino Island, Canada, 4210 km away, at the average speed of 686 kph. The slower speed in the first case is likely to be due to the fact that the tsunami had to bend considerably before heading towards its destination (cf. Helene & Yamashita, 2006).

Fig. 1 is a scatter plot of the wave amplitude versus the propagation distance. Since the maximum distance was under 8000 km, the flat space approximation holds (vide Appendix C) and Model B is applicable. The variation of the wave amplitude in this model is given by Eq. (16):

$$\psi_{ModelB} = \frac{A}{\sqrt{\rho}} \tag{21}$$

where A is the amplitude constant to be determined by regression analysis. Summing over the data points, one obtains:

$$A = \frac{\sum \psi_{ModelB}}{\sum \frac{1}{\sqrt{\rho}}} \tag{22}$$

The current tsunami data yield: A = 20.5265. The model equation is shown in Fig. 1. Also shown in the figure is the actual variation of the wave amplitude in accordance with the data. This is determined by first assuming a functional variation of the form

$$\psi_{observed} = A\rho^{\alpha} \tag{23}$$

where A and α are two constants to be determined from regression analysis. By taking natural logarithms of both sides first and then multiplying both sides by ρ, we get the two normal equations required to find the constants:

$$\log\psi = \log A + \alpha \log \rho \tag{24}$$

and

$$\rho \log\psi = \rho \log A + \alpha\rho \log \rho \tag{25}$$

By summing Eqs. (24) and (25) over the n data points, we get:

$$\sum \log\psi = n\log A + \alpha \sum \log \rho \tag{26}$$

and

$$\sum \rho \log\psi = \log A \sum \rho + \alpha \sum \rho \log \rho \tag{27}$$

By eliminating A between Eqs. (26) and (27), we have

$$\alpha = \frac{\sum \rho \sum \log\psi - n \sum \rho \log\psi}{\sum \rho \sum \log\rho - n \sum \rho \log\rho} \tag{28}$$

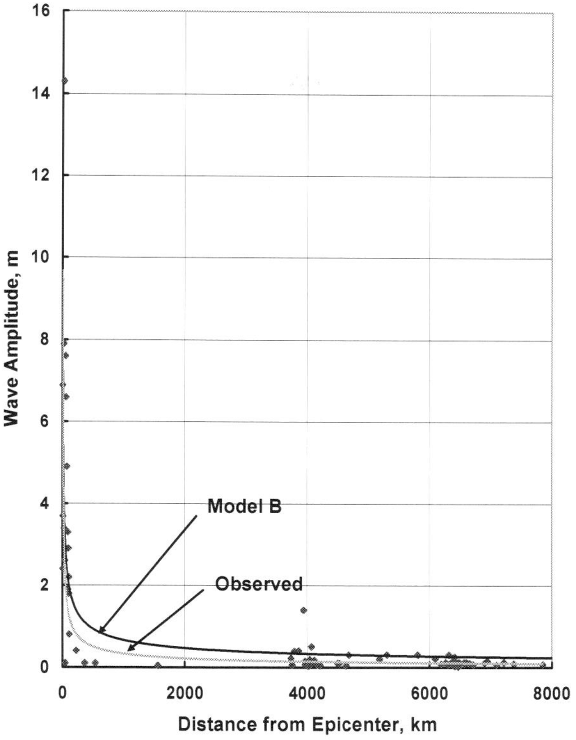

Figure 1. Observed and Model wave amplitudes of Hawaii tsunami of 1975.

A is then obtained from Eq. (23). The results give: A = 22.644; and α = -.6137. The actual variation of the wave amplitude with distance is then expressed as:

$$\psi_{Observed} = \frac{22.644}{\rho^{.6137}} \tag{29}$$

The actual wave amplitude falls off slightly faster than that predicted by Model B, which is based on conservation of energy. The difference may be assumed to represent the loss of energy due to as yet unidentified causes. The prime candidate appears to be the generation of atmospheric internal gravity waves by tsunamis, which can transport energy and momentum vertically through the atmosphere and produce travelling ionospheric disturbances (cf. Hickey, 2011). Assuming an exponential attenuation factor, we can write (vide Appendix B)

$$\psi_{Observed} = e^{-\beta\rho}\psi_{ModelB} \tag{30}$$

The attenuation factor then follows as

$$e^{-\beta\rho} = \frac{\psi_{Observed}}{\psi_{ModelB}} = \rho^{.5-\alpha} \tag{31}$$

The attenuation factor and the attenuation coefficient are calculated for this event as functions of the distance from the epicenter and entered in Table 1. It shows that the amplitude loss is most rapid near the epicenter with almost 50% of it dissipated in the first 400 km. The dissipation slows down dramatically thereafter and at 10,000 km, it is increases only to 65%. Consequently, the attenuation coefficient is not a constant but a function of the distance from the epicenter. By all accounts, the attenuation is small, suggesting the validity of models A and B.

Distance from epicenter ρ	Attenuation factor $\rho^{-.1137}$	Attenuation coefficient β
200 km	.5475	.00301/km
400 km	.5060	.00170/km
600 km	.4832	.00121/km
800 km	.4676	.000950/km
1,000 km	.4559	.000785/km
2,000 km	.4214	.000432/km
4,000 km	.3894	.000236/km
6,000 km	.3719	.000165/km
8,000 km	.3599	.000128/km
10,000 km	.3509	.000105/km

Table 1. Attenuation factor and attenuation coefficient as functions of distance from epicenter

4.2. The Great Japan tsunami of 2011

The Japan archipelago is one of the most earthquake-prone regions of the world. On 11 March 2011, a 9.0 magnitude earthquake struck on the east coast of Honshu. The epicenter was at 38.322°N latitude and 142.369°E longitude, 72 km east of Oshika peninsula with the hypocenter at a depth of 32 km below sea level (cf. http://www.tsunamiresearchcenter.com/news/earthquake-and-tsunami-strikes-japan; http://itic.ioc-unesco.org/index.php). It was the greatest earthquake to strike Japan and one of the greatest in recorded history. It was comparable to the 2004 Indian Ocean earthquake. There were an estimated 16,000 deaths, 27,000 injured and 3,000 missing (http://www.npa.go.jp/archive/keibi/higaijokya_e.pdf) with total property damage of $235 billion (according to World Bank reports), making it the costliest natural disaster of all time.

Fig. 2 provides a geometrical perspective of the 2011 Japan tsunami. The geodesic lines from the epicenter shown in the figure are great circles with a longitudinal separation of 90°, which define a 'lune' that covers one quarter of the Earth's surface area. Intersecting the great circles are 'circles of latitude' at angular distances of $\theta = \pi/4$, $\pi/2$ and $3\pi/4$ which translate to linear distances of 5,000 km, 10,000 km, and 15,000 km, respectively, from the epicenter. For constant propagation speeds, the circles of latitude define circular wavefronts. The 10,000 km great circle marks the 'equator' ($\theta = \pi/2$), past which the waves begin to converge according to Model C. A tsunami propagating in this lune does not encounter any continental landmass until after a distance of 17,000 km, which is 85% of the distance between the poles.

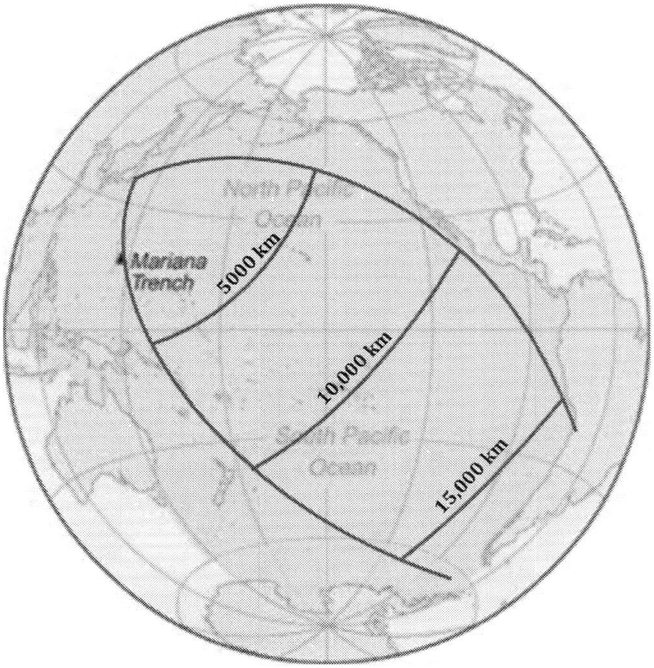

(adapted from worldatlas.com).

Figure 2. Propagation geometry of 2011 Japan tsunami in a lune of angle 90° with wavefronts at intervals of 5,000 km

The 2011 Japan tsunami was felt throughout the Pacific Ocean. Wave amplitudes were recorded at over 293 stations scattered in and around the Pacific. Fig. 3 is a scatter plot of the wave amplitude versus the distance from the epicenter. The highest amplitudes were recorded near the epicenter. Amidst a considerable scatter, a well-defined trend in the wave amplitudes emerges from the figure. The wave amplitude diminished rapidly as a function of the distance from the source, becoming nearly constant around the 10,000 km mark, and showing a discernible rise thereafter in accord with Model C. But for the intervention of the South American landmass, the waves would have converged at the anti-podal point in south Atlantic Ocean, and barring losses, the original wave amplitude restored.

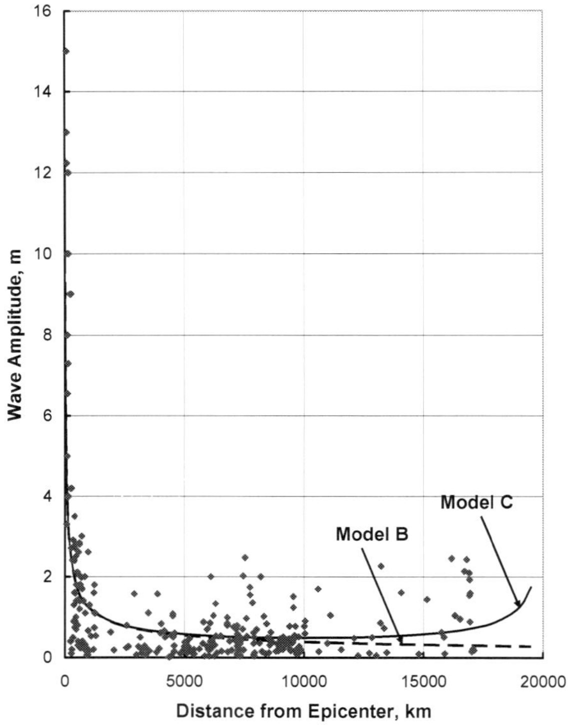

Figure 3. Observed wave amplitudes of 2011 Japan tsunami with least-squares error lines according to Models B and C.

The variation of the wave amplitude according to Model C can be written as [from Eq. (18)]:

$$\psi_{ModelC} = \frac{A}{\sqrt{\sin\theta}} \tag{32}$$

where A is a constant to be determined from the data. In terms of the radial distance ρ, we have

$$\psi_{ModelC} = \frac{A}{\sqrt{\alpha\sin\rho}} \tag{33}$$

where ρ is in kilometers and α = .009 deg/km or .00015708/km if θ is expressed in radian measure.

The constant A is determined from a single normal equation obtained by summing over the n observed data points

$$A = \sqrt{\frac{\sum \psi_{ModelC}{}^{2} \sin \alpha\rho}{n}} \tag{34}$$

giving A = .4896 m. The least-square error line according to Model C is then given by Eqs. (33) and (34). The same in Model B is obtained by Eq. (19):

$$\psi_{ModelB} = \sqrt{\frac{\sin \alpha\rho}{\alpha\rho}}\psi_{ModelC} \tag{35}$$

The regression lines according to Models B and C are superimposed on the data points in Fig. 3. As is shown in Appendix C, the difference between the two wave amplitudes is slight up to about the equator mark (10,000 km), after which the two curves begin to diverge. The predicted amplitude in Model B continues to fall according to Eq. (21), whereas that in Model C begins to rise according to Eqs. (32) and (33). The actual data clearly supports Model C and validates the convergence effect of the Earth's curvature on the wave amplitude. The same effect was earlier observed in the 8.3 magnitude Kuril islands earthquake of 2006 (Tan & Lyatskaya, 2009).

Finally, Eq. (9) of Model A furnishes a means to determine the average depth of the ocean along the travel path given the distance and the travel time to the destination:

$$d = \frac{v^2}{g} \tag{36}$$

From the observed travel times and the geodesic distances from the epicenter to 13 destinations along the coast of Chile, the average travel speeds were calculated from which the average depth of the ocean along the travel path determined (Table 2). The mean average speed of 739 kph yielded a mean average depth of 4303 m for the Pacific Ocean along these paths, which compares favorably with various estimates found in the literature: e.g., 4282 m (Herring & Clarke, 1971), 4190 m (Smith & Demopoulos, 2003), 4267 m (http://oceanservice.noaa.gov) , and 4080 m (britannica online encyclopedia).

The average depth of the Pacific Ocean (4300 m) is considerably grater than those of the Atlantic Ocean (3600 m) and Indian Ocean (3500 m) (cf. Herring & Clarke, 1971). Further, the smaller north-western half of the Pacific Ocean is substantially deeper than the larger south-eastern remainder, even though separate depth figures are hard to find in the literature. In order to estimate the average depth of north-western Pacific Ocean, we consider the travel times of the tsunami to reach various destinations on the coasts of the Hawaiian islands, which lie entirely in that region (Table 3). Travel time data from the epicenter to 8 destinations in the Hawaiian islands yield a mean average speed of 798 kph for a mean average depth of 5016 m for north-western Pacific Ocean along these paths. This confirms the fact that north-western Pacific Ocean is considerably deeper than the south-eastern remainder. These travel time studies further re-affirm the validity of the 'long wave in shallow water' approximation for tsunami propagation.

Latitude of Destination	Destination	Distance from Epicenter	Travel Time	Average Speed	Mean Average Speed	Average Depth	Mean Average Depth
18.467°S	Arica	16,166 km	21 h 26 m	754 kph		4,479 m	
20.217°S	Iquique	16,308 km	21 h 18 m	766 kph		4,615 m	
23.650°S	Antofagasta	16,522 km	21 h 33 m	767 kph		4,628 m	
27.067°S	Caldera	16,693 km	21 h 44 m	768 kph		4,645 m	
29.933°S	Coquimbo	16,799 km	22 h 04 m	761 kph		4,563 m	
33.033°S	Valparaiso	16,911 km	22 h 13 m	761 kph		4,562 m	
33.583°S	San Antonio	16,932 km	22 h 11 m	763 kph	739 kph	4,587 m	4,303 m
35.356°S	Constitucion	16,921 km	23 h 08 m	731 kph		4,213 m	
36.683°S	Talcahuano	16,905 km	22 h 57 m	737 kph		4,272 m	
39.867°S	Corral	16,946 km	22 h 54 m	740 kph		4,312 m	
41.483°S	P. Montt	17,006 km	25 h 20 m	671 kph		3,548 m	
45.467°S	72.339°W	17,065 km	25 h 06 m	680 kph		3,639 m	
54.933°S	P. Williams	17,115 km	24 h 05 m	711 kph		3,976 m	

Table 2. Travel Times, Average Speed and Average Depth of Pacific Ocean.

Latitude of Destination	Destination	Distance from Epicenter	Travel Time	Average Speed	Mean Average Speed	Average Depth	Mean Average Depth
21.960°N	Kauai	5,792 km	7 h 10 m	808 kph		5,143 m	
21.437°N	Mokuoloe	5,961 km	7 h 32 m	791 kph		4,930 m	
21.300°N	Honolulu	5,962 km	7 h 27 m	800 kph		5,042 m	
20.898°N	Maui	6,108 km	7 h 40 m	797 kph		4,998 m	
20.780°N	Lanai	6,071 km	7 h 48 m	778 kph	798 kph	4,770 m	5,016 m
20.036°N	Kawaihae	6,217 km	7 h 48 m	797 kph		5,002 m	
19.733°N	Hilo	6,302 km	7 h 56 m	794 kph		4,968 m	
19.634°N	156.507°W	6,182 km	7 h 33 m	819 kph		5,279 m	

Table 3. Travel Times, Average Speed and Average Depth of north-western Pacific Ocean.

4.3. The Great Chilean tsunami of 1960

On Sunday, May 22, 1960, at 19:11 GMT (15.11 LT), a super-massive earthquake occurred off the coast of south central Chile, with epicenter at 39.5°S latitude, 74.5°W longitude and focal depth of 33 km (cf. http://earthquake.usgs.gov/earthquakes/world/events/1960_05_22_tsunami.php). It happened when a piece of the Nazca Plate of the Pacific Ocean subducted beneath the South American Plate. The magnitude of the earthquake of 9.5 makes it the most powerful earthquake in recorded history (Kanamori, 2010). The tsunami generated by the earthquake, along with coastal subsidence and flooding, caused tremendous damage along the Chilean coast where an estimated 2,000 people lost their lives (http://neic.usgs.gov/neis/eq_depot/world/1960_05_22articles.html). The resulting tsunami raced across the Pacific Ocean causing the death of 61 people in Hawaii and 200 others in Japan and elsewhere (USGS reports). The estimated damage costs were near half a billion dollars.

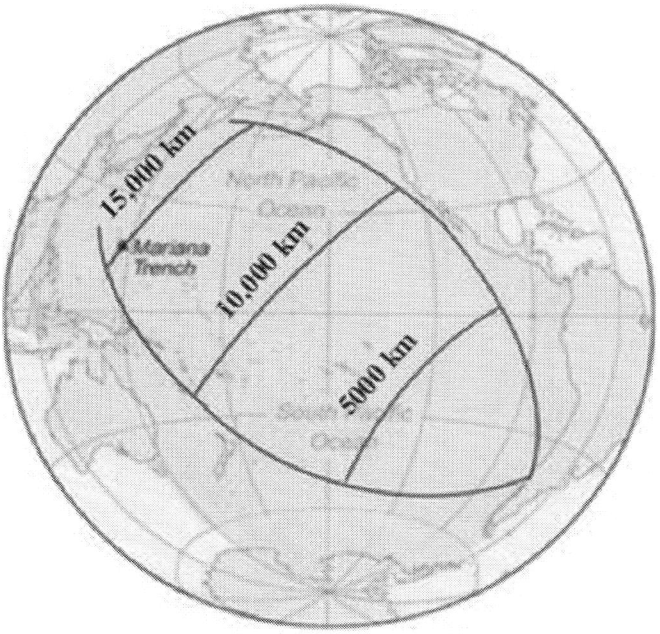

(adapted from worldatlas.com).

Figure 4. Propagation geometry of 1960 Chilean tsunami in a lune of angle 90° with wavefronts at intervals of 5,000 km

The Great Chilean tsunami of 1960 is similar to the Great Japan tsunami of 2011 coming from the opposite direction, only having greater amplitude. Both of these tsunamis, as well as many other analogous ones, traversed the longest stretch of continuous water covering over 85% of the distance between the poles. Thus the amplitude variations with distance of both of these tsunamis were similar. The strength of the Great Chilean tsunami was such that reflected waves from the Asian coasts were detectable (http://www.soest.hawaii.edu/GG/ASK/chile-tsunami.html).

Fig. 4 provides the geometrical perspective of the 1960 Chilean tsunami. As in the earlier example, the geodesic lines from the epicenter shown in the figure are great circles with a longitudinal separation of 90°, which define a 'lune' that covers one quarter of the Earth's surface area. Intersecting the great circles are 'circles of latitude' at angular distances of $\theta = \pi/4$, $\pi/2$ and $3\pi/4$ which translate to linear distances of 5,000 km, 10,000 km, and 15,000 km, respectively, from the epicenter. For constant propagation speeds, the circles of latitude define circular wavefronts. The 10,000 km great circle marks the 'equator' ($\theta = \pi/2$), past which the waves begin to converge according to Model C.

The Great Chilean tsunami of 1960 was felt throughout the Pacific Ocean. Over 1,000 measurements of wave amplitudes were recorded at 815 stations scattered in and around the Pacific. Fig. 5 is a scatter plot of the wave amplitude versus the distance from the epicenter. The highest amplitudes were recorded near the Chilean coast and at the diametrically opposite end, mostly on the Japanese coasts, where hundreds of data points were clustered. There is a second cluster past the 10,000 km mark at the Hawaiian Islands, where numerous measurements were taken. The over-all trend of the data points closely agrees with that predicted by the 'spherical ocean' Model C.

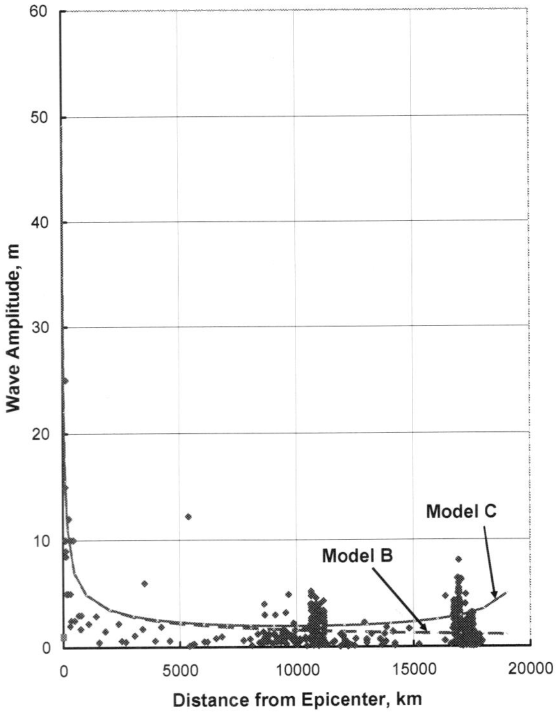

Figure 5. Observed wave amplitudes of 1960 Chilean tsunami with least-squares error lines according to Models B and C.

The least-squares regression lines of the data points as obtained from Eqs. (32) - (35) are shown in Fig. 5, with the amplitude constant $A = 1.9488$ m. This is 3.98 times that of the 2011 Japan tsunami constant. Since the energy density of the waves is proportional to A^2, this indicates that the energy density of the 1960 Chilean tsunami was 15.84 times that of the 2011 Japan tsunami, thus illustrating the power of the most powerful earthquake in recorded history. Normally, a 9.5 magnitude earthquake releases about 5.5 times the energy an a 9.0 earthquake (cf. Kanamori, 1977).

From the observed travel times and distances from the epicenter to destinations on the eastern coast of Japan, the average travel speeds were calculated and the average depth of the ocean determined along these paths (table 4). The mean average depth of 4,231 m compares favorably with the value of 4,303 m obtained from the 2011 Japan tsunami (Table 2). Also calculated were the travel speeds from the epicenter to locations on the Hawaiian Islands and the average depth of the ocean (Table 5). The mean average depth of 3,972 m reaffirms the fact that the north-western Pacific Ocean, at 5,016 m (Table 3) is substantially deeper than its south-eastern compliment.

Latitude of Destinatio n	Destination	Distance from Epicenter	Travel Time	Average Speed	Mean Average Speed	Average Depth	Mean Average Depth
41.783°N	Hakodate	17,058 km	23 h 27 m	727 kph	733 kph	4,166 m	4,231 m
39.267°N	Kamaishi	16,914 km	22 h 24 m	755 kph		4,489 m	
35.670°N	Tokyo	16,989 km	22 h 59 m	739 kph		4,302 m	
28.383°N	Nase	17,496 km	24 h 39 m	710 kph		3,967 m	

Table 4. Travel Times, Average Speed and Average Depth of Pacific Ocean

Latitude of Destinatio n	Destination	Distance from Epicenter	Travel Time	Average Speed	Mean Average Speed	Average Depth	Mean Average Depth
19.730°N	Coconut I.	10,621 km	15 h 30 m	685 kph	710 kph	3,697 m	3,972 m
19.733°N	Hilo	10,621 km	14 h 47 m	718 kph		4,064 m	
20.898°N	Kahului	10,817 km	15 h 07 m	716 kph		4,032 m	
21.300°N	Honolulu	10,957 km	15 h 22 m	713 kph		4,003 m	
28.960°N	Nawiliwili	11,124 km	15 h 29 m	718 kph		4,064 m	

Table 5. Travel Times, Average Speed and Average Depth of south-eastern Pacific Ocean

4.4. The Great Indian Ocean tsunami of 2004

The great 2004 Indian Ocean earthquake occurred off the west coast of Sumatra, Indonesia, on Boxing Day, December 26, 2004. Its revised magnitude of 9.2 makes it the second largest earthquake in recorded history, after the 9.5 magnitude Chilean earthquake of 1960 (cf. http://walrus.usgs.gov/tsunami/sumatraEQ/; Lay, et al., 2005). In terms of human casualties, however, it was the greatest natural disaster in recorded history, by far. The earthquake generated a super-massive tsunami that took the lives of an estimated 230,000 people in Indonesia, Sri Lanka, India, Thailand and elsewhere (Mörner, 2010). More than 1,000,000 people were displaced in the aftermath following the tsunami, which was eventually registered at every coast of the world ocean.

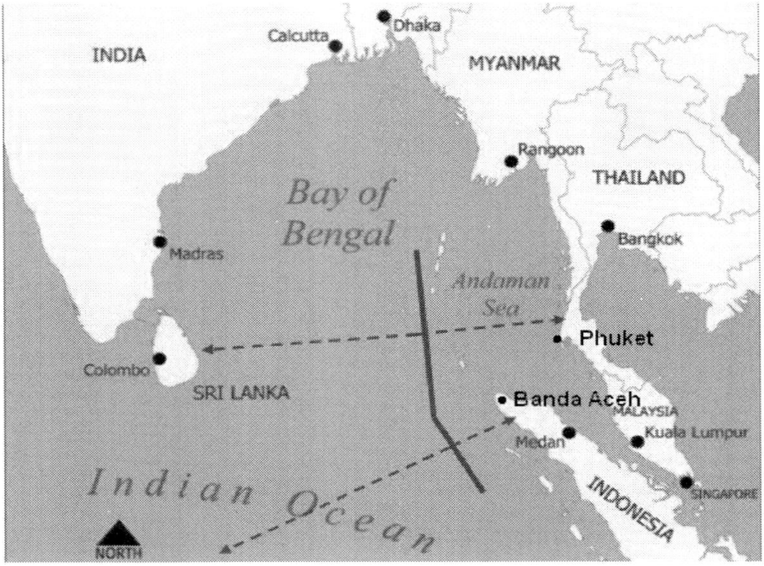

Figure 6. Propagation geometry of 2004 Indian Ocean tsunami showing the two major faultiness and propagation along their perpendicular bisectors (adapted from phuket.news.com).

This great earthquake was caused by the subduction of the Indo-Australian plate under the Eurasian plate near the Andaman and Nicobar Islands chain and its extension southwards under the Bay of Bengal (Fig. 6). Its hypocenter is listed at 3.295°N altitude and 95.982°W longitude. However, its fault-line was 1000 km long, which roughly consisted of two linear segments (cf. Kowalik, et al., 2005): (1) a 700 km long section off the west coasts of Andaman and Nicobar Islands; and (2) a 300 km section west of Aceh province of Sumatra, Indonesia (Fig. 6). The alignments of both the segments were generally north-south, with the southern segment titled slightly towards the south-easterly direction. The shorter southern segment had the more intense earthquake and generated the greater tsunami. Consequently, the epicenter lied on this segment of the fault-line. The earthquake is also variously referred to as the Sumatra-Andaman earthquake or the Boxing Day earthquake.

It is evident from Fig. 6 that the eastward tsunami from the southern segment of the fault-line (henceforth referred to as the Sumatra fault-line) had a direct impact on the Aceh province in northern Sumatra, where the highest waves of over 50 m were registered. More than half of all casualties were reported there. The westward tsunami from this fault-line, on the other hand, passed harmlessly over the open ocean, reaching the east coast of South

Africa, and entering the Atlantic Ocean. The northern segment of the fault-line (henceforth called the Andaman-Nicobar fault-line) produced tsunami which affected greater areas of landmass. Phuket lied near the perpendicular bisector of this fault and took a direct hit from the tsunami as did other coastal locations of Thailand and Myanmar. To the west, the east coasts of Sri Lanka and southern India were greatly affected. The tsunami rolled over the Maldive Islands and reached the eastern coast of Africa (Somalia, in particular), causing damage there. In consolation, Bangladesh, a densely populated area to the north, was spared the devastation.

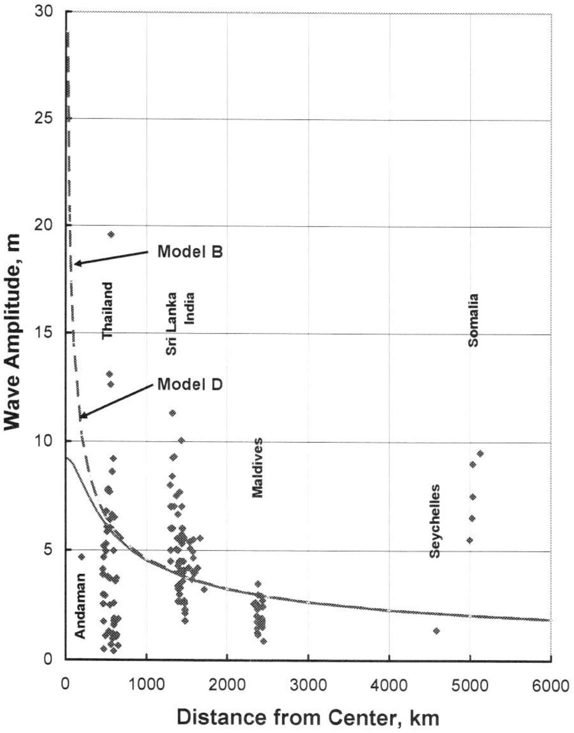

Figure 7. Observed and Model wave amplitudes of 2004 Indian Ocean earthquake due to the Andaman-Nicobar fault-line.

Assuming that the tsunami from the Sumatra fault-line was largely intercepted by the island (vide Fig. 6), we proceed to analyze the tsunami propagation from the Andaman-Nicobar fault-line based on Models B and D. For this, we have to consider the mid-point of the fault-line (approximately 9°N latitude and 92.5°E longitude) as the origin of the tsunami instead of the epicenter which lay on the Sumatra fault-line. Distances are now reckoned from this new center. If the coordinates (latitude, longitude) of the source and destination points be

(λ_s, ϕ_s) and (λ_d, ϕ_d), respectively, the angle subtended at the center of the Earth by the two locations is, from spherical trigonometry,

$$\theta = \cos^{-1}\left[\sin\lambda_s \sin\lambda_d + \cos\lambda_s \cos\lambda_d \cos\left(\phi_d - \phi_s\right)\right] \tag{37}$$

The linear distance between the two points on the surface of the Earth is then $d = \theta r_e$, where r_e = 6371 km, is the volumetric radius of the Earth prescribed by the International Union of Geodesy and Geophysics.

Fig. 7 is a scatter plot of the wave amplitudes as functions of the distance from the center of the Andaman-Nicobar fault-line. The data betray distinct clumps for Thailand, Sri Lanka, India and Maldives. At the far end are data for the Somalia coast. Also shown are the isolated data points for Andaman and Seychelles islands. Superimposed on the data points are the model amplitudes according to Models B and D. The model amplitudes are virtually indistinguishable for distances upwards of 1000 km (cf. Appendix D). In comparison with the Hawaii and Japan earthquakes (Figs. 1 and 3), the wave amplitudes are significantly higher, which is indicative of the magnitude of the earthquake. Unlike Hawaii and Japan earthquakes, however, there is a dearth of data points with high amplitudes near the origin. This is because the earthquake occurred beneath the ocean and there was no land close to it. Nevertheless, the lack of high wave amplitudes close to the fault-line appears to support the validity of Model D.

As before, the wave amplitudes according to Model B is given by (vide Eq. D.12):

$$\psi_{ModelB} = \frac{A}{\sqrt{\rho_b}} \tag{38}$$

where A is a constant to be determined from regression analysis of the data points. By summing over the data, we have

$$A = \frac{\sum \psi}{\sum \frac{1}{\sqrt{\rho_b}}} \tag{39}$$

The data yield a value of A = 146.296 m. The wave amplitudes in accordance with Model D then follows (vide Eq. D.8):

$$\psi_{ModelD} = \frac{\psi_{ModelB}}{\sqrt[4]{1 + \frac{c^2}{2\rho_b{}^2}}} \tag{40}$$

Here c = 350 km.

From the travel times and the travel distances from the epicenter on the Sumatra fault-line to five destinations on the South African coast and one on the Antarctic coast, the average speed of the tsunami was calculated, and the average depth of the Indian Ocean determined (Table 6). The average speed of 700 kph translates to an average depth of 3,840 m which is quite consistent with the reference figure of 3,963 m found in the literature (Herring & Clarke, 1971). Once again, the tsunami speed and depth of the ocean predicted by the 'long waves in shallow water' model turns out to be reliable.

Latitude of Destinatio n	Destination	Distance from Epicenter	Travel Time	Average Speed	Mean Average Speed	Average Depth	Mean Average Depth
28.800°S	Richards Bay	7,683 km	11 h 04 m	694 kph	700 kph	3,795 m	3,840 m
33.027°S	East London	8,193 km	11 h 29 m	713 kph		4,008 m	
33.958°S	Pt. Elizabeth	8,423 km	12 h 13 m	698 kph		3,743 m	
34.178°S	Mossel Bay	8,743 km	13 h 02 m	671 kph		3,543 m	
34.188°S	Simons Bay	9,078 km	12 h 53 m	705 kph		3,909 m	
69.007°S	39.584°E	9,088 km	12 h 41 m	717 kph		4,042 m	

Table 6. Travel Times, Average Speed and Average Depth of Central Indian Ocean

5. Conclusion

Tsunamis are complex geophysical phenomena not easily amenable to theoretical rendering. The formal theory based on solitary wave model is not easily accessible to the great majority of the scientific readers. This chapter has demonstrated that the 'long waves in shallow water' approximation of the tsunami explains many facets of tsunami propagation in the open ocean. The one-dimensional model provides accurate assessments of the general properties such as dispersion-less propagation, speed of propagation, bending of tsunamis around obstacles and depth of the ocean, among others. Two-dimensional models on flat and spherical ocean substantially account for the wave amplitudes for far-reaching tsunami propagation. Finally, the finite line-source model satisfactorily predicts the wave amplitudes near the source when the tsunami is caused by a long subduction zone.

Appendix

A. Propagation of water waves in one dimension

A wave is a transfer of energy from one part of a medium to another, the medium itself not being transported in which process (e.g., Coulson, 1955). The individual particles of the medium execute simple harmonic motions in one or two dimensions, depending upon the nature of the wave. For sound waves in air, the oscillations are parallel with the direction of propagation. Such waves are called *longitudinal waves*. For waves along a stretched string, the oscillations are perpendicular to the direction of propagation. Such waves are called *transverse waves*. Longitudinal waves can propagate through solids and fluid media (liquid or gas), whereas transverse waves can propagate through solids only. Longitudinal waves are sustained by compression forces, while transverse waves are due to shear forces. The sources of all waves are vibrations of some kinds.

In an earthquake, as many as four kinds of waves are produced, of which two are *body waves* and the other two are *surface waves*. The body waves are categorized as: (1) Longitudinal waves (called the *primary waves*); and (2) Transverse waves (called *secondary waves*). The surface waves, too, fall into two categories: (1) Transverse waves called *Love waves*, where the oscillations are parallel with the surface but perpendicular to the direction of propagation; and (2) *Rayleigh waves*, in which the oscillations take place in directions both parallel with and perpendicular to the direction of propagation, the latter taking place in the vertical direction. In Rayleigh waves, the particles below the surface undergo elliptical motions in the vertical plane.

Waves on the surface of water are similar to the Rayleigh waves. They are influenced by three physical factors: (1) *Gravitation*, which acts to return the disturbed surface back to the equilibrium configuration; (2) *Surface Tension*, since the pressure under a curved surface is different from that beneath a flat surface; and (3) *Viscosity*, which causes dissipation of energy. Of these, gravity is the dominant force except in the very short wavelength regions. Water waves controlled by gravity are called *gravity waves*. When surface tension dominates, the waves are called *capillary waves* or *ripples*. Such waves have wavelengths shorter than 1.7 cm and are quickly damped out by viscous forces (cf. Towne, 1967; Elmore & Heald, 1969).

The theory of gravity wave propagation in one dimension is well documented in the literature. A particularly elegant treatment is found in Elmore & Heald (1969). Conventionally, x is taken as the direction of propagation and y is taken to be the vertical direction. The conditions are assumed to be uniform in the z direction, which then disappears from the view of analysis. As water is an incompressible fluid, its density ρ is constant. The equation of continuity gives

$$\vec{\nabla} \circ \vec{v} = 0 \tag{A.1}$$

where $\vec{v} = v_x \hat{x} + v_y \hat{y}$ represents the velocity. Also, for irrotational motion, we have

$$\vec{\nabla} \times \vec{v} = \vec{0} \tag{A.2}$$

According to the potential theory, $\vec{v}$ is obtained from a velocity potential:

$$\vec{v} = -\vec{\nabla}\phi \tag{A.3}$$

where ϕ is a scalar function which satisfies Laplace's equation:

$$\nabla^2 \phi = \frac{\partial^2 \phi}{\partial x^2} + \frac{\partial^2 \phi}{\partial y^2} = 0 \tag{A.4}$$

If ξ and η represent the displacements in the horizontal and vertical directions, respectively, then

$$v_x = \frac{\partial \xi}{\partial t} = -\frac{\partial \phi}{\partial x} \tag{A.5}$$

and

$$v_y = \frac{\partial \eta}{\partial t} = -\frac{\partial \phi}{\partial y} \tag{A.6}$$

At the bottom, the vertical component of velocity must vanish:

$$v_y = -\frac{\partial \phi}{\partial y} = 0, \qquad y = 0 \tag{A.7}$$

At the surface of water, the pressure is atmospheric pressure P. Omitting the non-linear term $\frac{1}{2}v^2$ in Bernoulli's equation, we can write

$$P + \rho g(h + \eta) = \rho \frac{\partial \phi}{\partial t} + const., \qquad y = h \tag{A.8}$$

where h is the depth of water and g the acceleration due to gravity. Eq. (A.8) constitutes the relation between the variation of pressure due to the vertical displacement of water and the resulting change in the velocity potential. Differentiating Eq. (A.8) with respect to t, we get

$$v_y = \frac{\partial \eta}{\partial t} = \frac{1}{g}\frac{\partial^2 \phi}{\partial t^2} \tag{A.9}$$

Our task now is to find the velocity potential which satisfies Laplace's equation (A.4) and the two boundary conditions (A.7) and (A.9). We assume a trial solution

$$\phi(x,y,t) = X(x)Y(y)T(t) \tag{A.10}$$

and apply the method of separation of variables. From Eq. (C.4), we get

$$\frac{1}{X}\frac{d^2X}{dx^2} = -\frac{1}{Y}\frac{d^2Y}{dy^2} = -k^2 \tag{A.11}$$

where the separation constant k is chosen to insure that X is a periodic function of x. Written separately, Eq. (A.11) gives

$$\frac{d^2X}{dx^2} + k^2X = 0 \tag{A.12}$$

and

$$\frac{d^2Y}{dy^2} - k^2Y = 0 \tag{A.13}$$

The general solutions to Eqs. (A.12) and (A.13) are, respectively

$$X(x) = Ae^{ikx} + Be^{-ikx} \tag{A.14}$$

and

$$Y(y) = Ce^{kx} + De^{-kx} \tag{A.15}$$

For a wave travelling in the forward direction, $B = 0$. Further, from Eq. (A.10):

$$\frac{\partial \phi}{\partial y} = XT\frac{dY}{dy} \tag{A.16}$$

The boundary condition at the bottom (A.7) dictates that $C = D$. Thus

$$Y(y) = 2C\cosh ky \tag{A.17}$$

and

$$\phi(x,y,t) = A\cosh kye^{ikx}T(t) \tag{A.18}$$

where A is a new arbitrary constant. Now from Eqs. (A.6) and (A.9), we get:

$$v_y = -\frac{\partial \phi}{\partial y} = -kA\sinh kye^{ikn}T(t) \tag{A.19}$$

and

$$v_y = \frac{1}{g}\frac{\partial^2 \phi}{\partial t^2} = \frac{1}{g}A\cosh kye^{ikx}\frac{d^2T(t)}{dt^2} \tag{A.20}$$

Eqs. (A.19) and (A.20) yield:

$$\frac{d^2T}{dt^2} + (gk\tanh kh)T = 0 \tag{A.21}$$

Hence, the velocity potential has a simple harmonic time-dependence with angular frequency

$$\omega = \sqrt{gk\tanh kh} \tag{A.22}$$

If we choose $T = e^{-i\omega t}$, the velocity potential of a gravity wave propagating in the forward x direction over water of depth h becomes

$$\phi = A\cosh kye^{i(kx-\omega t)} \tag{A.23}$$

The wave velocity of the gravity wave is thus

$$v = \frac{\omega}{k} = \sqrt{\frac{g}{k}\tanh kh} \tag{A.24}$$

B. Wave propagation on two-dimensional flat surface

The equation of a wave emanating from a point source in a two-dimensional plane is conveniently expressed in plane polar coordinates (ρ, ϕ) with the source at the origin (e.g., Zatzkis, 1960):

$$\frac{\partial^2 \psi}{\partial \rho^2} + \frac{1}{\rho}\frac{\partial \psi}{\partial \rho} + \frac{1}{\rho^2}\frac{\partial^2 \psi}{\partial \phi^2} = \frac{1}{v^2}\frac{\partial^2 \psi}{\partial t^2} \tag{B.1}$$

where v is the velocity of the wave. Assuming circular symmetry (i.e., ψ independent of ϕ), we get:

$$\frac{\partial^2 \psi}{\partial \rho^2} + \frac{1}{\rho}\frac{\partial \psi}{\partial \rho} = \frac{1}{v^2}\frac{\partial^2 \psi}{\partial t^2} \tag{B.2}$$

To apply the method of separation of variables, let

$$\psi(\rho,t) = P(\rho)T(t) \tag{B.3}$$

Then Eq. (B.2) becomes

$$T\frac{d^2 P}{d\rho^2} + \frac{T}{\rho}\frac{dP}{d\rho} = \frac{P}{v^2}\frac{d^2 T}{dt^2} \tag{B.4}$$

Dividing both sides by ψ, separating the variables, and letting each side equal to a constant ($-k^2$), we get

$$\frac{1}{P}\frac{d^2 P}{d\rho^2} + \frac{1}{P\rho}\frac{dP}{d\rho} = \frac{1}{v^2 T}\frac{d^2 T}{dt^2} = -k^2 \tag{B.5}$$

The t-equation is

$$\frac{d^2 T}{dt^2} + k^2 v^2 T = 0 \tag{B.6}$$

whose solution is

$$T \propto e^{\pm ikvt} \tag{B.7}$$

The ρ-equation is

$$\rho^2 \frac{d^2P}{d\rho^2} + \rho \frac{dP}{d\rho} + k^2\rho^2 P = 0 \tag{B.8}$$

or

$$\rho \frac{d}{d\rho}\left(\rho \frac{dP}{d\rho} \right) + k^2\rho^2 P = 0 \tag{B.9}$$

This is Bessel's equation of the zeroth order. A novel technique to solve this equation is found in Irving & Mullineaux (1959). Let

$$P = \frac{R}{\sqrt{\rho}} \tag{B.10}$$

Then Eq. (B.9) assumes the form

$$\frac{d^2R}{d\rho^2} + k^2\left[1 + \frac{1}{4k^2\rho^2} \right] R = 0 \tag{B.11}$$

For large ρ (i.e., away from the source), the second term within the square bracket may be neglected. Thus

$$\frac{d^2R}{d\rho^2} + k^2 R = 0 \tag{B.12}$$

giving the solution

$$R \propto e^{ik\rho} \tag{B.13}$$

Hence, from Eq. (B.10):

$$P \propto \frac{e^{ik\rho}}{\sqrt{\rho}} \tag{B.14}$$

and

$$\psi(\rho,t) = P(\rho)T(t) = \psi_0 \frac{e^{ik(\rho \pm vt)}}{\sqrt{\rho}} \tag{B.15}$$

Since the wave is propagating outwards, we retain the + sign only, giving

$$\psi(\rho,t) = \psi_0 \frac{e^{ik(\rho - vt)}}{\sqrt{\rho}} \tag{B.16}$$

In terms of the wave number k,

$$\psi(\rho,t) = \psi_0 \frac{e^{i(k\rho-\omega t)}}{\sqrt{\rho}} \tag{B.17}$$

The amplitude of the wave falls off inversely as the square-root of the distance from the source:

$$\psi_{ModelB} \propto \frac{1}{\sqrt{\rho}} \tag{B.18}$$

The intensity of the wave (i.e., the energy density) is thus inversely proportional to the distance from the source:

$$I_{ModelB} \propto \psi_{ModelB}{}^2 \propto \frac{1}{\rho} \tag{B.19}$$

There is a simple alternative procedure to obtain Eq. (B.18) from Eq. (B.19) (Tan & Lyatskaya, 2009). Assuming conservation of energy, the energy spreads out in concentric circles of radius ρ, so that the intensity of the wave varies inversely as the radial distance whence the wave amplitude varies according to Eq. (B.18).

When k is complex, we can write $k = \kappa + i\beta$., Eq. (B.16) then takes the form

$$\psi(\rho,t) = \psi_0 \frac{e^{-\beta\rho} e^{i(k\rho-\omega t)}}{\sqrt{\rho}} \tag{B.20}$$

In Eq. (B.20), $e^{-\beta\rho}$ gives the attenuation factor when loss processes are present, with β representing the attenuation coefficient.

C. Wave propagation on two-dimensional spherical surface

For long-distance propagation of waves on a spherical surface, the curvature of the surface must be taken into account (Tan & Lyatskaya, 2009). Use spherical coordinates (r, θ, ϕ) with the center of the sphere as the origin and place the source of the wave at the north pole (θ = 0) (Fig. C.1). Assume azimuthal symmetry, i.e., ψ independent of ϕ. Then circular waves will propagate on the surface of the sphere ($r = a$) outwards in the direction of increasing zenith angle θ. The wave-fronts will be small circles on the sphere having radii $a\sin\theta$ and circumferences $2\pi a\sin\theta$. The energy conservation principle now requires that the intensity of the wave varies inversely as $\sin\theta$:

$$I_{ModelC} \propto \psi^2 \propto \frac{1}{\sin\theta} \tag{C.1}$$

Hence the amplitude of the wave varies inversely as the square-root of $\sin\theta$:

$$\psi_{ModelC} \propto \frac{1}{\sqrt{\sin\theta}} \tag{C.2}$$

The amplitude and intensity of the wave decrease from the north pole ($\theta = 0$) to the equator ($\theta = \pi/2$), but increase thereafter until the wave converges at the south pole ($\theta = \pi$), where the original amplitude and intensity are restored.

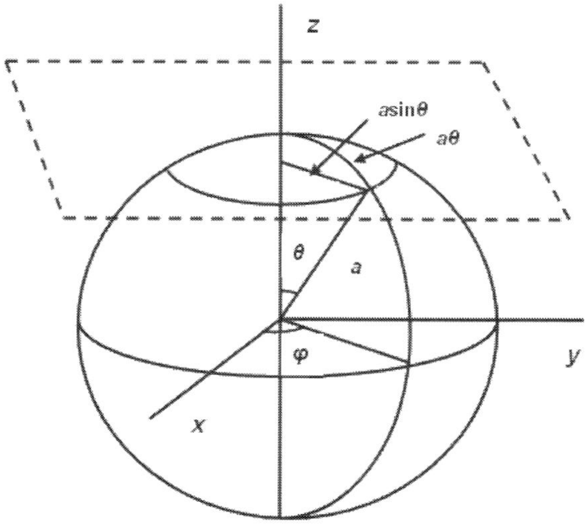

Figure C.1. Spherical surface with source at north pole ($\theta = 0$).

In order to compare the spherical surface solution with that of the flat surface solution (cf. Bhatnagar, et al., 2006), we notice that the corresponding circular wave-fronts on the flat surface will have radii of $a\theta$ and circumferences $2\pi a\theta$ (Fig. C.1). Thus the ratio of the intensities of the waves will be

$$\frac{I_{ModelC}}{I_{ModelB}} = \frac{I_{spherical}}{I_{flat}} = \frac{\theta}{\sin\theta} \tag{C.3}$$

Hence, the ratio of the amplitudes of the waves is

$$\frac{\psi_{ModelC}}{\psi_{ModelB}} = \frac{\psi_{spherical}}{\psi_{flat}} = \sqrt{\frac{\theta}{\sin\theta}} \tag{C.4}$$

Table C.1 shows the ratios of the wave intensities and amplitudes in the spherical and flat surface solutions. The values are independent of the radius of the sphere and are therefore applicable to all spherical surfaces. The departures of the spherical surface solutions from those of the flat surface solutions are slight for small values of θ. Even at the equator ($\theta = \pi/2$), the enhancements of the wave intensity and amplitude are merely 57% and 25% respectively. The enhancement becomes progressively greater until the south pole ($\theta = \pi$) is reached, where it becomes infinite.

θ, deg	θ, rad	$\sin\theta$	$\frac{\theta}{\sin\theta}$	$\sqrt{\frac{\theta}{\sin\theta}}$
0	0	0	1	1
15	.2618	.2588	1.0115	1.0057
30	.5236	.5000	1.0472	1.0233
45	.7854	.7071	1.1107	1.0539
60	1.0472	.8660	1.2092	1.0996
75	1.3090	.9659	1.3552	1.1641
90	1.5708	1.0000	1.5708	1.2533
105	1.8326	.9659	1.8972	1.3774
120	2.0944	.8660	2.4184	1.5551
135	2.3562	.7071	3.3322	1.8254
150	2.6180	.5000	5.2360	2.2882
165	2.8798	.2588	11.1267	3.3357
180	3.1426	0	∞	∞

Table C.1. Intensity and Amplitude Ratios in Spherical and Flat Surface Solutions.

D. Wave propagation from finite line source in two dimensions

Consider a finite line source of strength q and length $2c$ in the x - y plane, whose end-points are located at (- c, 0) and (c, 0) (Fig. D.1). The velocity potential ψ can be expressed as

$$\psi = \varphi(x,y)T(t) \tag{D.1}$$

where $\phi(x,y)$ satisfies the two-dimensional Laplace's equation

$$\frac{\partial^2\varphi}{\partial x^2}+\frac{\partial^2\varphi}{\partial y^2}=0 \tag{D.2}$$

The potential due to a line source constitutes a well-known problem in electrostatics (cf. Abraham & Becker, 1950). The equi-potential lines on which ϕ (x, y) is constant are con-focal ellipses with their foci located at the end-points of the line source. The lines of force are con-focal hyperbolas perpendicular to the ellipses (cf. Morse & Feshbach, 1953). In the case of a tsunami from a linear subduction zone, the ellipses represent the wave-fronts, whereas the hyperbolas are the lines along which the enrrgy is transferred (Fig. D.1).

The amplitude of the wave can be obtained by a practical approach similar to that of Tan & Lyatskaya (2009). An approximate expression for the perimeter of an ellipse with semi-major axis a and semi-minor axis b is found in the literature (cf. Weisstein, 2003):

$$L \approx \pi\sqrt{2\left(a^2+b^2\right)} \tag{D.3}$$

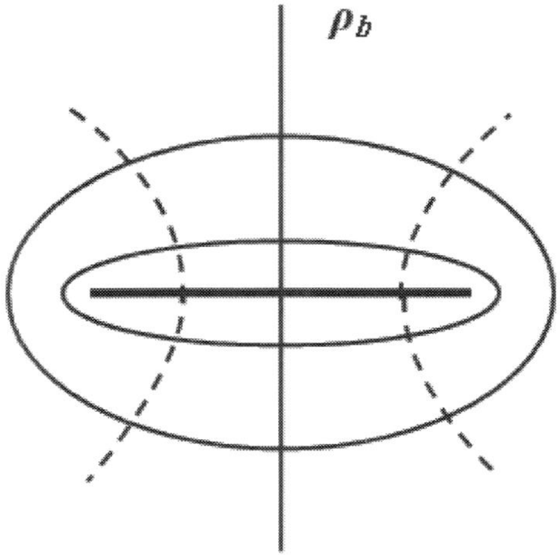

Figure D.1. Elliptical wave-fronts on which potential due to finite line source is constant.

From the property of the ellipse, $a^2 = b^2 + c^2$ (cf. Tan, 2008). Thus

$$L \approx \pi\sqrt{2b^2 + c^2} \tag{D.4}$$

As the elliptical wave-front propagates outwards, its intensity diminishes inversely as L, giving

$$I_{ModelD} \propto \frac{1}{\sqrt{b^2 + \frac{c^2}{2}}} \tag{D.5}$$

Hence, the wave amplitude varies as

$$\psi_{ModelD} \propto \frac{1}{\sqrt[4]{b^2 + \frac{c^2}{2}}} \tag{D.6}$$

Along the y-axis or the perpendicular bisector of the line source, we can equate b with the distance from the center of the source along that axis ρ_b:

$$\psi_{ModelD} \propto \frac{1}{\sqrt[4]{\rho_b^{\,2} + \frac{c^2}{2}}} \tag{D.7}$$

One can compare the wave amplitude in Model D with that of a point source of equivalent strength (Model B). Remembering Eq. (B.18), we can re-write Eq. (D.7) as:

$$\psi_{ModelD} \propto \frac{1}{\sqrt{\rho_b}} \frac{1}{\sqrt[4]{1+\frac{c^2}{2\rho_b^2}}} = \psi_{ModelB} \frac{1}{\sqrt[4]{1+\frac{c^2}{2\rho_b^2}}} \tag{D.8}$$

Thus

$$\frac{\psi_{ModelD}}{\psi_{ModelB}} = \frac{1}{\sqrt[4]{1+\frac{c^2}{2\rho_b^2}}} \tag{D.9}$$

Table D.1 shows the comparative values of the wave amplitudes in the line source (Model D) and point source (Model B) models. They differ greatly near the origin ($\rho_b = 0$), but the solutions begin to converge away from the origin. At $\rho_b = .2c$, the Model D value is only half that of Model B. For $\rho_b > c$, the difference is below 10%. For $\rho_b = 10c$, the two solutions are virtually indistinguishable.

$\frac{\rho_b}{c}$	$\frac{1}{\sqrt[4]{\rho_b^2+\frac{c^2}{2}}}$	$\frac{1}{\sqrt{\rho_b}}$	$\frac{1}{\sqrt[4]{1+\frac{c^2}{2\rho_b^2}}}$
0	1.189	∞	0
.1	1.183	3.162	.374
.2	1.167	2.236	.521
.3	1.141	1.826	.625
.4	1.109	1.581	.702
.5	1.075	1.414	.760
.6	1.038	1.291	.804
.7	1.003	1.195	.839
.8	.968	1.118	.866
.9	.935	1.054	.887
1.0	.904	1.000	.904
1.5	.777	.816	.951
2.0	.687	.707	.971
3.0	.570	.577	.987
4.0	.496	.500	.992
5.0	.445	.447	.995
6.0	.407	.408	.996
7.0	.377	.378	.997
8.0	.353	.354	.998
9.0	.333	.333	.999
10.0	.316	.316	1.000

Table D.2. Amplitude Ratios in Line Source and Point Source Solutions.

Author details

A. Tan, A.K. Chilvery and M. Dokhanian
Alabama A & M University, Normal, Alabama, USA

S.H. Crutcher
US Army RD&E Command, Redstone Arsenal, Alabama, USA

Acknowledgement

This study was partially supported by a grant from the National Science Foundation HBCU-UP HRD 0928904.

6. References

Abraham, M. & Becker, R. (1950). *The Classical Theory of Electricity and Magnetism*, Blackie and Sons, ISBN 545854, London, UK.

Bhatnagar, V.P., Tan, A. & Ramachandran, R. (2006). On response of the exospheric temperature to the auroral heating impulses during geomagnetic disturbances, *J. Atmos. Sol. Terr. Phys.*, Vol. 68, pp. 1237–1244, ISSN 1364-6826, UK.

Boussinesq, J. (1871). Théorie de l'intumescence liquide appelée onde solitaire ou de translation se propageant dans un canal rectangulaire, *Comptes Rendus*, Vol. 72, pp. 755-759, ISSN: 0567-7718, France.

Cecioni, C. & Bellotti, G. (2011). Generation and propagation of frequencydispersive tsunami, In *The tsunami threat – Research and technology* (2011). Mörner, N.-A., ed., InTech, ISBN 978-953-307-552-5, Croatia.

Coulson, C.A. (1955). *Waves*, Oliver and Boyd, ISBN 0582449545, Edinburhg, UK.

Dudley, W.C. & Lee, M. (1988). *Tsunami!*, University of Hawaii Press, ISBN: 0824819691, Honolulu, Hawaii, USA.

Elmore, W.C. & Heald, M.A. (1969). *Physics of Waves*, Dover, ISBN 978-0486649269, New York, USA.

Helene, O. & Yamashita, M.T. (2006). Understanding the tsunami with a simple model. *Eur. J. Phys.*, Vol. 27, pp. 855-863, ISSN 0 143-0807, UK.

Herring, P.J. & Clarke, M.R. (1971). *Deep Oceans*, Praeger Publishers, ISBN-10: 0213176165, New York, USA.

Hickey, M.P. in *The tsunami threat – Research and technology* (2011). Mörner, N.-A., ed., InTech, ISBN 978-953-307-552-5, Croatia.

Imteaz, M.A., Imamura, F. & Naser, J, Numerical model for multi-layered tsunami waves, Iin *The tsunami threat – Research and technology* (2011). Mörner, N.-A., ed., InTech, ISBN 978-953-307-552-5, Croatia.

Irving, J. & Mullineaux, N. (1959). *Mathematics in Physics and Engineering*, Dover, ISBN 9780123742506, New York, USA.

Kanamori, H. (1977). The energy release of great earthquakes, J. Geophys. Res., Vol. 82, pp. 2981-1987, ISSN: 2156-2202, USA.

Kanamori, H. (2010). *Revisiting the 1960 Chilean earthquake*, in USGS Open-File Report 2010-1152, U.S. Geological Survey, USA.
Karling, H.M. (2005). *Tsunamis*, Nova Science Publ. Inc., ISBN-10: 1594545189, Hauppauge, New York, USA.
Korteweg, D.J. & de Vries, G. (1895). On the change of form of long waves advancing in a rectangular canal and on a new type of long stationary waves, *Phil. Mag.*, Ser. 5, Vol. 39, pp. 422-443, ISSN: 13642812, UK.
Kowalik, Z., Knight, W., Logan, T., and Whitmore, P. (2005). Numerical modeling of the global tsunami: Indonesian tsunami of 26 December 2004, *Sci. Tsunami Hazards*, Vol. 23, pp. 40-56, ISSN 8755-6839, USA.
Lamb, H. (1993). *Hydrodynamics*, Cambridge University Press, ISBN: 0521458684, Cambridge, UK.
Lay, T., Kanamori, H., Ammon, C., Nettles, M., Ward, S., Aster, R., Beck, S., Brudzinski, M., Butler, R., DeShon, H., Ekström, G., Satake, S. & Sipkin, S. (2005). The great Sumatra-Andaman earthquake of 26 December 2004, *Science*, Vol. 308, pp. 1127-1133, ISSN: 0036-8075, USA.
Margaritondo, G. (2005). Explaining the physics of tsunamis to undergraduate and non-physics students, *Eur. J. Phys.*, Vol. 26, pp. 401-407, ISSN 0 143-0807, UK.
Mörner, N.-A. (2010). Natural, man-made and imagined disaster, *Disaster Advances*, 3(1), pp. 3-5, ISSN: 0974-262X, India.
Morse, P.M. & Feshbach, H. (1953). *Methods of Theoretical Physics*, McGraw-Hill, ISBN 9780070433168, New York, USA.
Pararas-Carayannis, G. (1976). *The Earthquake and Tsunami of 29 November 1975 in the Hawaiian Islands*, International Tsunami Information Center Report, USA.
Parker, B.B. (2012). *The Power of the Sea: Tsunamis, Storm Surges, Rogue Waves, and Our Quest to Predict Disasters*, Palgrave MacMillan, ISBN-10: 0230120741, New York, USA.
Rayleigh, Lord (1914). On the theory of long waves and bores, *Proc. Roy. Soc.*, Ser. A, Vol. 90, pp. 219-225, ISSN: 0950-1207, UK.
Russell, S. (1844). Report on waves, *British Association Repots*, UK.
Sharman, R.V. (1963). *Vibrations and Waves*, Butterworths, ISBN 9780408692007, London, UK.
Smith, C.R. & Demopoulos, A.W.J. (2003). In *Ecosystems of the Deep Ocean*, Tyler, P.A. ed., Elsevier, ISBN-10:044482619X, Amsterdam, Holland.
Stewart, G.B. (2005). *Catastrophe in Southern Asia: The Tsunami of 2004*, Lucent Books, ISBN-10: 1590188314, Farmington Hills, Michigan, USA.
Stoker, J.J. (1957). *Water Waves*, Interscience, ISBN 9780471570349, New York, USA.
Tan, A. (2008). *Theory of Orbital Motion*, World Scientific, ISBN 9789812709110, Singapore.
Tan, A & Lyatskaya, I. (2009). Alternative tsunami models. *Eur. J. Phys.*, Vol. 30, pp. 157-162, ISSN 0 143-0807, UK.
The Indian Ocean Tsunami: The Global Response to a Natural Disaster, P.P. Karan & S.P. Subbiah ed. (2011), University Press of Kentucky, ISBN: 978-0-8131-2652-4, Lexington, Kentucky, USA.
The tsunami threat – Research and technology (2011). Mörner, N.-A., ed., InTech, ISBN 978-953-307-552-5, Croatia.

Towne, D.H. (1967). *Wave Phenomena*, Dover, ISBN 9780201075854, New York, USA.

Tsunami (2012). Lopez, G.I. ed., InTech, ISBN 979-953-307-835-8, Croatia.

Weisstein, E.W. (2003). *CRC Concise Encyclopedia of Mathematics*, Chapman-Hall, ISBN 1584883472, Champaign, USA.

Zatzkis, H. in *Fundamental Formulas in Physics* Vol. 1 (1960). Menzel, D.H., ed., Dover, ISBN 9780486605951, New York, USA.

Chapter 7

A New Approach to Short-Term Tsunami Forecasting

Yury Korolev

Additional information is available at the end of the chapter

http://dx.doi.org/10.5772/51345

1. Introduction

A short-term tsunami forecast and effective tsunami warning is a key problem of the tsunami service. At present, the conventional method for short-term tsunami forecasting is based on seismological information only (earthquake magnitude, time of the main shock and epicentre location). An earthquake magnitude which exceeds the established threshold value, which is different for different tsunamigenic zones, typically results in the issuing of a tsunami warning (Gusiakov, 2011). This approach, based on the "magnitude-geographical principle", is straightforward and rather effective: at least it ensures few tsunami omissions. However, this method cannot provide sufficiently accurate estimates of tsunami wave heights, especially for specific coastal areas. Firstly, this is because the dependence of the tsunami intensity on the earthquake magnitude is far from being deterministic (Gusiakov, 2011): strong earthquakes can produce weak tsunamis (or no tsunamis at all) and, vice versa, sometimes destructive tsunamis are induced by relatively weak earthquakes. Secondly, because actual tsunami run-ups are strongly variable along the coast (c.f. MacInnes et al., 2009), the tsunami alarm can be justified for one coastal site, but false for another. Moreover, the magnitude-geographical warning method does not allow evaluation of the duration of the tsunami alarm or prediction of the arrival time of the largest tsunami wave (not necessarily the first, which is quite typical for tsunami events (c.f. Rabinovich et al., 2008)). Consequently, up to 75% of tsunami warnings turn out to be false (Gusiakov, 2011; Poplavsky et al., 1997). False tsunami warnings cause anxiety among emergency managers, government officials and the business community. These false alarms result in actual financial losses, sometimes significant, including losses due to production downtime and expenses for emergency evacuation procedures and navigation of vessels to the open sea (NOAA Tsunami, 2004).

According to the modern concept, a tsunami alarm should be declared in good time and only selectively for exposed coastal areas where the incoming tsunami will be a real threat.

Each tsunami alarm should be accompanied by trustworthy information on the expected tsunami arrival time, wave heights along the coast, their frequencies and the duration for dangerous oscillations to remain in effect. However, based merely on the seismological data such information cannot be obtained. Meanwhile, emergency managers and local officials are in urgent need the effective operational tools that can provide reliable tsunami forecasts as guidance for rapid, critical decisions in which lives and property are at stake (Titov et al., 2005).

From this point of view, direct tsunami monitoring in the open-ocean by DART[1] stations can significantly improve the situation (Titov et al., 2005; Titov, 2009). These deep-ocean real-time tsunameters have been developed by the Pacific Marine Environmental Laboratory (PMEL/NOAA) for the early detection, measurement and real-time reporting of far-field tsunamis. These stations have been deployed at strategic locations throughout the Pacific Ocean near the main seismically active regions with a history of tsunami generation (Mofjeld, 2009). DART tsunami monitoring systems were expected to become the backbone of the modern Tsunami Warning System (TWS) in the Pacific Ocean.

The principal methodology of the tsunami forecast based on open ocean sea level measurements was developed during 1987-2005 (Chubarov & Shokin, 1995; Chung, et al., 1995; Korolev & Poplavsky, 1995; Poplavsky et al., 1997; Satake, 1987; Titov et al., 2005; Voronina & Tcheverda, 1998; Wei et al., 2003). However, only recent advances in tsunami measurement and numerical modelling technology have made it possible to create effective tsunami forecasting systems. The tsunami forecasting technology elaborated at PMEL is based on the integration of real-time DART measurements and state-of-the-art numerical modelling (Titov et al., 2005). This technology was applied in an operative regime for several major events during 2005-2011 and was demonstrated to be highly efficient (Japan (East Coast of Honshu) Tsunami, 2011; Tang et al., 2008; Titov, 2009; Wei et al., 2008; Yamazaki et al., 2006). The maximum wave heights and other parameters of arriving tsunamis for several US sites were predicted and were later found to be in good agreement with the actual in situ observed waves (Titov, 2009). The method described in (Tsushima et al., 2011) is similar to the above method.

An inversion method proposed in (Satake, 1987) also gave good results (Fujii & Satake, 2008, 2011), however, determination of the area of a seismic centre on aftershock area for one day after the main shock is required for computation. The latter does not allow applying them in an operating mode.

It should be noted, however, that the PMEL technology is based on a *pre-computed Propagation Database* to compute a quick preliminary forecast of the ocean-wide propagation of the tsunami as a linear combination of unit sources selected to represent the initial earthquake parameters. Once the actual tsunami wave reaches a DART station, inversion is performed to adjust the slip distribution of the selected unit sources. The Propagation Database is a key component of the

[1] DART = Deep-ocean Assessment and Reporting of Tsunami is a deep-sea tsunami monitoring system by the US National Oceanic and Atmospheric Administration (NOAA).

operational tsunami forecasting system. At present, it includes a great number of pre-computed model runs for simulated earthquake sources in the Pacific, Atlantic and Indian Oceans (Titov, 2009; see also http://nctr.pmel.noaa.gov/propagation-database.html).

In the present study we suggest a simple method for short-term tsunami forecasting that does not require pre-computed tsunami waveforms. The method is based on a reciprocity principle (c.f. Rayleigh, 1945) and uses the only seismological information about co-ordinates of the earthquake epicentre and in situ tsunami data from one of the far-field open-ocean DART stations. This method can work in a real-time mode and, therefore, be effectively used for local tsunami warnings.

The method for a short-term tsunami forecast is based on Green's function for the wave equation possessing the fundamental property of symmetry. This property is well-known in acoustics and seismology as the reciprocity principle (Rayleigh, 1945). This principle is known as an asymptotic assumption.

The first use of this principle for long water waves was probably suggested by Loomis (Loomis, 1979). The reciprocity principle was found to be effective to optimize tsunami pre-computations. However, the effects of the ocean depth inhomogeneity had not been analysed previously. In view of large horizontal scales of tsunami sources, these effects can be significant. As a result of a study of the reciprocity principle and asymptotic conditions, the similarity conditions for reciprocal sources of long waves were obtained. Some applications of this principle to the tsunami research are considered in the current study. The numerical simulation was made using actual bathymetry. It was shown in numerical experiments that, with a quality sufficient for practical applications, the reciprocity principle works quite well, even when asymptotic conditions are not totally satisfied (i.e., when the characteristic time of the wave propagation between the reciprocity sources is only three times greater than the typical wave period and when the characteristic wave length is comparable with the size of the source area).

This paper presents a method of short-term tsunami forecasting using open-ocean sea level data from distant sites based on the reciprocity principle. The simple relationship (estimated transfer functions) enabled us to simulate tsunami waveforms for any selected oceanic point based only on the source location and sea level data from a remote reference site (Korolev, 2004, 2005). Information about the mechanism of the earthquake, the size of the centre or earthquake magnitude is not required.

This relationship (transfer function) consists of Laplace (Fourier) transforms. To obtain the target result, the inverse Laplace (Fourier) transform must be made. This problem is an ill-posed one. Two methods of inverse Laplace transform are described.

The method enables the transfer function to be evaluated during the event immediately after obtaining information about the earthquake epicentre location. The method enables us to compute expected tsunami waveforms in real-time mode and for any given ocean site. In contrast to the PMEL/NOAA technology, which is based on a pre-computed Propagation Database that needs updating for newly deployed open-ocean stations, the present method does not require pre-computed tsunami waveforms. The important advantage of this method is that it can be used irrespective of the actual source mechanism (seismic, submarine landslide or other phenomena).

The method was successfully applied to hindcast several recent tsunamis observed in the Pacific. The locations of the earthquake epicentres and the tsunami records from one of the NOAA DART sites were used as inputs for the modelling, while tsunami observations at other DART sites were used to verify and to estimate the effectiveness of this model.

The computed and observed tsunami waveforms for the regions of the Kuril Islands, Aleutian Islands and the US West Coast were in good agreement, satisfying the requirements of tsunami warning services (Korolev, 2011a, 2011b).

The good agreement is not only at points in open-ocean where DART stations are located, but also at points close to the coast of the Kuril Islands.

The proposed method can be considered as the basis for creating a program package that can be applied for early tsunami warnings for sites exposed to the threat of tsunamis. The method can be used for both regional and local tsunami warning services having access to the open-ocean (DART) data in a real-time mode. The result is a computational basis for decision making about issuing (confirmation) or cancelling a tsunami alarm. It is important that the method can be applied at newly developing tsunami warning centres.

2. Reciprocity principle

The reciprocity principle is a corollary of the properties of symmetry of Green's function for a linear wave equation. The Green's function is created, as the response of a medium on point-like (Dirac δ-function) disturbance. The reciprocity principle (relationship) expresses the same property of symmetry for sources with limited sizes. There are two aspects of this principle. A kinematic reciprocity principle says that the propagation time of a signal from point A to reciprocal point B is equal to the propagation time of a signal from point B to A and a dynamic reciprocity principle runs as follows: amplitudes of harmonic signals at points B (A) are identical, if the same source is located at A (B) (c.f. Rayleigh, 1945).

The validity of the reciprocity principle for an inhomogeneous medium with reflecting, absorbing and mixed (impedance) boundaries was proved later. It is valid for any area, including shadow zones and including the case of non-stationary waves (Chertock, 1970). The reciprocity principle is widely applied in hydroacoustics for hydrophone calibration (Uric, 1975).

Let us assume that a wave's process is described by the linear wave equation. Let, in the space limited by reflecting and/or absorbing boundaries, the harmonic Source **A** be located at point A with co-ordinates (x_A, y_A), while the analogous harmonic Source **B** be at point B (x_B, y_B). Let also the pressure amplitude on the radiating surface be p_A, $Q_A = p_A S_A$, where S_A is the square of the radiating surface and $p_A(B)$ is pressure amplitude in point B. The same is true for Source **B**: the pressure amplitude on the radiating surface is p_B, $Q_B=p_B S_B$, where S_B is the square of the radiating surface and $p_B(A)$ is pressure amplitude at point A.

So the following reciprocity relationship (asymptotic equality) is valid (Brekhovskikh, 1960; Landau & Lifshitz, 1975)

$$p_A(B)\cdot Q_B = p_B(A)\cdot Q_A. \tag{1}$$

The obtained relationship is fair, if the following asymptotic conditions are valid:

a. the distance between the sources has to be significantly longer than the wave lengths;
b. horizontal scales of both sources areas have to be significantly shorter than the wave lengths.

In the case of an inhomogeneous medium, these conditions may be formulated in time terms:

a' the time of wave propagation between the sources should be significantly longer than typical wave periods;

b' the wave period should be significantly longer than typical time in the source.

Any connections of a source's parameters with parameters of a medium are not analysed. If for the acoustic processes, for example at ocean, the medium inhomogeneities are insignificant (variations of wave speed do not exceed 7%), then for tsunami waves they can give significant effects, as the speed of long waves differ by more than ten times (from 10 m/s on water depth 10 m up to 200 m/s on depth 4000 м). Considering large horizontal sizes of sources of a tsunami, the influence of these effects can be significant. Besides, the condition (b) can be defaulted, as the characteristic lengths of tsunami waves do not always exceed characteristic sizes of the centre. The problem arises as to whether or not the reciprocity principle for tsunami waves is valid. If yes, what are the conditions of validity and do they differ from conditions in acoustics? The conditions on sources should be investigated.

2.1. The reciprocity principle for non-stationary long waves

The deducing of the reciprocity relationship for tsunami waves is indicated in this paper with the purpose of showing analogies and differences from a traditional relationship. We follow the algorithm described in (Brekhovskikh, 1960; Landau & Lifshitz, 1975).

Let us assume that a wave's process is described by the system of linear shallow-water equations, or wave equation, with variable water depth (Stoker, 1957)

$$\frac{\partial^2 \zeta}{\partial t^2} - g\vec{\nabla}\cdot(D(x,y)\cdot\vec{\nabla}\zeta) = 0 \tag{2}$$

where $\zeta=\zeta(t,x,y)$ – waveform, $D(x,y)$ – water depth, g – gravity acceleration, $\vec{\nabla}=(\frac{\partial}{\partial x},\frac{\partial}{\partial y})$ – differential operator.

Let Source **A** be located within area S_A with the centre at point A with co-ordinates x_A, y_A, while Source **B** be within area S_B with the epicentre at B (co-ordinates x_B, y_B). For simplicity, the initial sea surface elevation is used as the wave source. $\zeta(t,x,y)$ is a waveform from Source **A**, $\eta(t,x,y)$ is a waveform from Source **B**.

Initial conditions (t=0)

$$\zeta(0,x,y) = \zeta_0(x,y) \text{ in area } S_A \tag{3}$$

$$\eta(0,x,y) = \eta_0(x,y) \text{ in area } S_B \tag{4}$$

Boundary conditions

$$\begin{aligned} &\zeta(t,x,y) = 0 && \text{on } \Gamma_1 \text{ (absorbing boundary)}, \\ &\zeta(t,x,y) = 0 && \text{on infinity } \Gamma_\infty, \\ &u_n(t,x,y) = 0 && \text{on } \Gamma_2 \text{ (reflecting boundary)}, \\ &\alpha(x,y)u_n(t,x,y) + \beta(x,y)\zeta(t,x,y) && \text{on } \Gamma_3 \text{ (mixed, or impedance, condition)}. \end{aligned} \tag{5}$$

Here $u_n(t,x,y)$ – normal component of mass velocity.

The same boundary conditions are applied to waveform η and mass velocity w_n from Source **B**.

While tsunamis are non-stationary waves, Laplace transform is applied to equation (2) and boundary conditions (5) for a waveform from Source **A**

$$s^2\zeta(s,x,y) - g\vec{\nabla}(D(x,y)\cdot\vec{\nabla}\zeta(s,x,y)) = s\zeta_0(x,y).$$

The same is applied to waveform from Source **B**

$$s^2\eta(s,x,y) - g\vec{\nabla}(D(x,y)\cdot\vec{\nabla}\eta(s,x,y)) = s\eta_0(x,y)$$

Multiplying the last but one equation by $\eta(s,x,y)$ and the last equation by $\zeta(s,x,y)$ and subtracting one from another we obtain

$$\begin{aligned} &g\zeta(s,x,y)\cdot\vec{\nabla}(D(x,y)\cdot\vec{\nabla}\eta(s,x,y)) - g\eta(s,x,y)\cdot\vec{\nabla}(D(x,y)\cdot\vec{\nabla}\zeta(s,x,y)) = \\ &= s(\eta(s,x,y)\cdot\zeta_0(x,y) - \zeta(s,x,y)\cdot\eta_0(x,y)) \end{aligned}$$

This equation may be transformed to divergent form using relationship

$$f_1(x,y)\cdot\vec{\nabla}(D(x,y)\cdot\vec{\nabla}f_2(x,y)) = \vec{\nabla}(D(x,y)\cdot f_1(x,y)\cdot\vec{\nabla}f_2(x,y)) - D(x,y)\cdot\vec{\nabla}f_1(x,y)\cdot\vec{\nabla}f_2(x,y).$$

This yield

$$g\cdot\vec{\nabla}(D(x,y)\cdot\zeta(s,x,y)\cdot\vec{\nabla}\eta(s,x,y))-g\cdot\vec{\nabla}(D(x,y)\cdot\eta(s,x,y)\cdot\vec{\nabla}\zeta(s,x,y))= \\ =s\cdot(\eta(s,x,y)\cdot\zeta_0(x,y)-\zeta(s,x,y)\cdot\eta_0(x,y)) \quad (6)$$

Then the integral is taken over space with boundaries $\Gamma=\Gamma_\infty+\Gamma_1+\Gamma_2+\Gamma_3$. The integral of the left part of (6) is transformed into the integral along a contour $\Gamma=\Gamma_\infty+\Gamma_1+\Gamma_2+\Gamma_3$ (Korn & Korn, 1961):

$$g\cdot\oint_\Gamma D\cdot\left[\zeta\cdot\left(\frac{\partial\eta}{\partial x}dy-\frac{\partial\eta}{\partial y}dx\right)-\eta\cdot\left(\frac{\partial\zeta}{\partial x}dy-\frac{\partial\zeta}{\partial y}dx\right)\right]= \\ =-s\cdot\oint_\Gamma D\cdot\left(\zeta\cdot w_n-\eta\cdot u_n\right)\cdot dl$$

The obtained integral over Γ_∞ and Γ_1 is equal to 0 by virtue of the first and second conditions (5), the integral over Γ_2 is equal to 0 by virtue of the third condition (5). The integrand is transformed to impedance form and the integral over Γ_3 is equal to 0 by virtue of the fourth condition (5)

$$-s\cdot\int_{\Gamma_3} D\cdot(\zeta\cdot w_n-\eta\cdot u_n)\cdot dl=-s\cdot\int_{\Gamma_3}\frac{D}{\beta(x,y)}\cdot\left[\beta\cdot\zeta\cdot w_n+\alpha\cdot u_n\cdot w_n-\alpha\cdot w_n\cdot u_n-\beta\cdot\eta\cdot u_n\right]\cdot dl=$$

$$=-s\cdot\int_{\Gamma_3}\frac{D}{\beta}\cdot\left[\left(\alpha\cdot u_n+\beta\cdot\zeta\right)w_n-\left(\alpha\cdot w_n+\beta\cdot\eta\right)u_n\right]\cdot dl=0\,.$$

So, the integral of the left part of (6) is equal to 0. The integral of the right part (6) gives

$$\int_{S_A}\eta(s,x,y)\cdot\zeta_0(x,y)\cdot dS=\int_{S_B}\zeta(s,x,y)\cdot\eta_0(x,y)\cdot dS \quad (7)$$

Until now any suppositions, simplifying the problem, have not been done. Relationship (7) is the exact one.

The classical statement of the reciprocity principle requires the following conditions to be satisfied:

a. the distance between the sources has to be longer than the wave lengths (x, y scales of the source areas), i.e., the variation $\Delta\zeta$ should be significantly smaller than function $\zeta(s,x,y)$ within the area S_B : $\Delta\zeta<<\zeta(s,x,y)$. This should also be true for the function $\eta(s,x,y)$;

a' an alternative asymptotic condition may be formulated in the time domain: typical wave periods should be significantly smaller than the time of wave propagation between the sources;

b. the horizontal (x, y) scales of both source areas have to be shorter than the wave lengths of the generated tsunami waves.

If conditions (a), (a') and (b) are satisfied, functions $\zeta(s,x,y)$ and $\eta(s,x,y)$ in (7) may be removed from the integrals and replaced by mean values at points B and A. Thus, an approximate (asymptotic) expression that follows from (7) is

$$\eta(s,x_A,y_A)\cdot\int_{S_A}\zeta_0(x,y)\cdot dxdy \approx \zeta(s,x_B,y_B)\cdot\int_{S_T}\eta_0(x,y)\cdot dxdy\ .$$

Designating $\int_{S_A}\zeta_0(x,y)\cdot dxdy = Q_A$ and $\int_{S_B}\eta_0(x,y)\cdot dxdy = Q_B$, where Q_A and Q_B are the volumes of the initial disturbances, we can rewrite the last relationship as

$$\eta(s,x_A,y_A)\cdot Q_A \approx \zeta(s,x_B,y_B)\cdot Q_B \tag{8}$$

The form of relationship (8) matches with the form of traditional relation (1).

Due to the medium inhomogeneity (variable speed of wave propagation) and large horizontal source scales in comparison with the water depth, the initial free surface elevations should satisfy the following conditions based on common reasons of similarity and dimensions:

- the forms of the initial disturbances should be similar;
- the disturbance radii, R_A and R_B, are related as

$$\frac{R_A}{\sqrt{gD_A}} = \frac{R_B}{\sqrt{gD_B}}, \tag{9}$$

where D_A and D_B are the water depths at points A and B, and g is the gravitational acceleration.

The solution of one problem confirms this relationship (see Appendix A).

The approximate (asymptotic) relationship (8) and the condition of similarity (9) is the *dynamic reciprocity principle* (relationship) for non-stationary long water waves (Poplavsky et al., 1997).

Additionally, if the disturbance volumes are identical ($a_A R_A^2 = a_T R_T^2$, i.e. $Q_A = Q_B$), then after the inverse Laplace transform, the waveforms will also match each other

$$\zeta_T(t,x_A,y_A) = \zeta_A(t,x_T,y_T). \tag{10}$$

2.2. The analysis of the reciprocity principle for long waves

It is of interest to identify whether relationship (10) is fair if conditions (a) and (b) are not satisfied, therefore, this will be the subject of the following study.

To verify this approach we made several numerical experiments. The numerical simulations based on actual bathymetry were conducted for the region of the Southern Kuril Islands.

Source **1** was located in the Sea of Okhotsk northwest of Kunashir Island; Source **2** was located in the Yuzhno-Kurilsk Strait, between the Kunashir and Shikotan Islands; and Source **T** was located in the Pacific Ocean southeast of Shikotan Island. The computational domain for the numerical experiments indicating the location of the sources is shown in Fig. 1. The bathymetry grid for these experiments was chosen to have a spatial step $\Delta x = \Delta y =$ 1000 m.

The parameters of the initial free surface elevations were chosen taking into account the above-mentioned conditions. They were the following:

Source 1: Max ampl. a_1 = 0.4 m, radius R_1 = 25 km, water depth in epicentre D_1= 2418 m;

Source 2: Max ampl. a_2 = 15.6 m, radius R_2 = 4 km, water depth in epicentre D_2 63.6 m;

Source T: Max ampl. a_T = 0.49 m, radius R_T = 22.5 km, water depth in epicentre D_T= 2030 m.

The results of these simulations are shown in Fig. 2. The correlation coefficient, ρ, was chosen as a criterion for comparison of the reciprocal waveforms.

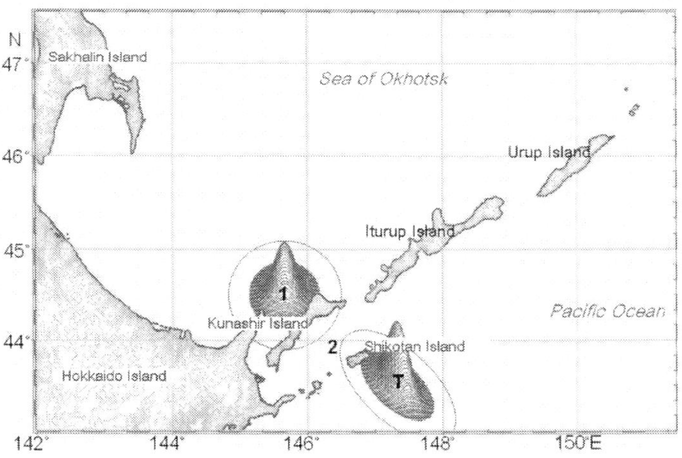

Figure 1. A map showing the region of the South Kuril Islands and a scheme of numerical experiments. Symbols 1, 2 and T indicate the epicentres of the sources. Initial sea surface elevations at Source 1 and Source T are pictured.

The coincidence of waveforms according to (10) is quite good despite wave sheltering by a chain of islands. The typical wave periods are about 10-18 min; the time of wave propagation is 57 min between Sources **1** and **T**, 50 min between Sources **1** and **2**, and 40 min between Sources **2** and **T**. It is possible to make certain that wave lengths in the reciprocity source areas slightly exceed their sizes. For example, the wave length of the wave from Source **1** (λ_{12}) at Site **2** for the period of T = 18 min and water depth D_2 = 63.6 m is $\lambda_{12} = T \cdot \sqrt{gD_2} \approx$ 27 km, while the diameter of Source **2** is 8 km.

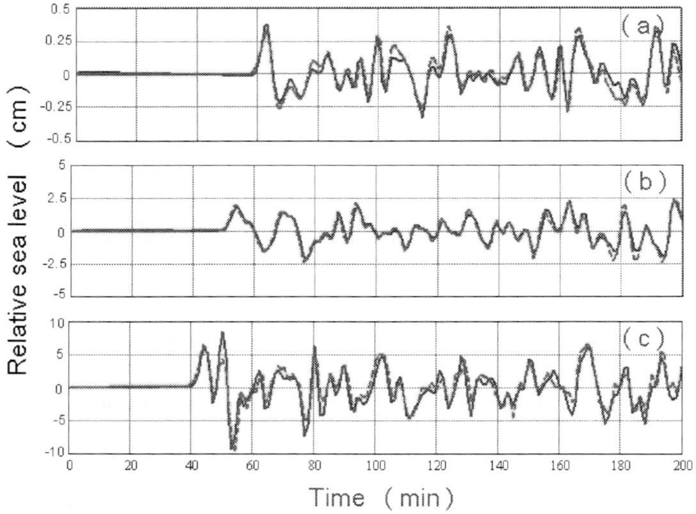

Figure 2. Illustration of the reciprocity relationships. (a) Waveforms at Site T due to Source 1 (blue solid line) and at Site 1 due to Source T (red dashed line); correlation coefficient $\rho_{1T} = 0.96$.
(b) Waveforms at Site 2 due to Source 1 (blue solid line) and at Site 1 due to Source 2 (red dashed line); correlation coefficient $\rho_{12} = 0.94$.
(c) Waveforms at Site T due to Source 2 (blue solid) and at Site 2 due to Source T (red dashed line) correlation coefficient $\rho_{2T} = 0.91$.

The wave length of the wave from Source **T** (λ_{T2}) at Site **2** for period of $T = 8$ min and water depth $D_2 = 63.6$ m is $\lambda_{T2} = T \cdot \sqrt{gD_2} \approx$ 12 km, while the diameter of Source **2** is 8 km.

Numerical experiments show that relationship (8) with condition (9) is true even if the time of wave propagation between sources is only 3 – 4 times longer than wave periods and if the wave lengths are comparable to or slightly longer than these sizes. As was found from our experiments, changing the radius of one of the sources by 25% does not significantly influence the output.

Additional experiments were done when Sources **1** and **2** were the same, but Source **T** was not a circle but an ellipse. The ellipse minor (b_2) and major (b_1) axes were chosen as (first) $b_1 = 63.6$ km and $b_2 = 31.8$ km ($b_2 / b_1 = 1/2$); and (second) $b_1 = 77.9$ km and $b_2 = 26.0$ km ($b_2 / b_1 = 1/3$); maximum amplitude $a_T = 0.49$ m and water depth at epicentre $D_T = 2030$ m (for both cases). We also changed the orientation of the major axis relative to true north. The condition of similarity of the source forms was not true, but it was accepted that $\frac{2R_{1(2)}}{\sqrt{gD_{1(2)}}} = \frac{\sqrt{b_1 \cdot b_2}}{\sqrt{gD_T}}$ and $4a_1R_1^2 = a_T b_1 b_2$. The respective numerical results are shown in Fig. 3.

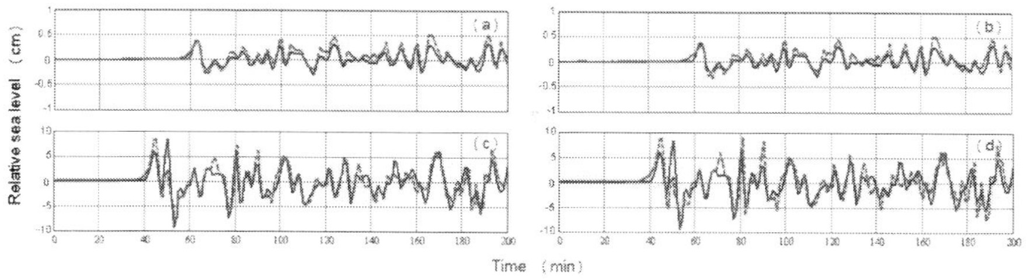

Figure 3. Fig. 3. Illustration of the reciprocity relationships. Elliptic Source T is elongated to the northwest (see Fig.1). The ratio of minor/major axes is equal to 1/2 (left column) and to 1/3 (right column). Upper row: waveforms (a) at Site T due to Source 1 (blue solid line), correlation coefficients are $\rho_{1T} = 0.90$ and (b) at Site 1 due to Source T (red dashed line), $\rho_{T1} = 0.87$. Lower row: waveforms (c) at Site T due to Source 2 (blue solid line), correlation coefficients are $\rho_{2T} = 0.86$ and (d) at Site 2 due to Source T (red dashed line), $\rho_{T2} = 0.78$.

The correlation coefficients in all experiments are in the range of 0.67 to 0.90. The best result is in the case when the major axis was directed toward the northwest. This agreement is good enough and sufficient for any practical applications. Relationship (10) was found to be fair not only for circular sources, but also for elliptic sources, at least for the moderate ratios of axes' lengths.

3. The method of short-term tsunami forecasting

The classical statement of the reciprocity principle requires conditions (a) and (b) (see the previous section), however, our numerical experiments demonstrate that it can still be applied without these strict contingencies. The results presented below could become the basis for the short-term tsunami forecasting.

3.1. Derivation of relationships for a short-term tsunami forecast

It was demonstrated (Korolev, 2004, 2005) that the derived reciprocity relationships could be efficiently applied for the problem of short-term (operative) tsunami forecasts based on sea level data from a remote open-ocean site. It is assumed that reciprocity relationship (8) with similarity condition (9) is true when one of the sources is natural, while the other source is artificial (numerical).

Let's tsunami Source **T**, the initial sea surface elevation, is located within the area S_T. The total volume of the disturbed uplift is Q_T. Let us also assume that M is the site of a sea level gauge (i.e., DART site) and A is the site of interest (a point for tsunami forecasting). The problem is to compute the tsunami waveform at site A ("target site") using the sea level data at site M ("reference site"). Mathematically, this problem does not have a unique solution, however, it can be considered as a "consumer problem". Specific methods should be developed for solving this problem.

In this section, "waveform" means the Laplace (Fourier) transform (spectra) of the waveform. Let the reciprocal sources be located within areas with epicentres at A and M. Their disturbed volumes are Q_A and Q_M, while $\eta_A(s,x_T,y_T)$ and $\eta_M(s,x_T,y_T)$ are the waveforms at site T from reciprocal sources S_A and S_M. For the natural tsunami waveform ζ_T from tsunami Source **T** and for waveforms from the reciprocal sources in S_A and S_M, relationships (8) are

$$\begin{aligned} \zeta_T(s,A)\cdot Q_A &= \eta_A(s,T)\cdot Q_T; \\ \zeta_T(s,M)\cdot Q_M &= \eta_M(s,T)\cdot Q_T. \end{aligned}$$

In these expressions we designated: $\zeta_T(s,A)\equiv\zeta_T(s,x_A,y_A)$, $\zeta_T(s,M)\equiv\zeta_T(s,x_M,y_M)$ and so on.

In addition, let's assume that the auxiliary sources are in the areas with the same epicentres at A, M and T (waveforms and disturbed volumes are designated by symbols with primes). For waveforms from these sources the reciprocity relationships are also true

$$\begin{aligned} \eta'_T(s,A)\cdot Q'_A &= \eta'_A(s,T)\cdot Q'_T \\ \eta'_T(s,M)\cdot Q'_M &= \eta'_M(s,T)\cdot Q'_T \end{aligned}$$

As the auxiliary sources may be selected rather arbitrarily, in particular, Q_A, Q_M and Q'_A, Q'_M may be selected in such a way that η'_A is equal to η_A and similarly for η'_M and η_M. It is true if $Q_A = Q'_A$ and the same for Source **M**.

Eliminating unknown Q_T from the above-mentioned two sets of equations yields the following relationship

$$\frac{\zeta_T(s,A)}{\zeta_T(s,M)} = \frac{\eta_T(s,A)}{\eta_T(s,M)} \tag{11}$$

Let us assume that functions on the left side of the equation relate to one tsunami source and functions on the right side relate to another source, and that both sources have the same epicentre. Relationship (11) supports the well-known fact that spectra of any two tsunamis from different sources are similar when recorded at the same site, whereas spectra of any two tsunamis from the same source are dissimilar at different sites (Takahasi & Aida, 1961; Miller, 1972).

From (11) we can get the following main relationship

$$\zeta_T(s,A) = \zeta_T(s,M)\frac{\eta_T(s,A)}{\eta_T(s,M)} \tag{12}$$

Here $\zeta_T(s,A)$ is the target function (tsunami waveform at A); $\zeta_T(s,M)$ is the reference function (tsunami waveform at M, i.e., open-ocean DART data); and $\eta_T(s.M)$ and $\eta_T(s,A)$ are the numerical waveforms at M and A from the auxiliary source located at S_T. All functions are Laplace transforms. The ratio on the right side of (12) plays the role of a transfer function. The transfer function allows computing a tsunami waveform at a target point on ocean level data at a reference site.

As the sea level data and computed data are digital ones, so the functions included in (12) are represented as numerical Laplace transformations or z-transformation, for example $\zeta(s,A)=\zeta(z,A)=\sum_{k=0}^{N}\zeta_k(A)e^{-ks\tau}=\sum_{k=0}^{N}\zeta_k(A)z^k$ (see Appendix B), where $\zeta_k(A)$ – is the value of function $\zeta(t,A)$ for a moment $t=k\tau$, τ – is a time sampling. Due to this, (12) may be rewritten as

$$\zeta(z,A)=\sum_{k=0}^{N}\zeta_k(A)\cdot z^k=\sum_{k=0}^{N_1}\zeta_k(M)\cdot z^k\,\frac{\sum_{k=0}^{N_2}\eta_k(A)\cdot z^k}{\sum_{k=0}^{N_3}\eta_k(M)\cdot z^k} \tag{13}$$

The arrival time of the forecasted tsunami T_A at target point A can be estimated as $T_A = T_1 + T_2 - T_3$, where T_1 - time of tsunami propagation to the point of registration M, T_2 - an arrival time of wave from an auxiliary source to point A, T_3 - an arrival time of wave from an auxiliary source to point M.

At the first stage, while the record obtained from the station DART has a short duration (for example, one half period of the first wave, N_1 = 10-20 minutes), the lengths of all power series in (13) should be equal: $N_1 = N_2 = N_3 = N$, otherwise the result will be inadequate. But the approximate forecast is possible if $N_3 = N_1$, but N_2 is significantly more than N_1. Such a forecast is approximate, but probably takes into account secondary waves. As the sea level information from the DART station at M is received, the duration and quality of the forecast is increased. The computations with variable duration of level data allow making the forecast in real-time mode (see Appendix B).

The practical interest is to predict tsunami waveforms of long duration adequately, which would take into account secondary waves reflected from island chains, submarine ridges and mainland coasts. These secondary waves have amplitudes which often significantly exceed amplitudes of first tsunami waves.

Such a tsunami forecast of long duration can be made on the limited information about a level. Often the tsunami in the ocean only has some first waves with significant amplitudes. If we neglect waves with small amplitude in a wave tail, the duration N_1 should be equal to the duration of the registered tsunami with significant waves and the duration N_2 could somehow be long, but durations N_3 and N_1 should be equal. The forecast duration N in this case is equal to N_2. In this way secondary waves can be taken into account.

The moment of the preliminary forecast completion is determined by the time of tsunami propagation up to a point of registration and the time necessary for tsunami identification (the first half period of the first wave).

The division and the multiplication of polynomials in the right part of (13) could give the required result. Unfortunately, the direct application of the offered algorithm, gives, as a rule, a diverging result. This problem is an ill-posed one. Two methods of the solving this problem (regularization methods) permit to obtain a converging result and are described in Appendix B.

It should be noted that the generation mechanism of the actual tsunami is not important here. The method can be applied for short-term forecasting irrespective of the tsunami source, in particular for tsunamis produced by earthquakes, volcanic eruptions, submarine landslides or other phenomena. However, for the practical realization of the proposed method it is important to assume that the transfer function for the corresponding event can be constructed as the ratio of the Laplace transforms of the waveforms at appropriate sites generated by an initial free surface elevation. This elevation was supposed to be circular with a diameter of 50-100 km which is the characteristic transversal size of the earthquake source. There is no requirement to know the detailed seismological information about the source; it is enough to know the co-ordinates of the earthquake (or another source) epicentre. This will enable us, with accuracy sufficient for practical use, to create a short-term tsunami forecast for specific coastal sites.

In fact, the proposed method is a generalization of the methods suggested earlier (Chubarov & Shokin, 1995; Chung et al., 1995).

3.2. Dependence of the hindcast quality on the diameter of the auxiliary source

There are questions about how the size of auxiliary sources affects the results of the computations.

Numerical experiments with auxiliary sources of different diameters (20, 30, 40, 50, 75 and 100 km) for sites in the ocean gave almost identical results.

3.3. Computing the short-term tsunami forecast

The algorithm for computing the tsunami waveform based on sea level data from remote sites includes the following steps:

1. deriving the co-ordinates of the earthquake (or other source) epicentre;
2. computing the tsunami waveforms for the reference site M and the target (forecasting) site A, i.e., functions $\eta_T(s,M)$ and $\eta_T(s,A)$. The auxiliary source is a circle with diameter of 50-100 km located at the centre of the actual tsunami source. The computation (evaluation of the transfer function) should be completed before the tsunami reaches the reference DART site (normally the site nearest to the source);
3. acquiring to the tsunami data from the appropriate reference DART station; eliminating tidal components from the corresponding record;
4. after recording and identifying tsunami waves at the reference DART station (Site M), the tsunami forecast can be made for the target site A based on relationship (12, 13). Similarly, in real-time mode, tsunami forecasts can be provided for other target sites (e.g., for particular sites along the coast). In fact, the preliminary forecast can be given after recording the first tsunami semi-wave. Hereafter, as new and more complete information is received, this forecast can be corrected and improved;
5. the forecast results can be used by TWSs to assist in making their decision about declaring or cancelling the tsunami warning (alarm).

4. Application of the method for short-term tsunami forecasting

Some preliminary research results published in 2005 (Korolev, 2005) demonstrated that the proposed method can be efficiently used for short-term tsunami forecasting. In particular, it was shown that the proposed method can work in a real-time mode and that the errors in the epicentre location only slightly affect the quality of the forecast.

Some new results were published recently by Korolev (2011a, 2011b) and some of these are represented below.

4.1. Preliminary information

In the present study we apply the method of short-term tsunami forecasting to hindcast tsunami waveforms for some tsunamigenic earthquakes. Earthquakes occurred in 2006, 2007 and 2009 in the region of the Central Kuril Islands, eastward of Simushir Island, near the Chilean coast in 2010 and eastward of Honshu Island in 2011. Associated tsunamis were recorded by some tide gauges on the coasts of Russia and by a number of DART stations located along the Kuril Islands, Japan, the Aleutian Islands and the US West Coast (NDBC DART, n.d.).

Tsunami waveforms were simulated at the locations of the DART stations and near to the coast of Kuril Islands. The actual in situ tsunami data from the DART stations closest to the earthquake epicentres were used for the hindcast. The data from other DART stations, located further from the epicentre, were used to compare with the predictions and to verify the method. The correlation coefficient was chosen as criterion of coincidence. In the present study we used deep-ocean stations not affected by coastal resonant effects and surf beats. All computations were made based on the shallow-water numerical model (Poplavsky et al., 1997). We used the bathymetric data that were interpolated from the ETOPO2 global dataset (Smith & Sandwell, 1994).

No seismological information about the earthquake, except the epicentre co-ordinates, was used in the following computations.

The scheme of the research experiments, indicating the earthquake epicentre and positions of the DART stations, is shown in Fig. 4. The hindcasting results are represented below.

4.2. Simushir tsunamis of 2006, 2007 and 2009

The auxiliary source with an initial circular sea surface uplift of 75 km in diameter and with a maximum height of 10 m was chosen for transfer function creation in the cases of the 2006 and 2007 events. The selected source centre coincided with the earthquake epicentre.

Results were published in (Korolev, 2011b).

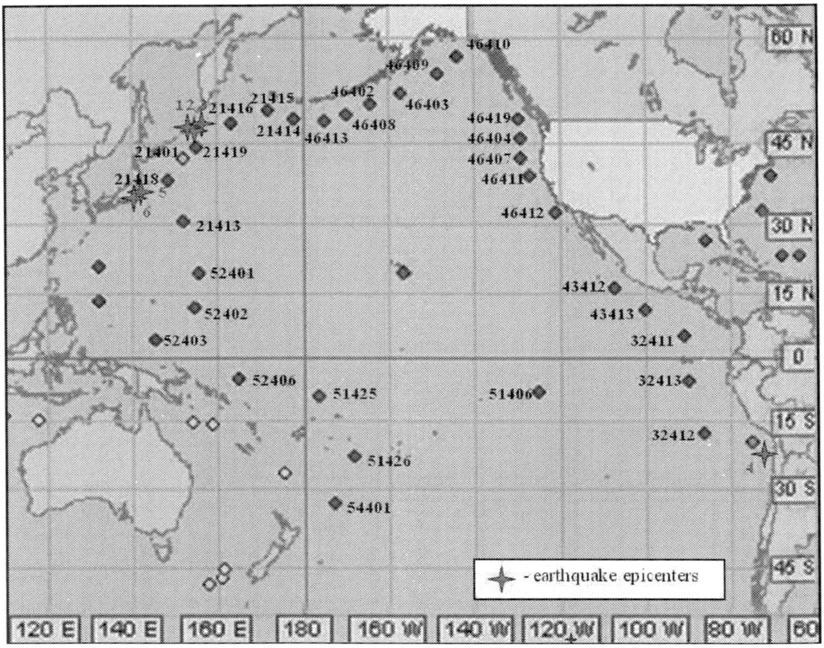

Figure 4. A map of the Pacific Ocean showing positions of DART stations (from NDBC DART, n.d.) and epicentres of some earthquakes.Digits denote: 1 – 2006 Simushir tsunami; 2 – 2007 Simushir tsunami; 3 – 2009 Simushir tsunami; 4 – 2010 Chile tsunami; 5 – 09.03.2011 tsunami; 6 – 2011 Great Tohoku tsunami.

4.2.1. The November 2006 Simushir tsunami

The tsunami was generated by a major earthquake (M_w = 8.3) on November 15, 2006 at 11:14 UTC with the epicentre at 46.592ºN, 153.266ºE near the Central Kuril Islands (NOAA/WDC, n.d). Significant wave heights were observed along the coasts of the Kuril Islands (c.f. Rabinovich et al., 2008; McInnes et al., 2009).

The November 15, 2006 Simushir tsunami was accompanied by a big run-up at islands nearest to the earthquake centre, but small amplitudes in the northern and southern Kuril Islands, and at the same time caused considerable damage in Crescent City on the US West Coast (Horrillo et al., 2008; Kowalik et al., 2008).

A tsunami alarm was declared at 11:30 UTC by Sakhalin Tsunami Warning Center on the coast of all Kuril Islands and an evacuation of people from dangerous areas was done. Vessels were ordered to sail out into open sea. The alarm was cancelled at 13:19 UTC.

The proposed method was applied to simulate the operational tsunami forecast in the North Pacific Ocean along the Aleutian Islands and the US West Coast. To compute the tsunami waveforms, the data of DART stations 21414 and 46408 were used as the basis (see Fig. 4 for the DART positions). The ocean depth at the epicentre, based on the bathymetry grid readings, was 2803 m. Spherical co-ordinates with a spatial grid step of 3.8 km at latitude 40ºN were used.

Figure 5 shows hindcast results and the waveforms of the observed tsunami waves.

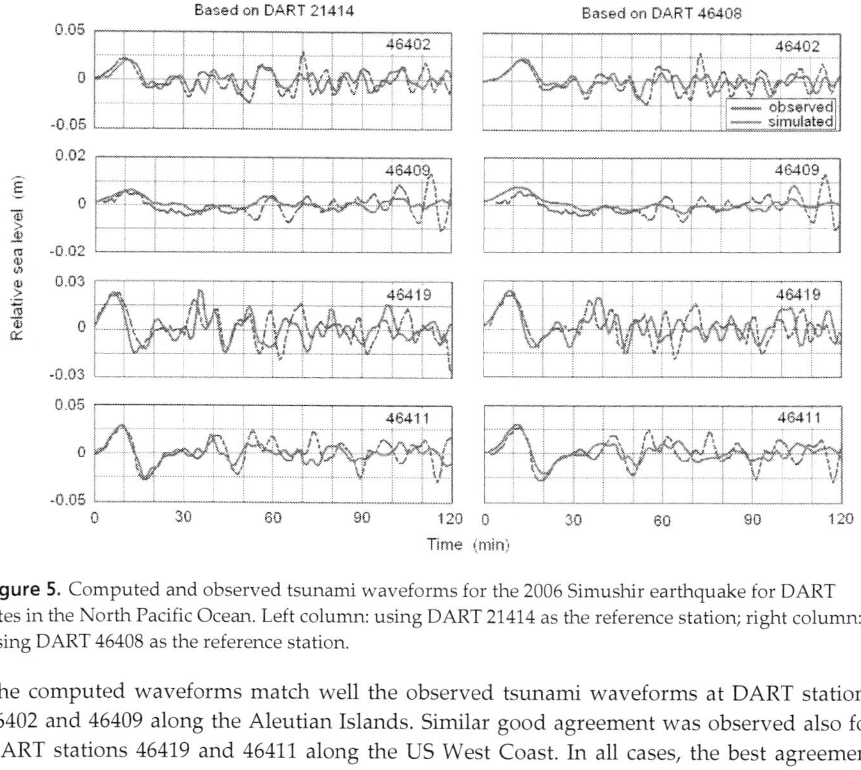

Figure 5. Computed and observed tsunami waveforms for the 2006 Simushir earthquake for DART sites in the North Pacific Ocean. Left column: using DART 21414 as the reference station; right column: using DART 46408 as the reference station.

The computed waveforms match well the observed tsunami waveforms at DART stations 46402 and 46409 along the Aleutian Islands. Similar good agreement was observed also for DART stations 46419 and 46411 along the US West Coast. In all cases, the best agreement was for a few leading waves; the correlation coefficients for these waves were within the range ρ = 0.70–0.85.

4.2.2. The January 2007 Simushir tsunami

The tsunami was generated by a major earthquake (M_w = 8.1) on January 13, 2007 at 4:23 UTC with the epicentre at 46.243ºN, 154.524ºE located on the oceanic slope of the Central Kuril Trench (NOAA/WDC, n.d.). This tsunami was weaker than the 2006 Simushir tsunami, but was locally quite intensive (Rabinovich et al., 2008).

The tsunami alarm was declared on the coast of the Kuril Islands at 4:36 UTC.

Tsunami waves were recorded with amplitudes at Yuzhno-Kurilsk 5 - 6 cm and at Malokurilskoe 36 cm. Due to the weak manifestation of the tsunami on the coasts of the Kuril Islands, Kamchatka and the Aleutian Islands, the tsunami alarm was cancelled at 7:49 UTC.

This tsunami was recorded by a smaller number of DART stations than the 2006 tsunami (c.f. Laverov et al., 2009), however, one of these stations was DART 21413, which was not in operation during the previous (2006) event. This station was located southward from the 2007 epicentre (690 nautical miles southeast from Tokyo). During the 2006 event the tsunami was only recorded by DART stations located to the east of the source. It is of additional interest to compare the results for open-ocean stations located in various directions from the source. Simulation of the operational tsunami forecast was done. The data of DART stations 21414 (east of the source) and 21413 (south of the source) were used as the basis (see Fig. 4 for the DART positions). The ocean depth at the epicentre was 6877 m. Spherical co-ordinates with a spatial grid step of 3.8 km at latitude 40ºN were used. The corresponding results are shown in Fig. 6.

In general, the agreement between computed and observed waveforms is satisfactory both for DART 46408 near the Aleutian Islands and for DARTs 46419 and 46412 along the US West Coast. Hindcasts based on data from both stations 21414 (eastward from the source) and 21413 (southward from the source) are of almost identical quality. For all cases, the computations show correctly the tsunami manifestation during the initial (negative) phase. The correlation coefficients were within the range ρ = 0.50-0.72.

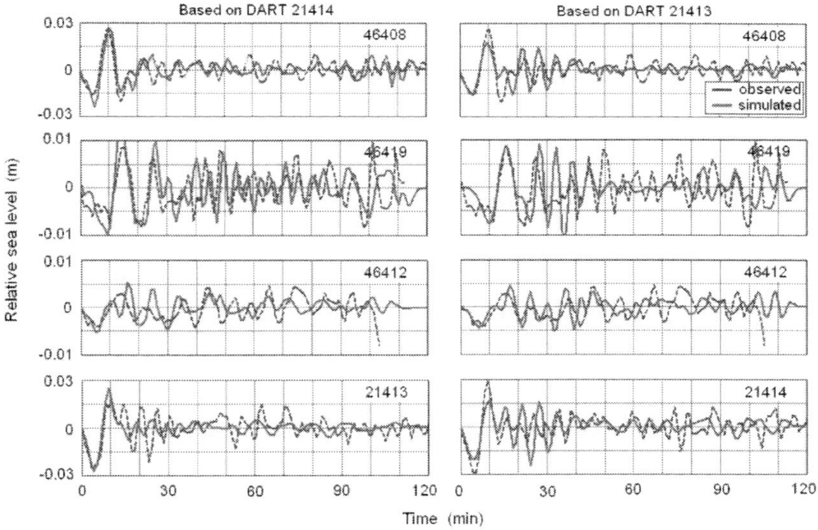

Figure 6. The same as in Fig. 5, but for the 2007 Simushir tsunami. Left column is for DART 21414 as the reference station; right column is for DART 21413 as the reference station.

4.2.3. The January 2009 Simushir tsunami

This earthquake (M_w = 7.4) occurred on January 15, 2009 at 17:49 UTC approximately in the same region of the Central Kuril Islands as the 2006 and 2007 earthquakes, but it was weaker than the two previous earthquakes. The epicentre of the earthquake was located at 46.857ºN, 155.154ºE (NOAA/WDC, n.d.). The earthquake generated a tsunami that was recorded on the coast of the Kuril Islands and by a few DART stations, in particular by DART 21416 located 240 miles from the Kamchatka Peninsula. The latter station was the nearest to the epicentre. A tsunami alarm was not announced during this event.

Figure 7 shows a more detailed map than in Fig. 4, indicating epicentres of earthquakes, DART positions in the northwest Pacific and target points on the coast of the Kuril Islands.

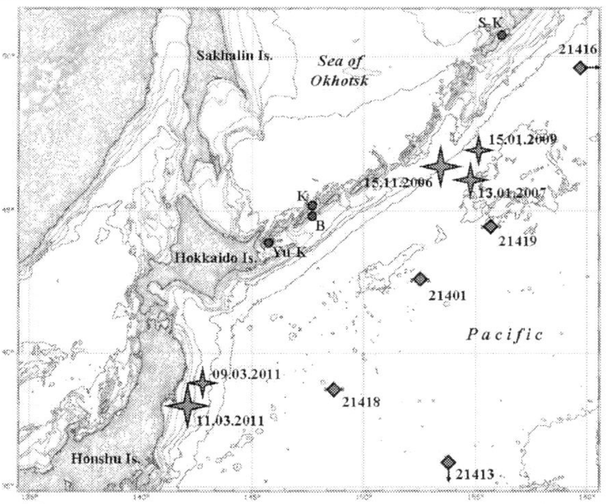

Figure 7. Map fragment of northwest Pacific. Conventional signs: S-K – Severo-Kurilsk, K - Kurilsk, B – Burevestnik (Kasatka Bay), Yu-K – Yuzhno-Kurilsk; asterisks – locations of earthquake epicentres (and dates), rhombuses – locations of DART stations.

The location of DART station 21416 is out of the scheme, the true position is to the east of that marked on the map at longitude 163° 30'E, the true site of DART station 21413 is to the south of that marked on the map at a point with latitude 30° 31'N.

The auxiliary source was taken as a circular uplift of the sea surface 50 km in diameter and maximum height of 8 m; the source centre coincided with the earthquake epicentre. This source was used as the input for the model. The ocean depth at the epicentre was 6645 m.

The computed and observed tsunami waveforms are shown on Fig. 8. The data from DART 21416 were used to hindcast the waveforms for DART 46408 and 46413, located eastward from the source, and for DART 21413, located southward from the source. Similar computations were also made for DARTs 46408, 46413 and 21416 based on the data from DART 21413.

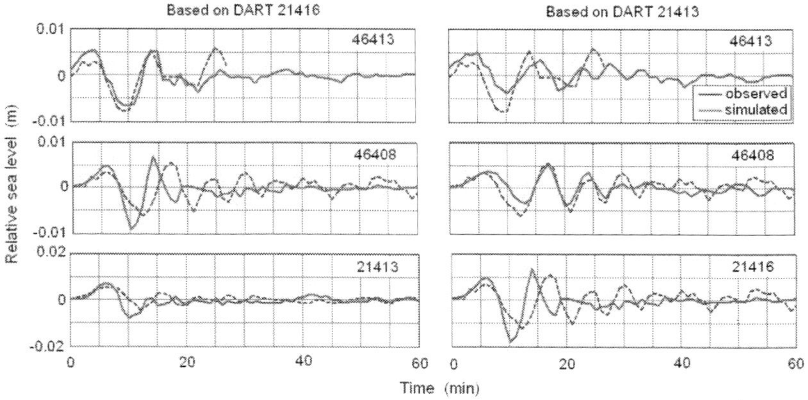

Figure 8. The same as in Fig. 5, but for the 2009 Simushir tsunami. Left column is for DART 21416 as the reference station; right column is for DART 21413 as the reference station.

The computed waveforms are in good agreement with the observed tsunami records for both the area of the Kamchatka Peninsula and the area of the Aleutian Islands, especially for the leading waves. The correlation coefficients were within the range ρ = 0.75–0.80.

4.2.4. The 2009 Simushir tsunami on the Kuril Islands

Previous results demonstrated the possibilities of the proposed method. More interesting is the answers the following questions. How can the proposed method produce a tsunami forecast near to the coast? Can the proposed method predict secondary waves of large amplitudes? How well can the proposed method estimate the duration of the alarm period?

The auxiliary source to create a transfer function was an axisymmetric free surface elevation with a diameter of 50 km and a maximum amplitude of 8 m. Water depth at the earthquake centre based on the bathymetry grid readings was equal to 6887 m. Mercator projection with a spatial grid step 1.1 km was used with latitude of scale 52º.

The data of the closest to the earthquake centre DART station 21416, located north-east of the source, were used for the simulation of the tsunami forecast on the coast of the Kuril Islands. The sea level recorded by this station is shown in Fig. 9.

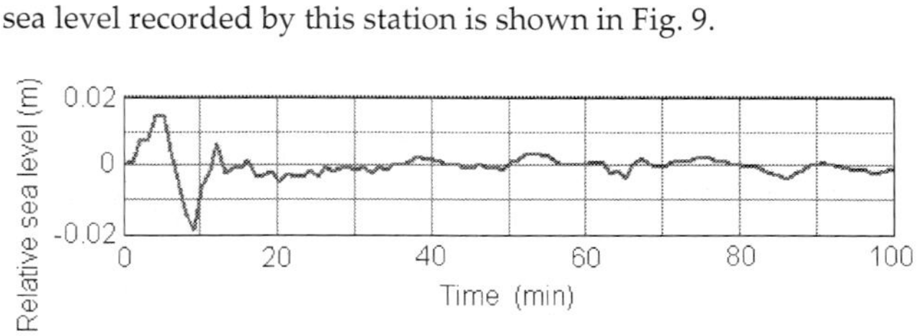

Figure 9. The sea level recorded by DART station 21416.

The January 15 tsunami was recorded by a tide gauge at Severo-Kurilsk, located north of the source. The tsunami waveform was computed at the point near Severo-Kurilsk in the strait at a distance of about 1.5 km offshore and at a depth of 20 m.

Results of simulation are represented in Fig. 10. This figure shows a tide gauge record at Severo-Kurilsk with a duration of 17 hours. The specific natural oscillations of the strait up to 19 hours on the sea level record on January 15 can be seen. Then the nature of the oscillations changed due to the tsunami. After 6 - 7 hours the specific mode of natural oscillations was restored, but with increased amplitude.

The blue line designates the recorded form of oscillations of a sea level at the port of Severo-Kurilsk, red - hindcast:

a. preliminary forecast at the point S-K (see Fig. 8) based on data of DART station 21416 with a duration of 20 min;
b. updated forecast at the same point based on data of DART station 21416 with a duration of 82 min.

On Fig. 10 the vertical solid line shows the time of the main shock (17:49 UTC), the vertical dashed line shows the moment of the forecast.

The preliminary estimated tsunami waveform based on 20 minute data of DART station 21416 is shown in Fig. 10 (a). The updated forecast using 82 minute level data of DART station21416 is shown in Fig. 10 (b).

Forecast, preliminary and adjusted, shows that the tsunami arrival is expected at 19:20, 90 min after the main shock. Waves with a greater amplitude followed the first wave are expected 1.5 hours later. The duration of the tsunami is estimated to be 6 - 7 hours.

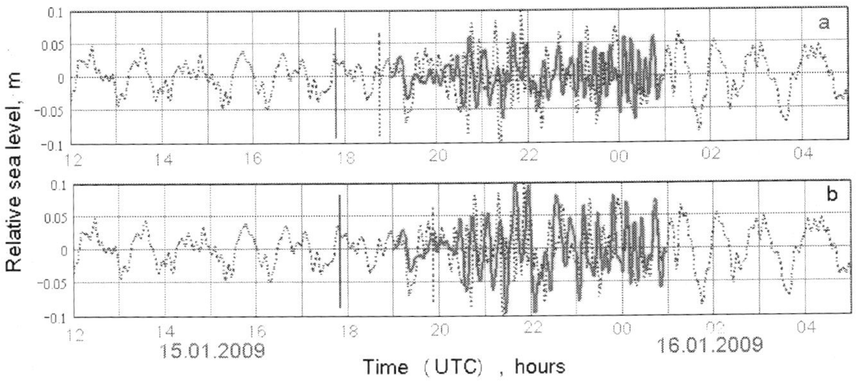

Figure 10. Results of the 2009 Simushir tsunami hindcast close to Severo-Kurilsk.

The computations necessary to create a transfer function are performed for 15 min, while the tsunami travel time to the point 21416 equals to 41 min, to Severo-Kurilsk - 75 min. Taking into account the time of tsunami identification (20 min), the preliminary tsunami forecast at Severo-Kurilsk may be given 30 minutes before the first wave arrives at this point.

The above example shows the principal possibility of tsunami forecast near to the coast of the Kuril Islands based on data of level stations, located on the ocean side of the Kuril-Kamchatka Trench.

4.3. 2010 Chile tsunami

The tsunami was generated by earthquake (M_w = 8.8) which occurred on February 27, 2010 at 6:34 UTC near the Chilean coast. The co-ordinates of the earthquake epicentre are 36.122ºS, 72.898ºW (NOAA/WDC, n.d.). The tsunami was recorded by a number of DART stations in the Pacific.

4.3.1. Hindcast of 2010 Chile tsunami in ocean

Post-event computations simulating the short-term tsunami forecast are based on data closest to the epicentre of the earthquake recorded at DART station 32412 over 30 minute duration (Fig. 11). An auxiliary source to create the transfer function was defined as a circular initial free surface elevation with a diameter of 100 km and a maximum height of 10 m with the centre coincident with the epicentre of the earthquake. The computations were performed in spherical co-ordinates with a spatial grid step 5 km at the equator.

The results of hindcasting the tsunami waveforms in the Pacific (see Fig. 4 for the positions of DART stations and tsunami epicentre) are represented in Fig. 12.

The results of the simulation show a good agreement between hindcasted and observed tsunami waveforms in the north and the west directions from the earthquake centre.

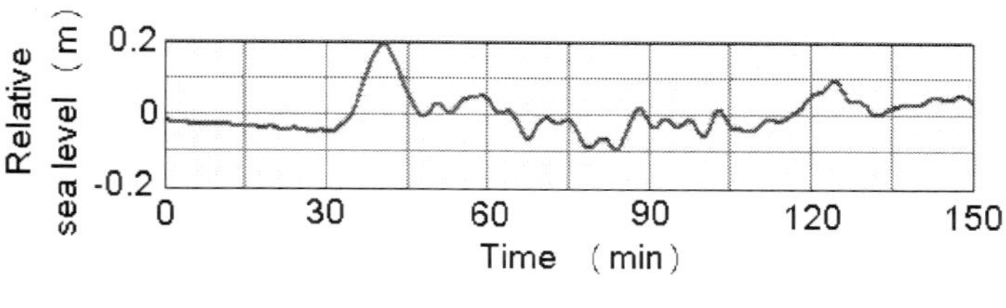

Figure 11. Sea level recorded by DART station 32412.

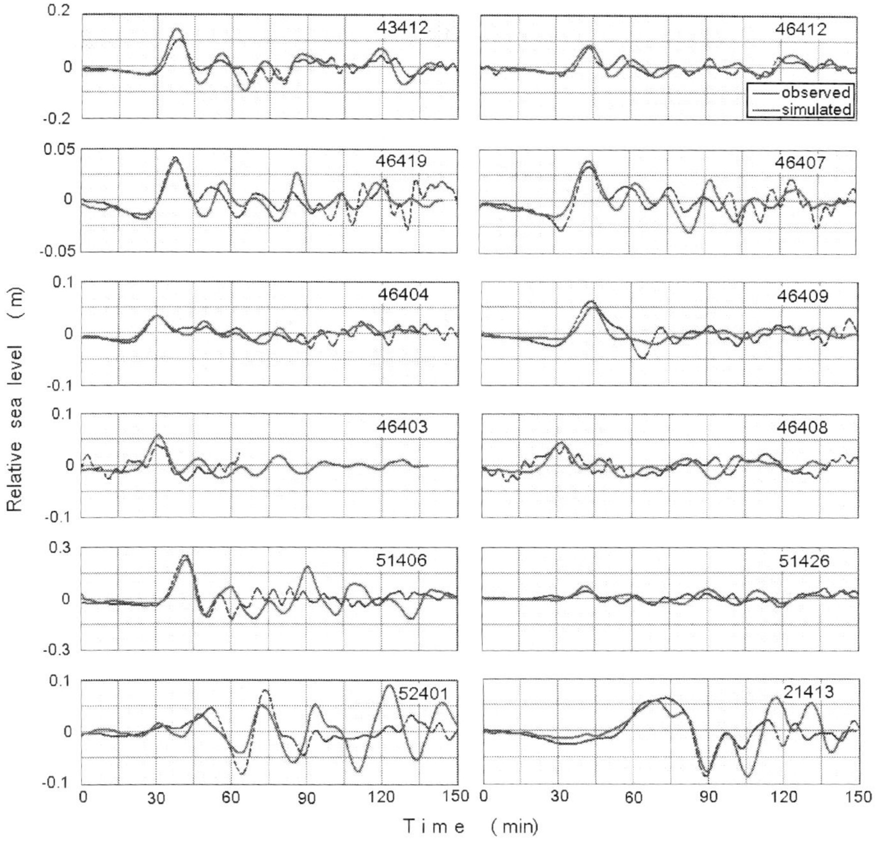

Figure 12. Results of hindcasting the 2010 Chile tsunami in the Pacific based on DART 32412 data.

4.3.2. Hindcast of 2010 Chile tsunami near the coast of the Kuril Islands

The earthquake occurred on February 27 at 06:34 UTC. The tsunami alarm was declared by the Sakhalin Tsunami Warning Center on all the Kurile Islands on February 28 at 00:50 UTC (10:50 local time) 3 h 22 min before the first wave arrived at Severo-Kurilsk (Paramushir Island). The population of all settlements of the Kuril Islands was evacuated and ships were withdrawn to the open sea. The wave amplitudes for 4 hours did not exceed 0.3 m. The duration of the alarm was about 4 hours. Shortly after the alarm was cancelled (4 hours 8 minutes after the first wave arrival) a wave with an amplitude of 1 m was recorded. Once again the tsunami alarm had not been announced.

For tsunami prediction near the coast of the Kuril Islands, computations have been made from the source of the tsunami off the coast of Chile, initially using spherical co-ordinates with a spatial grid step 5 km at the equator. Then computations were made using a more detailed grid (left bottom corner 37° 51'N, 139° 26'E, right top corner 45° N, 180° E). The spatial grid step was 1 km on scale latitude 45°N. Part of the scheme is shown in Fig. 7.

The tsunami hindcasting is based on data of DART station 32412 with a 30 minute duration The hindcast result is represented in Fig. 13.

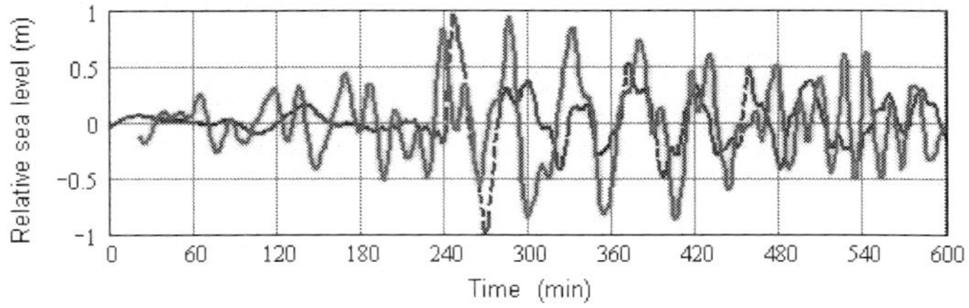

Figure 13. 2010 Chile tsunami hindcast at Severo-Kurilsk (point S-K on Fig. 7) using data of DART station 32412.

The simulated tsunami waveform coincides well with the form of the observed tsunami. The wave amplitudes during the first 4 hours do not exceed 0.5 m. After 4 hours the appearance of waves with increased amplitude up to 1 m is correctly predicted.

4.4. Tohoku foreshock 2011

The weak tsunami M_w=7.2 which arose off east coast of Honshu Island on March 09, 2011 (epicentre co-ordinates 38.510°N, 142.792°E (NOAA/WDC, n.d.)) was recorded by a small number of DART stations. The tsunami hindcast using data of the 21418 DART station, nearest to the centre, was made in the area eastward of the Kuril Islands at points 21401, 21419 and in the Pacific Ocean at a point 21413 1300 km to the southeast of Tokyo (see Fig.7 for DART positions).

The auxiliary source was taken as a circular uplift of the sea surface 100 km in diameter and maximum height of 10 m; the source centre coincided with the earthquake epicentre. The ocean depth at the epicentre was 1374 m. We used the Mercator projection with a spatial grid step 5 km at latitude 45°N.

Represented in Fig. 14, the result shows a good coincidence between computed and actual tsunami waveforms.

The computations were executed for points in the open-ocean. The various edge, resonant effects do not influence tsunamis in the open-ocean. This result demonstrates the possibilities of the proposed method. Based on data of DART station 21418, located to the east of the earthquake epicentre, the tsunami waveforms are obtained to the northeast of the epicentre (points 21401 and 21419) and to the southeast (point 21413).

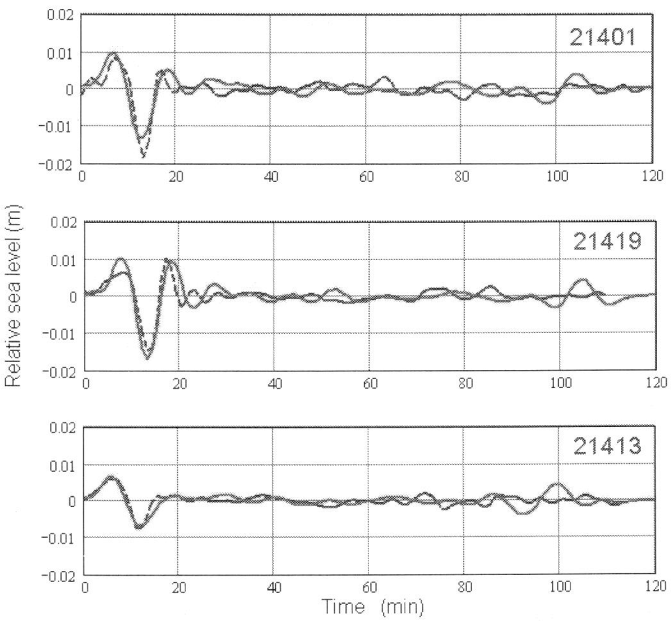

Figure 14. Hindcast of the March 9, 2011 tsunami in the Pacific.

4.5. 2011 Great Tohoku tsunami

4.5.1. 2011 Great Tohoku earthquake and tsunami

This earthquake with magnitude M_w = 9.0 off the northeast coast of Honshu Island occurred on March 11 at 5:46 UTC. The co-ordinates of the epicentre are 38.322ºN, 142.369ºE (NOAA/WDC, n.d.).

As a result of the earthquake a tsunami occurred. It was anomalously high and very destructive on the coast of Honshu Island. Against the backdrop of a long rise in sea level with a height of 2 metres there was an additional shorter rise, with increased amplitude up to 5 m 11 minutes after (Fig. 15).

The great tsunami resulted from a combination of crustal deformations of the ocean floor due to up-thrust tectonic motions, augmented by additional uplift due to the quake's slow and long rupturing process, as well as large coseismic lateral movements which compressed and deformed the compacted sediments. This mechanism is discussed in (Pararas-Carayannis, 2011).

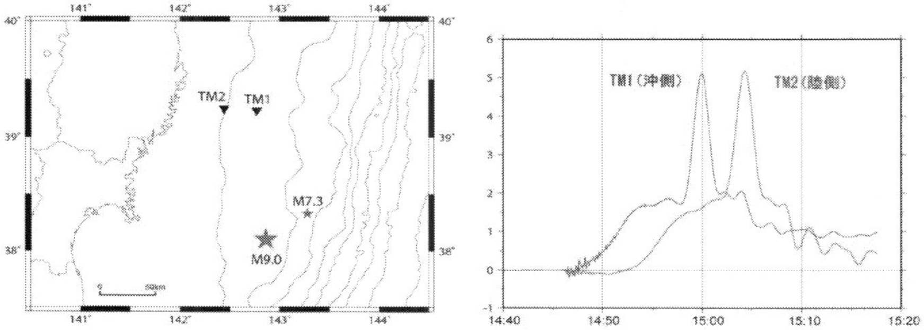

Figure 15. Registration of the 11.03.2011 tsunami near to the centre. Positions of earthquake epicentre with M = 9.0 and sea level stations TM1 and TM2 - at the left. The form of tsunami registered by stations - on the right (see http://outreach.eri.u-tokyo.ac.jp/wordpress/wp-content/uploads/2011/03/TsunamiKamaishiMeter.jpg).

4.5.2. The hindcast of 2011 Great Tohoku tsunami in the ocean

A simulation of operative tsunami forecasting for March 11, 2011 was made for points in the open-ocean. The data of DART station 21401 were used in the computations. The relative level of the ocean registered by station 21401 is shown in Fig. 16.

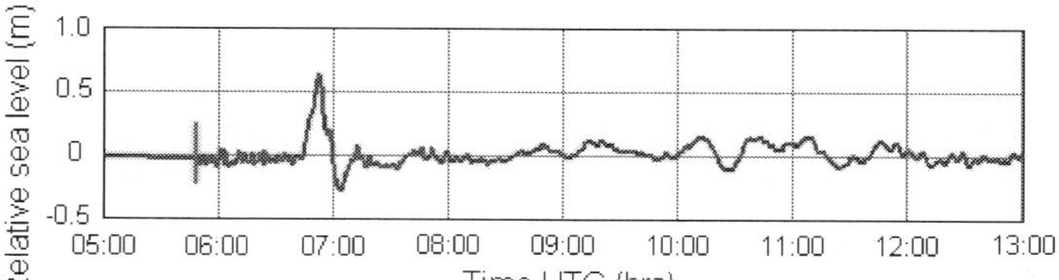

Figure 16. Sea level recorded by DART station 21401 on March 11, 2011 (the tidal components are deleted), the vertical red line – the moment of the main shock.

To create the transfer function the auxiliary computations of the waveform at the corresponding points from the source in the form of an initial circular elevation of the free-surface with diameter 100 km and an amplitude 10 m are made. The centre of the source coincides with the epicentre of the earthquake, the depth of the ocean according to the bathymetric grid equals to 887 m. Computations were made using spherical co-ordinates with the spatial grid step 5 km at a scale latitude 0°. The computations were based on a 30 minute tsunami duration of the data from 21401 DART station.

Figure 17 demonstrates the high quality of the coincidence of simulated and observed tsunami waveforms both in the northeastern and southern directions from the source. Durations exceeding 30 min in obtained data from DART station 21401 used in the computations do not affect the quality of the forecast.

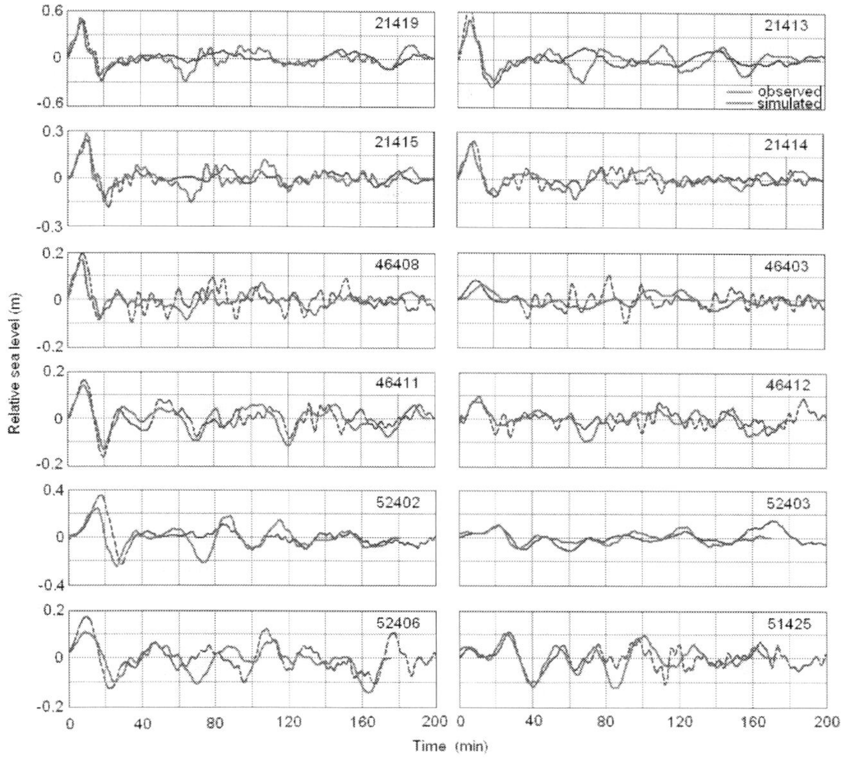

Figure 17. Hindcast of the March 11, 2011 Great Tohoku tsunami in the Pacific.

Analogous simulations have been performed using other means by Titov, Fujii and Satake (Japan (East Coast of Honshu) Tsunami, 2011; Fujii & Satake, 2011). The quality of the coincidence of simulated and observed tsunami waveforms in these works as well as in the presented work is comparable.

4.5.3. The hindcast of 2011 Great Tohoku tsunami near the coast of the Kuril Islands

This strong earthquake off the northeast coast of Honshu Island occurred on March 11 at 5:46 UTC.

The tsunami alarm was declared by the Sakhalin Tsunami Warning Center at 5:58 UTC (local time 15:58) on all Kuril Islands. The population deemed to be in the threatened areas was evacuated to safe places and vessels were recommended to sail out into the open sea.

At 6:43 deep-sea DART station 21401 (located southeast of Iturup Island) registered a tsunami wave with amplitude 67 cm.

The oscillations of sea level, caused by the tsunami, with amplitude of about 1 metre lasted until March 12, therefore, the duration of the tsunami alarm was about 20 hours. The tsunami alarm on all Kuril Islands was cancelled on March 12 at 2:20 UTC (local time 12:20).

The tsunami arrived at Severo-Kurilsk at 9:00. From information from vessels in the sea near Severo-Kurilsk the sea level under keel changed from 4.2 m up to 2.6 m, the swing was 1.6 m. Recorded wave amplitudes were about 0.3 m in Kurilsk. According to visual observations the amplitudes of the tsunami in Burevestnik were approximately 1 m. The tsunami reached Yuzhno-Kurilsk at 7:44, the recorded wave amplitudes were about 1 m. The tsunami alarm was declared 1.5 to 3 hours before the tsunami arrival at different points of the Kuril Islands.

To create the transfer function the auxiliary computations of the waveform at the corresponding points from the source in the form of an initial circular elevation of the free-surface with diameter 100 km and an amplitude 10 m were made. The sea depth at the epicentre is equal to 970.1 m. The centre of the source coincides with the epicentre of the earthquake. We used the Mercator projection with a spatial grid step 0.9 km at latitude 45°N. Computations based on data from DART station 21401 with tsunami duration 30 min were made.

The hindcast was made near the above-mentioned points S-K, K, B and Yu-K. The points are located at distances from the coast, respectively, 1.1, 0.8, 2.1 and 5.4 km. Water depths at these points are 16, 3.2, 30 and 29 m. Locations of these points are shown in Fig. 7.

The results of a hindcast of tsunami waveforms with a duration of 6 - 7 hours made for the above-mentioned points at the Kuril Islands are represented in Fig. 18.

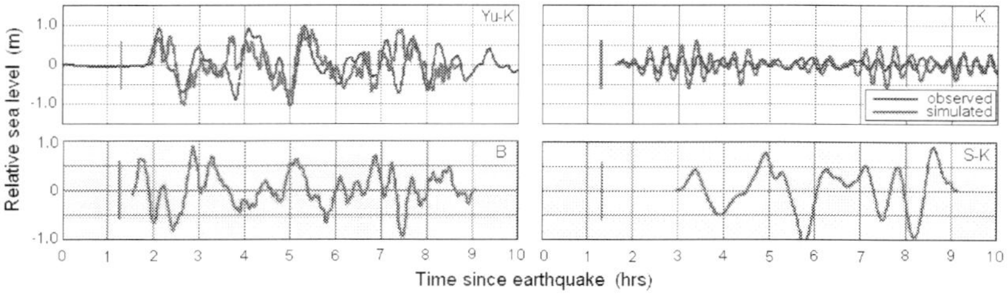

Figure 18. Tsunami hindcast at the points of the Kuril Islands. The vertical red line – the moment of forecast producing.

The simulated waveform at point Yu-K and the observed tsunami waveform at Yuzhno-Kurilsk are in good agreement. The structure of the tsunami, arrival times of all four waves and wave amplitudes coincided well. According to computations, maximal wave amplitudes should not exceed 1 m; that is in agreement with the observations.

The simulated tsunami waveform at Kurilsk (point K) demonstrates a brightly expressed batch wave structure. The computed wave periods coincide well with recorded periods. According to the forecast, the wave amplitudes should not exceed 0.5 metres.

According to the forecast for Burevestnik (point B), the wave amplitudes should not exceed 1 m. This result agrees with observational data: the wave amplitudes on visual observations were about 1 metre.

At Severo-Kurilsk (point S-K), according to the forecast, increase of amplitude of consequent waves almost up to 1 m is possible. The range of oscillations of a level should not exceed 1.8 - 2 metres. This result agrees well with observational data at Severo-Kurilsk. Based on information from vessels, the swing of sea level was equal to 1.6 m.

Unfortunately, the registration of a tsunami at Burevestnik and Severo-Kurilsk was not conducted. Nevertheless, the results of the computation are confirmed by visual observations and information from vessels in these points.

The time of auxiliary computations for transfer function creation did not exceed 25 min, while the tsunami travel time up to DART station 21401 is equal to 58 min. The auxiliary computations are completed before registration of a tsunami by station 21401. Taking into account the time of tsunami recognition at point 21401 (the first period with a duration of 19 min) the preliminary forecast could be given 78 - 80 min after the main shock of the earthquake. That is 2 hours prior to arrival of the first wave of a tsunami at point S-K, 46-48 minutes prior to arrival of a wave at Yu-K, 40 minutes before tsunami arrival at K and 20 minutes before arrival of the first wave of the tsunami at B. In the latter case the offered hydrophysical method appears to be ineffective.

5. Conclusion

The reciprocity principle, known in acoustics as an asymptotic assumption, is applied to describe non-stationary long ocean waves. The similarity conditions for reciprocal sources, not previously studied, are obtained. It is shown in numerical experiments that, with a quality sufficient for practical applications, the reciprocity principle works quite well, even when asymptotic conditions are not totally satisfied (i.e., when the characteristic time of the wave propagation between the sources is only three to five times greater than the typical wave period or when the characteristic wave length is only two to three times greater than the size of the source area).

The proposed method of computing the tsunami waveform from an unknown source using ocean level data at some points demonstrated its efficiency.

There were questions how the size of auxiliary sources affects the results of computations. Numerical experiments with auxiliary sources of different diameters (with a 20 – 100 km range) gave almost identical results.

There were also questions as to whether it is possible to effectively apply auxiliary axially symmetric sources to create a transfer function. The form of the source influences the directivity of the tsunami. The relief of ocean bottom has a major influence on the evolution of the waves. The use the observational data from the 2009 Simushir, 2010 Chile, 2011 weak Honshu and 2011 Great Honshu tsunamis gives very encouraging results. Despite the fact that the auxiliary source was axially symmetric, the tsunami forecast gives almost identical results based either on the data from reference DART stations located in the same direction as the target sites or from the stations located in other directions.

Despite its approximate character, the proposed method can provide a forecast of tsunami wave parameters for any ocean site with sufficiently high quality and provide real-time operative information for tsunami warning services. The results of this paper demonstrate that the proposed method satisfies the requirements of the warning services and can be effectively used for tsunami warnings.

The numerical experiments described in the paper were conducted for open-ocean target sites. For these sites the method works pretty well, however, the main purpose of TWSs is to provide reliable tsunami forecasts for coastal areas. For this purpose we made auxiliary computations (to evaluate transfer functions) using the detailed bathymetry grid (detailed enough to resolve small-scale resonant features of coastal topography). This allows us to provide the adequate tsunami forecast for the entire coastal zone exposed to arriving tsunami waves. This ensures a straightforward forecast, not only of the leading tsunami waves, but also the following tsunami waves which sometimes are much more intensive than the head waves. The practical interest is to predict tsunami waveforms of long duration, which would take into account secondary waves reflected from island chains, submarine ridges and mainland coasts. Examples of tsunami hindcasts of 2009, 2010, 2011 have shown that the forecast of long duration based on short information from DART stations is possible.

The proposed method of short-term tsunami forecasting uses only seismological information about the location of the earthquake epicentre and works independently of the mechanism of tsunami generation. Numerical experiments show that, despite the anomalous character of the 2011 Great Tohoku tsunami, the hindcast using only level data from DART 21401 without the additional seismological data gives an adequate result. The same example shows that the forecast, based on preliminary computations, is not always effective.

The advantage of the proposed method is that it does not require a pre-computed database of synthetic tsunami waveforms. This is especially important for areas where tsunami early warning systems are just being established and there are no pre-computed databases.

The time of the forecast producing is determined by the time of tsunami propagation up to a point of registration by the DART station and the time of tsunami identification (the first half-cycle, about 15 - 20 minutes). Such a forecast is effective for points for which the difference between the time of tsunami attack and the time of forecast producing is enough for an evacuation of the population and vessels from a threatened zone. For the Kuril Islands this time is equal, conditionally, to 30 minutes.

The comparison of results presented in this paper with some other results based on assimilation of DART data (Japan (East Coast of Honshu) Tsunami, 2011; Tang et al., 2008; Titov, 2009; Wei et al., 2008) shows that the forecast quality obtained by different methods is comparable.

Currently, post-event computations are performed using four non-related software, thence routine work takes a long time. This, unfortunately, does not allow forecasting in real-time.

Appendix A

To verify the condition (9) the problem of wave propagation in a boundless basin is solved. One half of the basin (area A, $x < 0$) has a constant depth D_A, the other (area B, $x > 0$) has a constant depth D_B.

Let's the axial–symmetrical initial elevation of a free surface $\zeta_{A0}(x-X_A,y-Y_A)$ with radius R_A is given in area A with centre at a point A. The waveform of a free surface in a point with co-ordinates X_B, Y_B in a right half-space B is sought.

The evolution of waves is described by wave equation with variable water depth (2) with the initial condition $\zeta(t,x,y)/_{t=0} = \zeta_{A0}(x-X_A,\ y-Y_A)$ in area S_A and also with conditions on infinity: $\zeta(t,x,y) \rightarrow 0$ when $|x|,\ |y| \rightarrow \infty$.

The Laplace transform on time and Fourier transforms on both co-ordinates x, y are applied to the equation (2) and conditions (3) - (5)

$$f(s,k_x,k_y) = \int_0^{\infty}\int_{-\infty}^{\infty}\int_{-\infty}^{\infty} f(t,x,y)\cdot \exp(ik_x x)\cdot \exp(ik_y y)\cdot \exp(-st)\cdot dydxdt\,.$$

So the equation (2) became

$$\left(\frac{s^2}{c_A^2} + k_x^2 + k_y^2\right)\cdot \zeta(s,k_x,k_y) = \frac{s}{c_A^2}\cdot \zeta_{A0}(k_x,k_y), \tag{14}$$

where s, k_x, k_y – are parameters of Laplace and Fourier transforms,

c_A – long wave speed on water depth D_A: $c_A=(gD_A)^{1/2}$,

$$\zeta_{A0}(k_x,k_y) = \exp(ik_x X_A + ik_y Y_A)\cdot \int_{S_A} \zeta_{A0}(x',y')\cdot \exp(ik_x x' + ik_y y')\cdot dx'dy',$$

x', y' – are Cartesian co-ordinates with origin in point A.

The inverse Fourier transform on k_x of function $\zeta(s,k_x,k_y)$ from (14):

$$\zeta_1(s,x,k_y) = \frac{1}{2\pi}\cdot\frac{s}{c_A^2}\int_{-\infty}^{\infty} e^{ik_x X_A + ik_y Y_A} \int_{S_A} \zeta_{A0}(x',y')\cdot e^{ik_x x' + ik_y y'} dx'dy' \frac{e^{-ik_x x}}{k_x^2 + \chi_A^2} dk_x,$$

where $\chi_A^2 = \frac{s^2}{c_A^2} + k_y^2$, is obtained by taking residue. The residue is taken in one pole $k_x = -i\chi_A$, considering the boundedness of the required function.

The result is

$$\zeta_1(s,x,k_y) = \frac{s}{2c_A^2\chi_A} e^{ik_yY_A} \cdot e^{-\chi_A(x-X_A)} \cdot \int \zeta_{A0}(x',y') \cdot e^{\chi_A x' + ik_y y'} dx'dy'$$

The obtained result is a particular solution of the inhomogeneous equation. By virtue of the fact that the solution is sought in an area with a saltus of depth, it is necessary to take into account a general solution for the homogeneous equation. It is $\zeta_2(s,x,k_y) = C_2 exp(\chi_A x)$. Only one fundamental solution is taken satisfying the condition $\zeta_2(s,x,k_y) \to 0$ when $x \to -\infty$.

In a right half-space B a general solution is $\zeta_3(s,x,k_y) = C_3 exp(-\chi_B x)$ satisfying the condition $\zeta_3 \to 0$ when $x \to \infty$, where $\chi_B^2 = \frac{s^2}{c_B^2} + k_y^2$, $c_B = (gD_B)^{1/2}$.

Indeterminate coefficients C_2 and C_3 are sought from the following conditions when $x=0$

$\zeta_1(s,0,k_y) + \zeta_2(s,0,k_y) = \zeta_3(s,0,k_y)$ - continuity of sea level,

$D_A \cdot \frac{d}{dx}\left(\zeta_1(s,x,k_y) + \zeta_2(s,x,k_y)\right) /_{x=0} = D_B \cdot \frac{d}{dx}\zeta_3(s,x,k_y) /_{x=0}$ - continuity of stream.

From the two last equations

$$C_3 = \frac{2\chi_A D_A}{\chi_A D_A + \chi_B D_B} \cdot \frac{s}{2c_A^2\chi_A} \cdot e^{\chi_A X_A + ik_y Y_A} \cdot \int_{S_A} \zeta_{A0}(x',y') \cdot e^{\chi_A x' + ik_y y'} dx'dy'.$$

So, the solution (water level) in the point $B(X_B, Y_B)$ from source A is

$$\zeta_A(s, X_B, k_y) = \frac{s}{g \cdot (\chi_A D_A + \chi_B D_B)} \cdot e^{\chi_A X_A + ik_y Y_A} \cdot \int_{S_A} \zeta_{A0}(x',y') \cdot e^{\chi_A x' + ik_y y'} dx'dy' \cdot e^{-\chi_B X_B}.$$

Inverse Fourier transform on k_y

$$\zeta_A(s, X_B, Y_B) = \frac{1}{2\pi} \int_{-\infty}^{\infty} \frac{s}{g \cdot (\chi_A D_A + \chi_B D_B)} e^{\chi_A X_A - \chi_B X_B + ik_y Y_A} \int_{S_A} \zeta_{A0}(x',y') \cdot e^{\chi_A x' + ik_y y'} dx'dy' e^{-ik_y Y_B} dk_y.$$

The regular wave reflection (refraction) from the line of depth changing is hereinafter considered.

It is supposed that $|X_A|$, $|X_B|$, $|Y_A|$, $|Y_B| >> R_A$. The last integral may be estimated by the steepest descent method (Lavrentiev & Shabat, 1987). The exponent of power $\Psi(k_y) = \chi_A X_A - \chi_B X_B + ik_y Y_A - ik_y Y_B$ should be expanded in a power series of k_y near the saddle point. This point is sought from the following condition

$$\Psi'_{k_y}(k_y) = \frac{k_y X_A}{\chi_A} - \frac{k_y X_B}{\chi_B} + i \cdot (Y_A - Y_B) = 0 .$$

Taking into account X_A, $Y_A<0$ appropriate root of this equation is $k_y = k_+ = -i \cdot \frac{s}{c_A} \sin\theta_A = -i \cdot \frac{s}{c_B} \sin\theta_B$, where θ_A, θ_B – angles of incidence and refraction of a ray, connecting points A and B, in some point on the line $x=0$.

So, the function $\Psi(k_y)$ and the second derivative $\Psi''(k_y)$ in the saddle point $k_y=k_+$ are

$$\Psi(k_+) = \frac{s}{c_A}\cos\theta_A \cdot X_A - \frac{s}{c_B}\cos\theta_B \cdot X_B + \frac{s}{c_A}\sin\theta_A \cdot Y_A - \frac{s}{c_B}\sin\theta_B \cdot Y_B = -s \cdot T_{AB},$$

where T_{AB} – propagation time from A to B along a ray and

$$\Psi''_{k_y}(k_+) = \frac{c_A X_A}{s\cos^3\theta_A} - \frac{c_B X_B}{s\cos^3\theta_B}.$$

Let us assume that $\zeta_{A0}(x',y')$ is an axial symmetric function in circle area S_A with radius R_A. It is convenient to replace the Cartesian co-ordinates on polar ($x'=r\cos\phi$, $y'=r\sin\phi$) and to enter dimensionless co-ordinates and function ($r=r'R_A$, $\zeta_{A0}(x',y')=a_A\,\zeta'_{A0}(r')$).

The integrand is substituted by its value in a saddle point and function $\Psi(k_y)$ - by an expansion in a power series near to this point. Thus, the sought-for solution (Laplace transform) is described by expression

$$\begin{aligned}
&\zeta_A(s,X_B,Y_B) = \\
&= \frac{a_A R_A^2}{2\pi} \frac{e^{-s\cdot T_{AB}}}{c_A\cos\theta_A + c_B\cos\theta_B} \int_0^1 \zeta'_{A0}(r') \cdot \int_0^{2\pi} e^{\frac{sR_A}{c_A} r'\cos(\phi-\theta_A)} d\phi \cdot r'dr' \cdot \\
&\cdot \int_{-\infty}^{\infty} \exp\left(\frac{1}{2}\frac{c_A X_A}{s\cos^3\theta_A}\left(k_y + i\frac{s}{c_A}\sin\theta_A\right)^2 - \frac{1}{2}\frac{c_B X_B}{s\cos^3\theta_B}\left(k_y + i\frac{s}{c_B}\sin\theta_B\right)^2\right) dk_y
\end{aligned} \tag{15}$$

If s is imaginary value: $s=i\sigma$,- then integral over φ is $2\pi \cdot J_0(\frac{\sigma R_A}{c_A} r')$ - Bessel function of order 0. So, expression (15) may be rewritten

$$\begin{aligned}
&\zeta_A(\sigma,X_B,Y_B) = \\
&= a_A R_A^2 \frac{\exp(-i\sigma \cdot T_{AB})}{c_A\cos\theta_A + c_B\cos\theta_B} \int_0^1 \zeta'_{A0}(r') J_0\left(\frac{\sigma R_A}{c_A} r'\right) \cdot r'dr' \int_{-\infty}^{\infty} \exp\left(-i\frac{1}{2}\frac{c_A X_A}{\sigma\cos^3\theta_A}\left(k_y - \frac{\sigma}{c_A}\sin\theta_A\right)^2 + \right. \\
&\left. + i\frac{1}{2}\frac{c_B X_B}{\sigma\cos^3\theta_B}\left(k_y - \frac{\sigma}{c_B}\sin\theta_B\right)^2\right) dk_y
\end{aligned} \tag{16}$$

Analogously, the solution in point A from circle source $\zeta_{B0}(x-X_B,y-Y_B)$ with radius R_B is obtained

$$\zeta_B(\sigma, X_A, Y_A) =$$
$$= \frac{a_B R_B^2}{2\pi} \frac{\exp(-i\sigma \cdot T_{BA})}{c_B \cos\theta_B + c_A \cos\theta_A} \int_0^1 \zeta'_{B0}(r') J_0\left(\frac{\sigma R_B}{c_B} r'\right) \cdot r' dr' \int_{-\infty}^{\infty} \exp(i \frac{1}{2} \frac{c_B X_B}{\sigma \cos^3\theta_B} (k_y + \frac{\sigma}{c_B} \sin\theta_B)^2 - \quad (17)$$
$$-i \frac{1}{2} \frac{c_A X_A}{\sigma \cos^3\theta_A} (k_y - \frac{\sigma}{c_A} \sin\theta_A)^2) dk_y$$

Expressions (16) and (17) are enough for a comparison.

The evaluations of integrals over k_y, obtained with allowance for deformation of an integration contour, are identical.

If the disturbances (sources) are similar, dimensionless function $\zeta'_{A0}(r') = \zeta'_{B0}(r')$.

Then, if volumes of initial elevations are equalled ($a_A R^2_A = a_B R^2_B$), functions $\zeta_A(\sigma, X_B, Y_B)$ and $\zeta_B(\sigma, X_A, Y_A)$ will be identical, if the arguments of Bessel functions will be equal: $\frac{\sigma R_A}{c_A} = \frac{\sigma R_B}{c_B}$.

The similarity condition $\frac{R_A}{\sqrt{gD_A}} = \frac{R_B}{\sqrt{gD_B}}$ (9) follows from here.

Appendix B. Methods of inverse Laplace transform

For a final tsunami evaluation in a specific point A it is necessary to execute the inverse Laplace transforms of (12) or (13). The problems of inverse transform for such expressions are related to ill-posed problems and some regularization methods are required.

The initial functions in (12, 13) are discrete ones, therefore, their Laplace images can be represented by discrete Laplace transforms as a power series with a time step τ, for example, (Doetsch, 1967)

$$\zeta(s, M) = \frac{1 - e^{-s\tau}}{s} \sum_{k=0}^{N} \zeta_k(M) \cdot e^{-ks\tau},$$

where $\zeta_k(M)$ – sea level in the point M in moments $t=k\tau$, $k=0\ldots N$, τ – time sample. $N\tau$ is duration of $\zeta(t,M)$.

After representing all functions in the form of time series and replacing $z=e^{-s\tau}$, (13) will be represented as z-transformation

$$\zeta(z, A) = \sum_{k=0}^{N} \zeta_k(A) \cdot z^k = \sum_{k=0}^{N_1} \zeta_k(M) \cdot z^k \frac{\sum_{k=0}^{N_2} \eta_k(A) \cdot z^k}{\sum_{k=0}^{N_3} \eta_k(M) \cdot z^k}. \quad (18)$$

The factor $\frac{1 - e^{-s\tau}}{s}$, being an image of square impulse and being out of interest, hereinafter is omitted.

B.1. The empirical method of inverse Laplace transform: The root-displacement method (Method 1)

After division and multiplication of polynomials in the right part of (18) the polynomial will be derived $\zeta(z,A)=\sum_{k=0}^{N} f_k z^k = \sum_{k=0}^{N} f_k e^{-ks\tau}$. The result of inverse Laplace transform of this polynomial is the sequence consisting of δ-functions with factors equal to coefficients of a polynomial. These factors f_k are values of a tsunami level being sought at moments $t = k\tau$, $k = 0 \ldots N$, after an expected tsunami arrival at point A.

In expression (18) the upper limits of the sums should be identical. Otherwise, in the case of various limits on the result of inverse transforms, there would be doubtful summands.

Unfortunately, the power series obtained in such a way diverges. Two methods are offered below to avoid this effect.

The power series obtained after division and multiplication of polynomials in the right part of (18) is diverging if roots of a polynomial in a denominator have absolute values which are less than 1. The following procedure is applied to avoid a divergence. In a polynomial of a denominator the roots having absolute values less than 1 are substituted by roots with absolute values equalled to 1, but with the same arguments

$$\sum_{n=0}^{N} \eta_n(M)\cdot z^n \rightarrow \sum_{n=0}^{N} \eta_n(M)\cdot z^n \cdot \frac{\prod_{k=0}^{K}(z-\frac{z_k}{|z_k|})}{\prod_{k=0}^{K}(z-z_k)}$$

where z_k – root with $|z_k|<1$, K – the number of such roots.

As a result of this procedure, and multiplication and division of power series, the converging power series $\sum_{n=0}^{N} f_n z^n$ is obtained, though distorted by a procedure of root-displacement.

For reduction of these distortions, similar operation applied to a denominator is used to obtained series

$$\sum_{n=0}^{N} f_n \cdot z^n \rightarrow \sum_{n=0}^{N} f_n \cdot z^n \cdot \frac{\prod_{k=0}^{K}\left(z-\frac{z_k}{|z_k|}\right)}{\prod_{k=0}^{K}\left(z-z_k\right)} = \sum_{n=0}^{N} d_n \cdot z^n$$

The resulting series $\sum_{n=0}^{N} d_n z^n$ is converging and the coefficients d_n are approximate values of the required function.

So the required tsunami waveform in moments $t=n\tau$, $n=0\ldots N$, is $\zeta_n(A)=d_n$.

B.2. Method of inverse Laplace transform based on the solution of an ill-posed system of linear equations (Method 2)

Representing required function $\zeta(z,A)$ as the sum $\zeta(z,A)=\sum_{k=0}^{N}\zeta_k(A)\cdot z^k=\sum\zeta_k\cdot z^k$, the expression (18) can be written as

$$\sum_{k=0}^{N}\zeta_k\cdot z^k\cdot\sum_{k=0}^{N}\eta_k(M)\cdot z^k=\sum_{k=0}^{N}\zeta_k(M)\cdot z^k\cdot\sum_{k=0}^{N}\eta_k(A)\cdot z^k,$$

where the coefficients $\zeta_k=\zeta_k(A)$ are required values.

After a multiplication of the sums and the equating of coefficients on identical powers z (in products only $N+1$ summands are kept) the set of equations can be composed as

$$U\,\zeta = b, \tag{19}$$

where U ($U_{ij}=\eta_{i-j}(M)$ $i\geq j$, $i=0\ldots N$) – trigonal matrix of order N, b – column matrix ($b_i=\sum_{k=0}^{i}\zeta_k(M)\cdot\eta_{i-k}(A)$, $i=0\ldots N$),

ζ - column matrix of unknowns.

Determinant of matrix U in numerical simulations $\det U=\eta_0^{N+1}\sim10^{-200}$ – 10^{-100}, when $\eta_0(M)\approx 0.01$ -0.1, $N=100$. It is obvious that system (19) is ill-posed one (Tikhonov & Arsenin, 1977).

System (19) can be solved using the following method described in (Tikhonov & Arsenin, 1977).

The solution of system (19) is sought from the condition of minimum smoothing functional

$$\|U\cdot\zeta-b\|^2+\alpha\cdot\|\zeta\|^2\rightarrow min,\alpha>0$$

Norms of matrices are $\|U\|=\sqrt{\sum_{i,j=0}^{N}U_{ij}^2}$, $\|\zeta\|=\sqrt{\sum_{i=0}^{N}\zeta_i^2}$ and $\|b\|=\sqrt{\sum_{i=0}^{N}b_i^2}$.

An approximate solution based on the above condition of minimum is a solution of the following system of linear algebraic equations

$$U'\cdot U\cdot\zeta+\alpha\cdot E\cdot\zeta=U'\cdot b,$$

where U' – transposed matrix, E – unit matrix, α – regularization parameter. This regularization parameter α may be estimated by the method (Tikhonov & Arsenin, 1977).

B.3. Comparison of the results of two methods of inverse Laplace transform

Figure 19 shows one of the results of application of both methods of the inverse Laplace transform to the computation the tsunami waveform for event of March 9, 2011

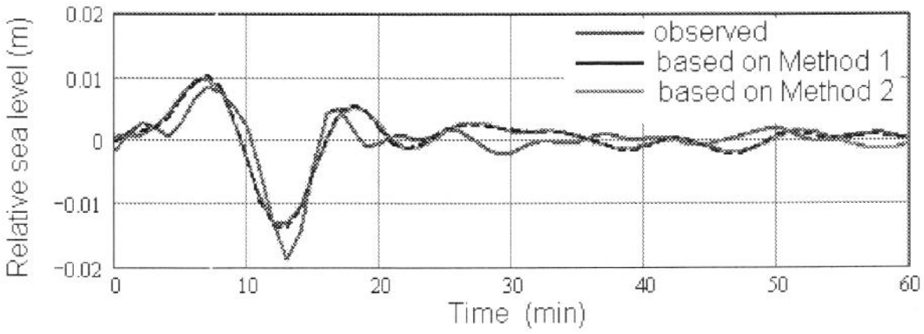

Figure 19. Comparison two methods of inverse Laplace transform.

As seen in Fig. 19, the coincidence of the computed waveforms obtained by different methods (black and red lines), except for the last 5 minutes, is excellent. The real tsunami recording is shown in blue. Further comment on Fig. 19 is not required.

All hindcastings in the present paper were obtained based on Method 2.

Author details

Yury Korolev
Institute of Marine Geology and Geophysics, Russia

6. References

Brekhovskikh, L.M. (1960). *Waves in Layered Media,* Academic Press, New York.

Chertock, G. (1970). Transient flexural vibrations on ship-like structures exposed to underwater explosions. *J. Acoust. Soc. America,* Vol. 48, No. 2, pp. 170-180

Chubarov, L.B. & Shokin, Yu.I. (1995). Mathematical modeling in mitigating the hazardous effect of tsunami waves in the ocean. A priori analysis and timely on-line forecast. *Sci. Tsunami Hazards,* Vol. 13, N0. 1, pp. 27-44

Chung, J.Y., Kim, S.D. & Ivanov, V.V. (1995). Tsunami wave hindcasting in the Japan Sea, In: *Tsunami: Progress in Prediction, Disaster Prevention and Warning,* edited by Y. Tsuchiya and N. Shuto, pp. , 85–98, Kluwer Academic Publishers, Dordrecht, Netherlands

Doetsch, G. (1967). *Anleitung zum praktischen gebrauch der Laplace-transformation und der z-transformation.* R. Oldenbourg, Munchen, Wien

Fujii, Y. & Satake, K. (2008). Tsunami sources of the November 2006 and January 2007 Great Kuril Earthquakes. *Bull. Seism. Soc. Amer.,* Vol. 98, No. 3, pp. 1559-1571

Fujii, Y. & Satake, K. (2011). Off Tohoku-Pacific Tsunami on March 11, 2011. Available from: http://iisee.kenken.go.jp/staff/fujii/OffTohokuPacific2011/tsunami.html.

Gusiakov, V. K. (2011). Relationship of tsunami intensity to source earthquake magnitude as retrieved from historical data. *Pure Appl. Geophys.*, Vol. 168, pp. 2033-2041

Horrillo, J., Knight, W. & Kowalik, Z. (2008). The Kuril Islands Tsunami of November 2006. Part II: Impact at Crescent City by local enhancement. *J. Geophys. Res.*, Vol. 113, C01021

Japan (East Coast of Honshu) Tsunami, March 11, 2011. Main Event Page. (2011). In: *NOAA Center for Tsunami Research*, Currently accessed, Available from: http://nctr.pmel.noaa.gov/honshu20110311/images/comp_plots/20110311_darts_comp.png

Korn, G.A. & Korn, T.M. (1961). *Mathematical handbook for scientists and engineers. Definitions, theorems and formulas for reference and review*. McGraw-Hill Book Company, Inc., New York, Toronto, London

Korolev, Y. P. (2004). Tsunami numerical modeling for short-term forecasting using data of remote level gauges. *Oceanology*, Vol. 44, pp. 346-352

Korolev, Y. (2005). The method of the early tsunami warning using data of remote level gauges, Proc. of 22nd Int. Tsunami Symp., Chania, Crete Island, Greece, 27-29 June 2005, pp. 113-119

Korolev, Y. P. (2011a). Retrospective Short-Term Forecast of the 1996 Andreanov (Aleutian Islands) Tsunami. *Oceanology*, Vol. 51, pp. 385–393

Korolev, Y. P. (2011b). An approximate method of short-term tsunami forecast and the hindcasting of some recent events. *Nat. Hazards Earth Syst. Sci.*, Vol. 11, pp. 3081–3091, doi: 10.5194/nhess-11-3081-2011. Available at: www.nat-hazards-earth-syst-sci.net/11/3081/2011/

Korolev, Y. P. & Poplavsky, A.A. (1995). A method of tsunami estimation for short-range forecast service, In: *IUGG, XXI General Assembly*, Boulder, Colorado, A335

Kowalik, Z., Horrillo, J., Knight, W. & Logan, T. (2008). The Kuril Islands Tsunami of November 2006. Part I: Impact at Crescent City by distant scattering. *J. Geophys. Res.*, Vol. 113, C01020

Landau, L.D. & Lifshitz, E.M. (1975). *Fluid Mechanics: Vol. 6 (Course of Theoretical Physics)*. Pergamon Press, Oxford, New York, Toronto, Sydney, Paris, Braunschweig

Laverov, N.P., Lobkovsky, L.I., Levin, B.W., Rabinovich, A.B., Kulikov, E.A., Fine, I.V., & Thomson, R.E. (2009). The Kuril tsunamis of November 15, 2006, and January 13, 2007: Two trans-Pacific events. *Transactions (Doklady) of the Russian Academy of Sciences, Earth Science Section*, Vol. 426, No. 3, pp. 386-392

Lavrentiev, M.A. & Shabat B.V. (1987). *Methods of the theory of the functions of complex variables*, (in Russian), Nauka, Moscow

Loomis, H.G. (1979). Tsunami prediction using the reciprocal property of Green's functions. *Marine Geodesy*, Vol. 2, pp. 27-39

MacInnes, B.T., Pinegina, T.K., Bourgeois, J., Razhigaeva, N.G., Kaistrenko, V.M. & Kravchunovskaya, E.A. (2009). Field survey and geological effects of the 15 November 2006 Kuril tsunami in the middle Kuril Islands. , *Pure Appl. Geophys.*, Vol. 166, pp. 9-36

Miller, G.R. (1972). Relative spectra of tsunamis. *Hawaii Inst. Geophys. HIG-72-8*, Honolulu, 7 p.

Mofjeld, H.O. (2009). Tsunami measurements, In: *The Sea*, Vol. 15, edited by E.N. Bernard and A.R. Robinson, pp. 201-235, Harvard University Press, Cambridge

NDBC DART Program. (n.d.). In: National Data Buoy Center, Currently accessed, Available from: http://www.ndbc.noaa.gov/dart.shtml

NOAA/WDC Global Historical Tsunami Database at NGDC. (n.d.). In: *NOAA National Geophysical Data Center*, Currently accessed, Available from: http://www.ngdc.noaa.gov/hazard/tsu_db.shtml

NOAA Tsunami Warning System Receives High Marks. (2004). In: *NOAA Magazine on-line*, 30.03.2012, Available from: http://www.magazine.noaa.gov/stories/mag153.htm

Pararas-Carayannis, G. (2011). Tsunamigenic source mechanism and efficiency of the March 11, 2011 Sanriku Earthquake in Japan. *Science of Tsunami Hazards*, Vol. 30, No. 2, pp. 126-152

Poplavsky, A.A., Khramushin, V.N., Nepop, K.I. & Korolev, Y.P. (1997). *The Operative Tsunami Prediction on the Sea Coasts of the Far East*, (in Russian), DVO RAN, Yuzhno-Sakhalinsk, Russia

Rabinovich, A.B., Lobkovsky, L.I., Fine, I.V., Thomson, R.E., Ivelskaya, T.N., & Kulikov, E.A. (2008). Near-source observations and modeling of the Kuril Islands tsunamis of 15 November 2006 and 13 January 2007. *Advances in Geosciences*, Vol. 14, No. 1, pp. 105-116

Rayleigh Lord (J.W. Strutt). (1945). *The Theory of Sound*, Vol. 2. Dover Publications, New York

Satake, K. (1987). Inversion of tsunami waveform for the estimation of a fault heterogeneity: method and numerical experiments. *J. Phys. Earth*, Vol. 35, No. 3, pp. 241-254

Smith, W.H.F. & Sandwell, D.T. (1994). Bathymetric prediction from dense satellite altimetry and sparse shipboard bathymetry. *J. Geophys. Res.*, Vol. 99, No. 21, pp. 803-824

Stoker, J.J. (1957). *Water waves. The mathematical theory with applications*. Interscience Publishers, New York, London

Takahasi, R. & Aida, I. (1961). Studies on the spectrum of tsunami. *Bull. Earthq. Res. Inst.*, Vol. 39, pp. 523–535

Tang, L., Titov, V.V., Wei, Y., Mofjeld, H.O., Spillane, M., Arcas, D., Bernard, E.N., Chamberlin, C., Gica, E., & Newman J. (2008). Tsunami forecast analysis for the May 2006 Tonga tsunami. *J. Geophys. Res.*, Vol. 113, C12015, doi: 10.1029/2008JC004922

Tikhonov, A. N. & Arsenin, V. Y. (1977). *Solutions of ill-posed problems* (F. John, translation editor), V. H. Winston and Sons (distributed by Wiley, New York)

Titov, V.V., González, F.I., Bernard, E.N., Eble, M.C., Mofjeld, H.O., Newman, J.C. & Venturato, A.J. (2005). Real-time tsunami forecasting: challenges and solutions. *Natural Hazards*, Vol. 35, No. 1, pp. 35–41

Titov, V.V. (2009). Tsunami forecasting, In: *The Sea*, Vol. 15, edited by E.N. Bernard and A.R. Robinson, pp. 367-396, Harvard University Press, Cambridge

Tsushima, H., Hirata, K., Hayashi, Y., Tanioka, Y., Kimura, K., Sakai, S., Shinohara, M., Kanazawa, T., Hino, R., & Maeda, K. (2011). Near-field tsunami forecasting using offshore tsunami data from the 2011 off the Pacific coast of Tohoku Earthquake. *Earth Planets Space*, Vol. 63, pp. 821-826

Uric, R.J. (1975). *Principles of Underwater Sound*, McGraw-Hill Book Company, New York

Voronina, T.A. & Tcheverda, V.A. (1998). Reconstruction of tsunami initial form via level oscillation. *Bull. Novosibirsk Comp. Center,* Ser. Mathematical Methods in Geophysics, Vol. 4, pp. 127-136

Wei, Y., Cheung, K.F., Curtis. G.D., & McCreery, C.S. (2003). Inverse algorithm for tsunami forecasts. *J. Waterw., Ports, Coastal Ocean Eng.*, ASCE, Vol. 129, No. 2, pp. 60–69

Wei, Y., Bernard, E., Tang, L., Weiss, R., Titov, V., Moore, C., Spillane, M., Hopkins, M., & Kânoğlu, U. (2008). Real-time experimental forecast of the Peruvian tsunami of August 2007 for U.S. coastlines. *Geophys. Res. Lett.*, Vol. 35, L04609, doi: 10.1029/2007GL032250.

Yamazaki, Y., Wei, Y., Cheung, K.F., & Curtis, G.D. (2006). Forecast of tsunamis from the Japan-Kuril-Kamchatka source region. *Natural Hazards*, Vol. 38, No. 3, pp. 411-435

Chapter 8

A Warning System in Taiwan for South China Sea Tsunamis Using *W* Phase Inversion and Unit Tsunami Methods

Po-Fei Chen, Wen-Tzong Liang and Bor-Yaw Lin

Additional information is available at the end of the chapter

http://dx.doi.org/10.5772/52271

1. Introduction

Ocean-wide tsunamis are almost always triggered by mega-earthquakes, rupturing substantial segments on the zones of interplate boundary between subducting slab and overriding plate. Whether, or specifically when, a mega-earthquake will occur on a subduction zone is crucial for preparedness of tsunami hazard mitigations. However, conditions controlling occurrences of mega-earthquakes are not well understood, as the 2004 Mw9.3 Sumatra earthquake contradicted the previous paradigm that the maximum-size earthquake in a given subduction zone can be predicted from its simple tectonic parameters such as slab age and convergence rates (Ruff & Kanamori, 1980). On top of that, the 100 years or so sampling periods of earthquake catalogue since instrumental era are not long enough to reveal the repeating periods, if any, of mega-earthquakes. Consequently, the fault dimensions necessary for a mega-earthquake to rupture in general become the most reliable constraint on its potential - for any subduction zone with length greater than 1,000 km, one cannot rule out the possibility that it is currently accumulating the strain energy of a mega-earthquake, which will release eventually to trigger ocean-wide tsunamis.

In the South China Sea (SCS) region, the Manila subduction zone, stretching over 1,000 km northward from Mindoro, Philippines to offshore SW Taiwan, poses a potential zone for the occurrence of a mega-earthquake and subsequently, the threat of widespread SCS tsunami hazards. Indeed, it is a reasonable doubt that the historical tsunami hazards on coasts of SW Taiwan in 1661 and 1782 (Soloviev & Go, 1984) might be caused by SCS tsunamis, amplified by SCS shelf when approaching Taiwan straits. Although it remains an open question as to when (or even whether) a mega-earthquake will occur in the Manila subduction zone, we endeavor to establish a warning system in Taiwan for SCS tsunamis.

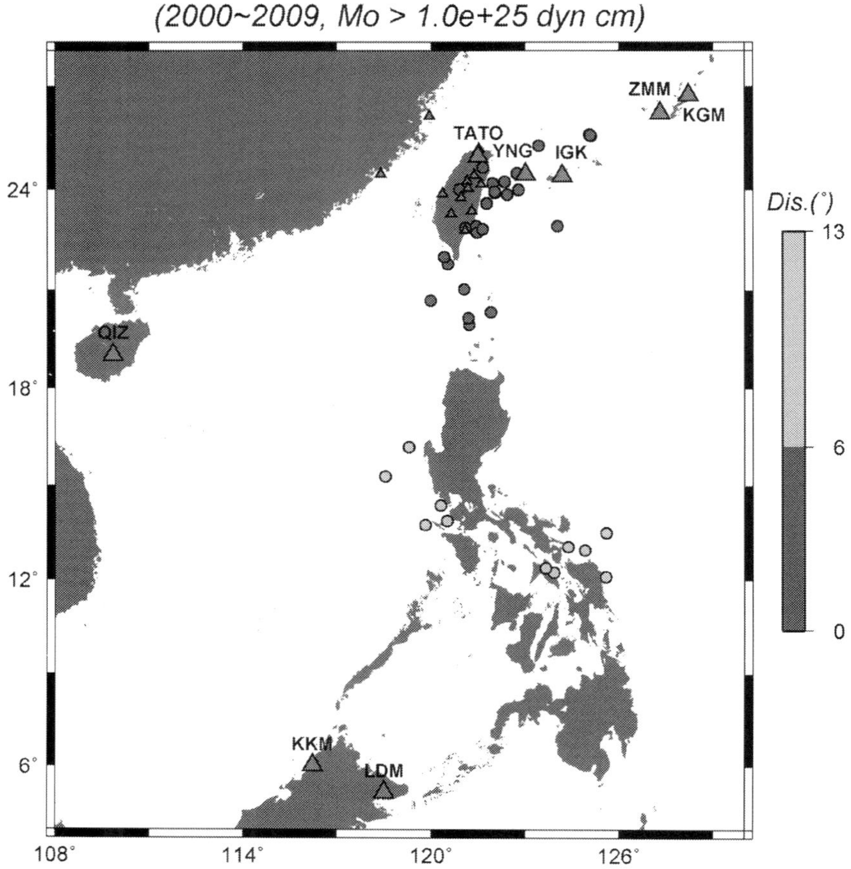

Figure 1. Distribution of SCS earthquakes occurring from January 2000 to July 2009 (circles) with color-keyed epicentral distances relative to central Taiwan (blue for near group and green for far group). Triangles represent distributions of virtual network that mimic those of extended BATS. The non-BATS stations are labeled with greater symbols.

Being able to predict the arrival times and amplitudes of the approaching tsunamis promptly after the occurrence of a tsunamigenic earthquake is key to the success of a tsunami warning system, which in turn hinges on modeling the evolution of earthquake-generated tsunami waves. There are three distinctive stages involved: generation, propagation, and run-up (Titov, 1997). For the generation stage, rapid determination of reliable earthquake source parameters (lon., lat., depth, moment, and focal mechanism) is crucial to infer the seafloor vertical displacements as initial conditions for tsunami wave propagations. Here, we adopt the source inversion using seismic *W* phase, which carries the long period information of earthquake source at high speeds and thus is suitable for tsunami warning purposes (Kanamori & Rivera, 2008). We propose using data of "extended BATS" for source inversion of SCS earthquakes. "Extended BATS" refers to the ongoing expansion of the Broadband Array in Taiwan for Seismology (BATS; Kao et al., 1998) network, leaded by the Institute of Earth Sciences Academia Sinica (IESAS) and incorporated with the Vietnamese Academy of Science and Technology (VAST) and the Philippines Institute of Volcanology and Seismology (PHIVOLCS) to construct a real-time network that encompasses SCS [see Figure 10 of Huang *et al.* (2009)]. In this study, we test the applicability of *W* phase inversion for past large SCS earthquakes, using data of stations that mimic distributions of "extended BATS" – the bulk of BATS network plus eight other stations surrounding SCS (Figure 1).

Given seismic source parameters, the vertical static seafloor displacements, as initial conditions of tsunami wave propagations, can be determined immediately by solving the surface deformations of an inclined fault in a half-space elastic medium, such as those described by Mansinha & Smylie (1971) and Okada (1985). However, the propagation stage that followed is the most time consuming stage and for the warning purposes of a regional scale, it is hardly practical for real-time calculations. Here, we employ the unit tsunami methods (Lee *et al.*, 2005) to solve the dilemma. The methods divide all potential source regions into pixels and calculate the propagations of an initial unit amplitude uniformly assigned on the pixel. The resulting waves to a target station are stored in database and referred to unit tsunamis of that station-source pair. In the case of a real event, the predicted tsunamis of one station are synthesized by linear combinations of those unit tsunamis that are paired the station with all source pixels. The weightings of unit tsunamis in linear combination are determined by the vertical seafloor displacements averaged over the region of the corresponding pixel. We will elaborate the methods further in next section when building the database. Unit tsunamis are conceptually analogous to Green's functions in Seismology and are applicable solely for linear systems. Therefore, in our warning system, we strategically focus on predicting the offshore arrival times and amplitudes of approaching tsunamis that is prior to the run-up stage and subsequently, the non-linear effects are minor and can be ignored.

2. Data and methods

2.1. *W* phase inversion

We refer readers to Kanamori & Rivera (2008) regarding theory, modeling, and source inversion of *W* phase. We recapitulate here that *W* phase can be interpreted as superposition of the fundamental mode, first, second and third overtones of spheroidal modes at long period and can be synthesized by normal-node summation. The Green's functions of six moment tensor elements are pre-computed for a distance range of $0° \leqq \Delta \leqq 90°$ with an interval of 0.1° and for a depth range of 0-760 km. The synthetic waveforms of an earthquake are derived by convolving the Green's function with its moment rate function, which is a triangular function defined by two parameters, half duration, t_h, and the centroid delay, t_d (Figure 2). The broadband seismic data are deconvolved to displacement with instrument response removed and a band-pass filtered. A time period of 15Δs (Δ epicentral distance in degree) from the beginning of *P* wave is windowed to extract *W* phase. A time domain recursive method is used not only for real time operation but also for using available data to the point where it gets clipped at the large amplitude *S* or surface waves. The same procedures are applied on synthetic waveform and, together with data, a linear inversion is performed on concatenated time series using a given hypocenter location and origin time Kanamori & Rivera (2008).

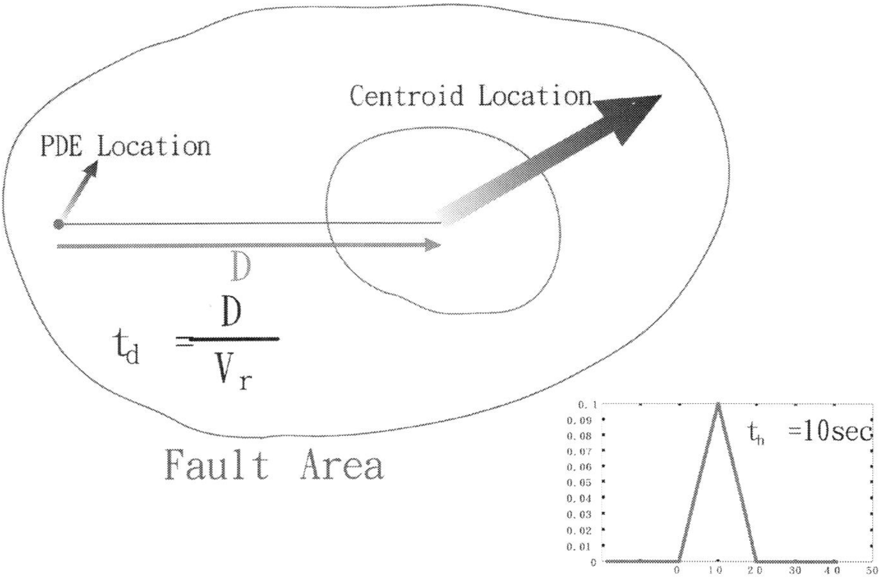

Figure 2. Schematic illustration of the half duration (t_h) and centroid delay (t_d). The t_h value is the half width of the triangular moment rate function (10 sec in this example). The t_d value is the temporal position of the triangle central measured from the origin time, which is the distance (D) from the initial rupture to the centroid location divided by the rupture velocity (V_r).

We sorted out earthquakes from the GCMT catalog (Dziewonski *et al.*, 1981; Ekström *et al.*, 2005) that occurred between January 2000 and July 2009, were bounded by 10°N/26°N and 115°E/135°E, and had moments greater than 10^{25} dyn-cm (Figure 1). Table 1 listed the PDE (Preliminary Determination of Epicenters) locations, t_d, t_h, and M_w values of the GCMT solutions for the 38 selected earthquakes. We divided the earthquakes into near and far groups relative to central Taiwan (Figure 1). A virtual regional array, including the bulk of current BATS and a few stations from F-NET, Malaysian National Seismic Network, etc., was set up to mimic the distributions of extended BATS (Figure 1). The three component

(ZNE) LH channel data (1 sample-per-second) were collected through Web site for *W* phase source inversions. Some stations were eliminated with their data continually exhibiting low quality (e.g., LYUB). We tested for six scenarios of two groups: group I using vertical component only and group II using all three ZNE components. In each group, three scenarios are tried with different level of knowledge on earthquake parameters: (1) using the GCMT centroid parameters (impractical in real events), (2) using the hypocenter parameters (lon., lat., depth, origin time) reported by the PDE catalogue, and a centroid time, t_d, determined by grid search; the source half duration, t_h, was set equal to t_d, (3) the same as (2) except the centroid location (lon., lat.) determined by a 2-dimensional grid search (depth fixed at PDE's). We indicated the three scenarios as gCMT location, t_d location, and [t_d+xy] location, respectively. The cutting distance – within which data are removed – was determined to be 1.5° by examining results as a function of cutting distance. The inversion was iterated three times with an increasingly assigned threshold to discard stations of bad fitting data from previous iteration. An example of final fitting between synthetic and observed waveforms of vertical component for the first shock of the M_w7.0 Dec. 26, 2006, Pingtung earthquake is shown (Figure 3). The frequency band used to filter seismic waveforms basically followed Table 1 of Hayes *et al.* (2009) except that for a few M_w◎6.0 earthquakes, the band was fine-tuned to have better results.

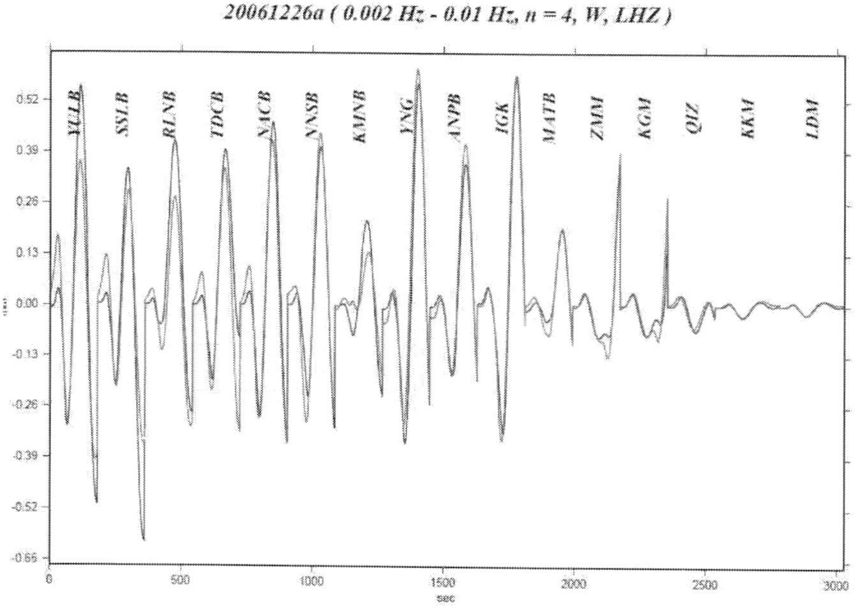

Figure 3. Observed (black) and synthetic (red) *W* phase for the first shock of the M_w7.0 Dec. 26, 2006, Pingtung earthquake. The *W* phase time window of each station was trimmed and concatenated for inversion. Labels are names of stations.

Event number	PDE Location					GCMT parameters			W phase results				
	Date	Origin time	Lat.(°)	Long.(°)	Depth (km)	t_d(s)	t_h(s)	M_w	t_d(s)	M_w	Frequency Band(Hz)	# of stations	# of channels
1	1/6/2000	21:31:6.20	16.09	119.48	33.0	3.40	3.00	6.18	6.0	6.28	0.0067-0.015*	7	12
2	6/10/2000	18:23:29.30	23.84	121.22	33.0	4.60	4.30	6.42	3.0	6.53	0.002-0.010	3	4
3	7/16/2000	3:21:45.50	20.25	122.04	33.0	1.10	3.80	6.36	4.0	6.42	0.0067-0.010*	10	24
4	12/18/2001	4:2:58.30	23.95	122.73	14.0	2.00	2.00	6.81	1.0	6.73	0.0067-0.010*	8	19
5	3/26/2002	3:45:48.70	23.35	124.09	33.0	2.40	4.40	6.41	4.0	6.34	0.0067-0.010*	9	11
6	3/31/2002	6:52:50.50	24.28	122.18	32.8	9.50	10.10	7.09	9.0	7.07	0.00167-0.010	5	11
7	5/15/2002	3:46:5.80	24.64	121.92	10.0	5.20	3.00	6.12	7.2	6.04	0.0067-0.015*	6	11
8	5/28/2002	16:45:17.10	24.07	122.26	33.0	1.50	2.50	6.05	5.0	5.87	0.0067-0.020	5	8
9	10/12/2002	23:43:12.90	15.10	118.50	33.0	-0.30	2.40	6.00	5.0	6.03	0.010-0.015*	15	34
10	2/15/2003	11:1:59.90	12.17	124.08	10.0	3.50	3.20	6.24	6.0	6.25	0.0067-0.010*	14	27
11	6/10/2003	8:40:30.80	23.52	121.63	44.5	3.28	2.30	5.95	7.0	5.94	0.0067-0.015*	6	10
12	6/26/2003	14:11:44.60	12.25	123.85	10.0	6.12	2.30	5.94	9.0	6.02	0.0067-0.015*	15	31
13	7/1/2003	17:0:22.80	12.80	124.90	33.0	3.45	2.20	5.94	6.0	5.92	0.0067-0.020	12	22
14	10/9/2003	22:19:13.90	13.76	119.94	33.0	2.13	2.40	5.96	8.0	6.03	0.0067-0.020	14	24
15	11/18/2003	17:14:22.60	12.02	125.42	35.0	3.46	4.30	6.49	6.0	6.50	0.0067-0.020	16	48
16	12/10/2003	4:38:11.60	23.04	121.36	10.0	9.15	6.20	6.80	13.0	6.82	0.002-0.020*	8	23
17	5/19/2004	7:4:11.70	22.66	121.50	20.0	2.54	3.10	6.21	6.0	6.28	0.0067-0.020	7	16
18	9/15/2004	19:10:50.60	14.22	120.41	115.4	0.59	2.30	5.94	5.0	6.03	0.0067-0.020	15	28
19	10/8/2004	14:36:6.10	13.93	120.53	105.0	3.75	4.00	6.41	8.0	6.47	0.002-0.020*	13	26
20	10/15/2004	4:8:50.20	24.53	122.69	94.0	4.71	4.70	6.56	4.7	6.59	0.0067-0.020	11	26
21	11/8/2004	15:55:1.10	24.10	122.54	29.0	4.60	3.30	6.27	5.0	6.23	0.0067-0.020	9	18
22	10/15/2005	15:51:7.20	25.32	123.36	183.4	3.98	4.00	6.42	5.0	6.45	0.0067-0.020	17	41
23	4/1/2006	10:2:19.60	22.87	121.28	9.0	2.75	3.00	6.13	3.8	6.15	0.0067-0.020	9	23
24	10/9/2006	10:1:46.70	20.65	120.02	10.0	4.04	3.40	6.28	7.0	6.34	0.0067-0.020	17	47
25	12/26/2006	12:26:21.10	21.80	120.55	10.0	7.91	7.60	6.98	7.9	6.97	0.002-0.010	15	38
26	12/26/2006	12:34:13.80	21.97	120.49	10.0	8.53	6.90	6.91	11.0	6.95	0.0067-0.020*	17	42
27	4/20/2007	0:26:40.60	25.72	125.09	10.0	4.32	2.80	6.13	7.0	6.15	0.0067-0.020	16	34
28	4/20/2007	1:45:56.10	25.71	125.11	9.0	3.89	3.30	6.26	6.0	6.32	0.0067-0.020	15	31
29	4/20/2007	2:23:34.00	25.62	125.04	11.1	3.65	2.30	5.95	5.0	5.94	0.0067-0.020	18	43
30	9/6/2007	17:51:26.20	24.34	122.22	53.0	2.81	3.20	6.25	3.8	6.24	0.0067-0.020	14	25
31	3/3/2008	13:49:40.40	19.91	121.33	10.0	4.60	2.40	5.98	7.0	6.00	0.0067-0.020	21	50
32	3/3/2008	14:11:14.60	13.35	125.63	24.0	6.96	6.70	6.87	9.0	6.89	0.002-0.010	17	37
33	4/23/2008	18:28:41.90	22.88	121.62	10.0	3.18	2.40	5.97	3.2	5.96	0.0067-0.020	10	22
34	6/1/2008	1:57:24.30	20.13	121.37	35.0	2.88	3.60	6.33	6.0	6.35	0.0067-0.020	18	46
35	7/13/2008	14:58:32.30	21.01	121.14	10.0	4.61	3.00	6.19	6.0	6.23	0.0067-0.020	19	35
36	8/15/2008	10:25:16.50	12.90	124.32	10.0	2.51	2.40	5.99	6.0	6.07	0.0067-0.015*	22	48
37	7/13/2009	18:5:1.40	24.06	122.20	17.0	5.22	3.50	6.32	5.2	6.25	0.0067-0.015*	13	21
38	1/25/2007	10:59:17.60	22.56	121.93	36.2	-0.23	2.50	6.02	*.*	*.**	*.****-*.***	**	**

Table 1. List of the earthqaukes studied. Origin time (UT), latitude, longitude, and depth are taken from PDE catalog. GCMT parameters are from the GCMT catalog. W Phase results indicate frequency band, number of stations, and channels used for the final solutions. The solutions are the scenario of using PDE location and grid searching For t_d. The derived t_d and M_w are shown.

2.2. Unit tsunami methods

The unit tsunami methods divide all potential source regions into pixels and assign an initial unit amplitude uniformly on each pixel, called unit source (Figure 4). The propagations of unit sources are pre-calculated with the resulting tsunamis at observing stations recorded as unit tsunamis to be stored in database. Each and every unit source has his unique unit tsunamis to a specific station stored and is pulled out to synthesize the composite tsunamis of that station in the case of a real event. The synthetics are done by linear combination with weighting factors determined by vertical seafloor displacements averaged over the region of corresponding unit source. In this way, the time consuming propagation stage is performed in advance, which makes prompt issuing of tsunami warnings more efficient. The principle of unit tsunami methods is conceptually analogous to Green's function in Seismology and only works in linear systems. As compliance, we aim at predicting the offshore – prior to the run-up stage - arrival times and amplitudes of approaching tsunami waves. Under the circumstances, tsunami amplitudes are in general at least an order of magnitude less than the ocean depths and the nonlinear effects, such as convections and bottom frictions, are negligible. Furthermore, at open seas, the wavelengths (tens to hundreds of kilometers) of ocean-wide tsunamis are always much greater than ocean depths (a few kilometers). The vertical components of accelerations are thus neglected and horizontal motions of water mass are taken as uniform from top to bottom (Satake, 2002). Finally, we end up with the simplest form of hydrodynamic equations to govern the motions of incompressible fluids, i.e. the linear shallow water wave equations.

The potential source region of the Manila subduction zone was divided into 14×10 square pixels, each with 0.5° ×0.5° in size and an initial vertical seafloor displacement of 1 m was assigned to each pixel to constitute the group of unit sources (Figure 4). We employed Cornell Multigrid Coupled Tsunami Model (COMCOT; Liu *et al.*, 1998) to simulate the propagations of each unit source. COMCOT is a finite difference scheme and in our case, we solved for linear shallow water wave equations in spherical coordinates with 1 minute (~1.8 km) and 1 sec in space and time, respectively. The boundary conditions were total reflection for ocean-land interfaces and radiation for map boundaries. We set up 32 virtual stations representing existing tidal stations of the Central Weather Bureau (CWB; Figure 5). The wave fields of one station from one unit source were referred to the unit tsunamis corresponding to the station-unit source pair. For each unit source, we simulated for four-hour propagations with the resulting wave fields as a function of time at the 32 stations to be stored as 32 unit tsunamis in database. In the end, a total of 14×10×32 unit tsunamis were stored in database for synthetics of tsunami waves in the case of a real event. The stored unit tsunamis were readily available for arrival time predictions even without the occurrence of real events. We applied Short Time Average over Long Time Average (STA/LTA; Allen, 1982), a conventional scheme for picking seismic P and S phases, to automatically pick the arrival times of unit tsunamis with results stored in database for arrival time predictions.

(a)

(b)

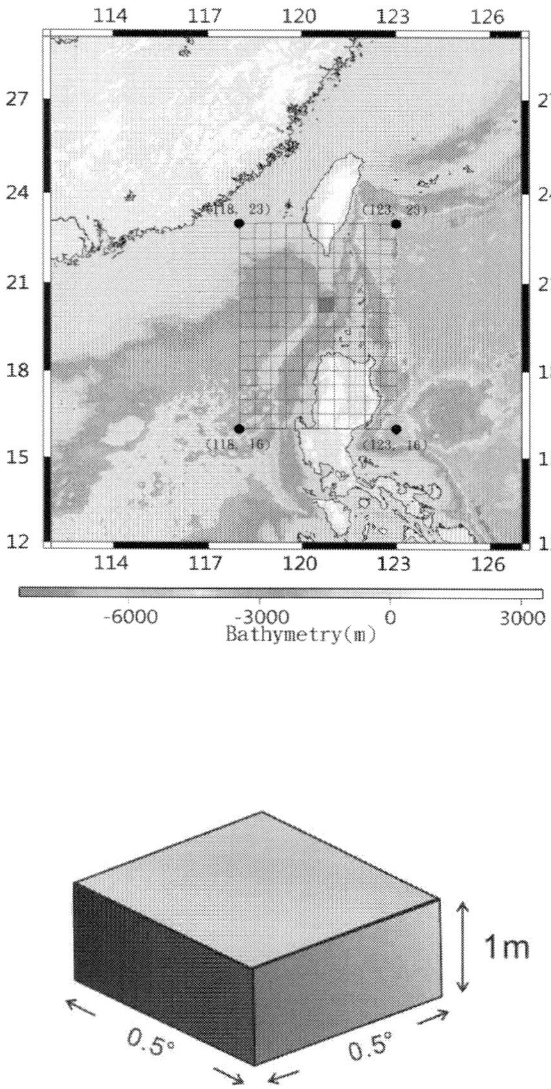

Figure 4. (a) Division of the source region of the Manila subduction zone into 14 (latitudinal) times 10 (longitudinal) subregions of pixels. The red pixel is the ninth from bottom and the sixth from left, indicated with 09_06. (b) The vertical displacement assigned to the pixel as unit source.

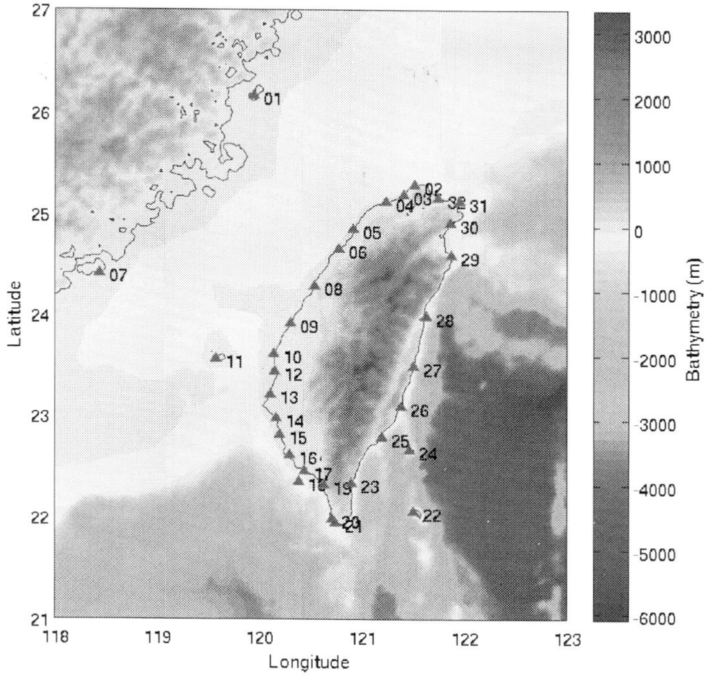

Figure 5. Locations of 32 CWB tidal stations where we set up virtual stations to record unit tsunamis of the 14×10 unit sources. The stations are numbered from north to south in a counterclockwise sense surrounding Taiwan.

In the case of a real tsunamigenic event, we calculated the vertical seafloor displacements caused by the earthquake (Okada, 1985), using source parameters inverted by fitting seismic *W* phase. The weighting factors of each unit source were derived by averaging the vertical displacements over area of the corresponding unit source. The tsunami waves at one station were then synthesized by summing the 14×10 unit tsunamis of the station with corresponding weighting factors.

3. Results

3.1. *W* phase inversion

We judge the qualities of solutions based on the discrepancies between those of *W* phase inversion and those of GCMT solutions of the same event, in terms of both moment magnitudes (M_w) and focal mechanisms. The comparisons of M_w for six scenarios are presented in Figure 6 with group I (Z component only) in the first column and group II (ZNE components) in the second column. The numbers in each box (scenario) indicate the

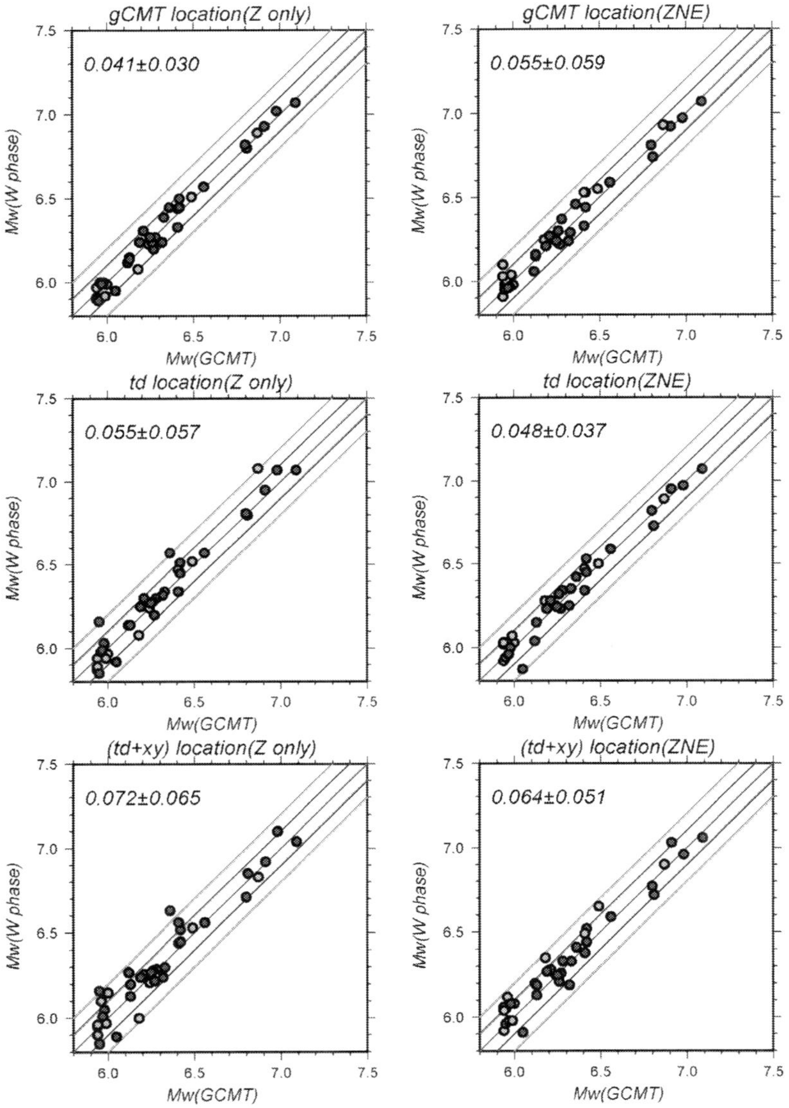

Figure 6. Comparison of M_w (GCMT) and M_w (*W* phase) for earthquakes used in this study with blue for near group and green for far group. The left column is results of Z component only (group I) and the right one ZNE components (group II). Top to bottom represents three sets of scenario for different level of knowledge on earthquake parameters (see text). Numbers are mean values for differential M_w and standard deviation. The best result is those of group I GCMT location (top left) and the best one among practicable scenarios is group II t_d location.

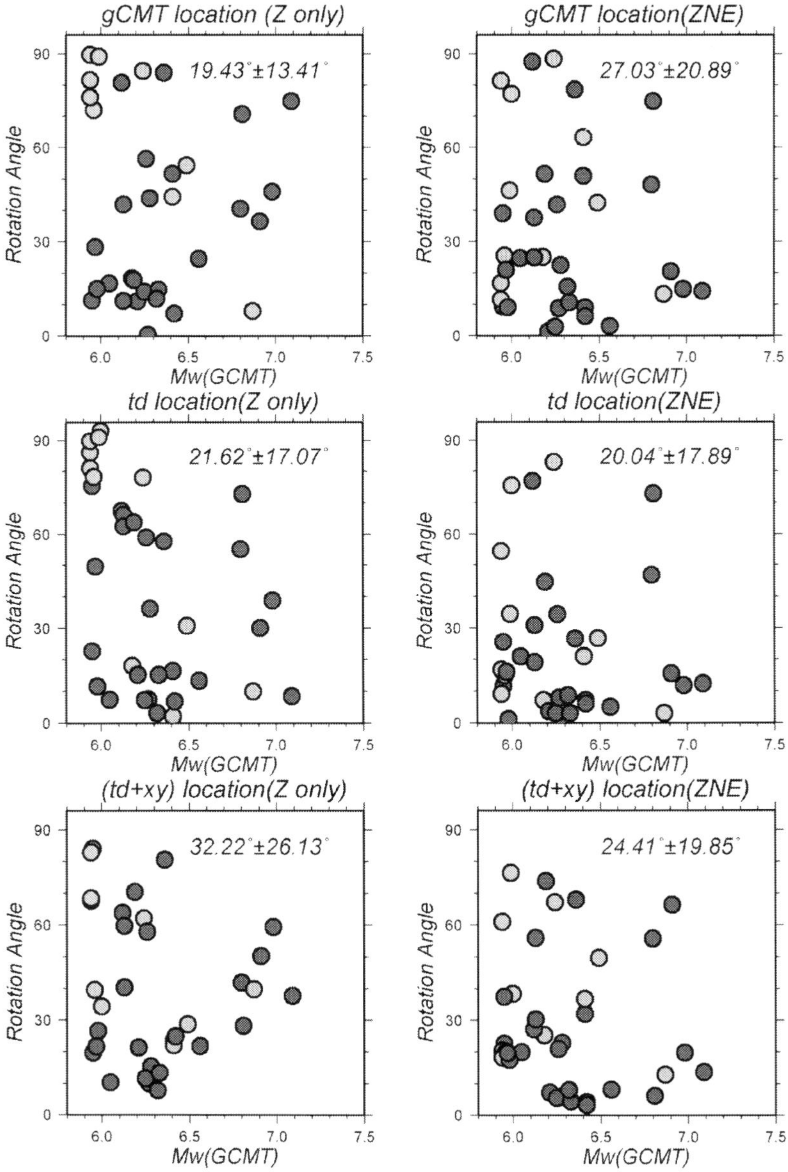

Figure 7. Rotation angle of source geometries between *W* phase and GCMT solutions, presented in the same fashion as in Figure 6. Again, the best result is those of group I GCMT location (top left) and the best one among practicable scenarios is group II t_d location.

absolute means of magnitude differences and corresponding standard deviations. The absolute means of all six scenarios are less than 0.1 unit, validating the application of *W* phase inversion using data of regional network to determine source parameters of SCS earthquakes greater than M_w5.9. Among the three tried scenarios (gCMT location, t_d location, and [t_d+xy] location), only the last two are practicable in real-time *W* phase inversion, among which the group II t_d location is the best scenario (least discrepancies). The Kagan rotation angle refers to the solid angle rotating from one double couple to another (Kagan, 1991) and thus is a measurement of focal mechanism discrepancies. We present the Kagan angles - relative to the GCMT solutions of the same events - of the six scenarios in Figure 7 with mean and standard deviation indicated, using the same fashion as Figure 6. Again, the group II t_d location is the one with the minimum mean among all real-time practicable scenarios.

3.2. Unit tsunamis

Characteristics of tsunami wave propagations in SCS and around Taiwan depend on bathymetric features and can be learned from simulated propagations of unit sources. Figure 8 shows the propagating waves at different time frames of an exemplary unit source, from which we learned that tsunamis triggered by earthquake offshore north Luzon will hit south Taiwan at about 20 min., that the propagation speeds along east Taiwan coasts are much faster than those along west Taiwan coasts, meaning more warning time for the west coasts, and that almost all coasts around Taiwan will be attacked by SCS tsunamis within three hours after the generation of tsunamis. Figure 9a shows the resulting 32 unit tsunamis of the exemplary unit source, which demonstrate that stations 14 to 26 are the most affected ones with southern tip of Taiwan (stations 20, 21, and 23) being the most vulnerable (Figure 6). The STA/LTA scheme works well in determining the arrival times of unit tsunamis (Figure 9b) and we compile the data to produce one arrival-time map for each unit source (Figure 10), which will also be stored in database and are readily for prompt arrival-time predictions.

4. Discussion

4.1. *W* phase inversion

Although the *W* phase source inversion was initially designed using data of global network and is now routinely operated in this fashion by the U.S. Geological Survey (USGS) National Earthquake Information Center (NEIC; Hayes *et al.*, 2009), the *W* phase inversion is also applicable using exclusively regional data (e.g., the Full Range Seismograph Network of Japan) provided there is dense broadband network at regional distances (epicentral distances < 12°)[Kanamori H., personal comm.]. In this study, we have attested that, using data of regional network (the mimic network of extended BATS), the *W* phase is equally applicable for inverting source parameters of SCS earthquakes greater than M_w5.9. We expect the applicability can be even improved as a result of better azimuthal coverage once the construction of the extended BATS is finished.

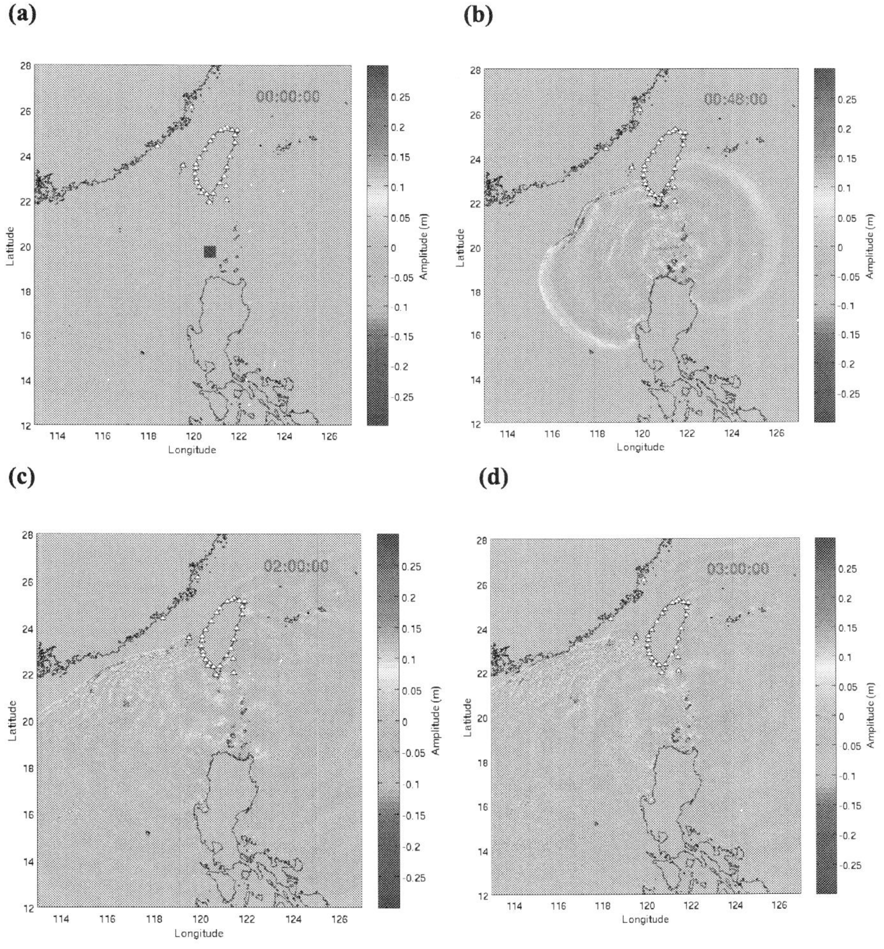

Figure 8. Propagation of tsunami waves of an exemplary unit source. The triangles surrounding Taiwan are the existing CWB tidal stations where we set up virtual stations to record the unit tsunamis. Color bars represent wave heights in meter. (a) Location of the unit source. (b) After 48 min., note that the wave has reached Hualien on east Taiwan coast, but only Kaohsiung on that of western Taiwan due to significantly bathymetric differences. (c) At 2 hours, the tsunami wave along eastern Taiwan has reached the majority of north Taiwan and note that northwest Taiwan coasts have the most warning time. (d) After three hours, almost all Taiwan coasts have been attacked by tsunamis.

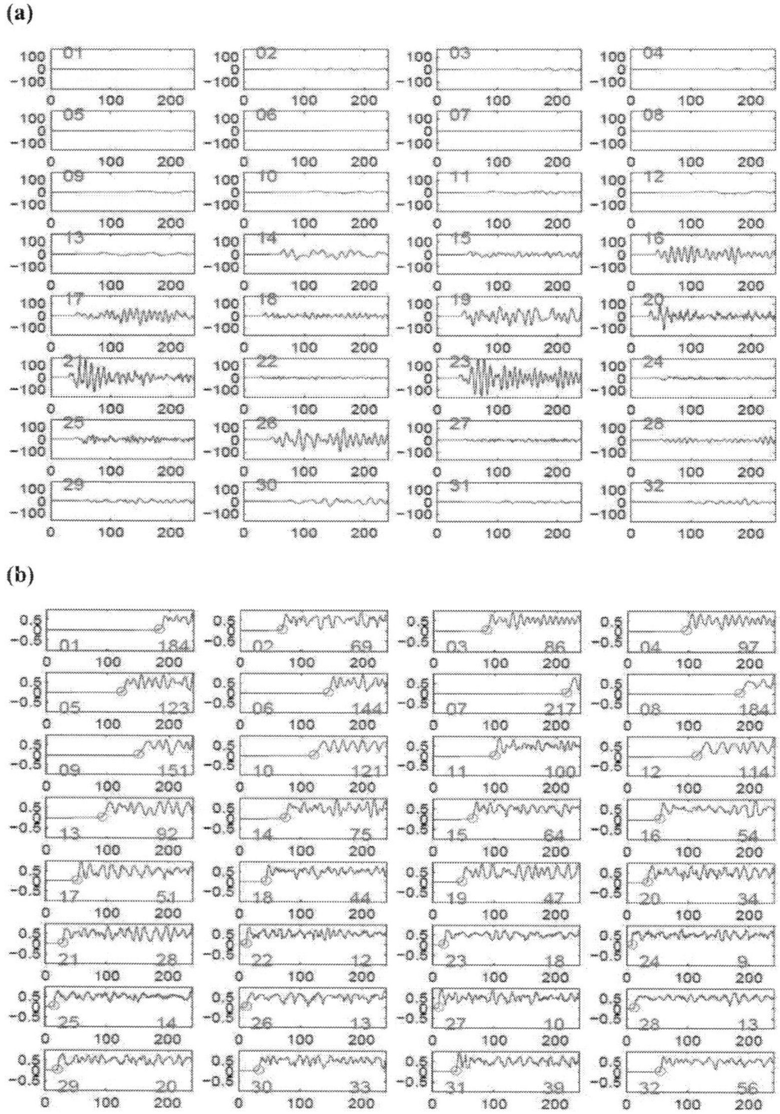

Figure 9. (a) Unit tsunamis of the 32 virtual stations for the unit source in Figure 8. Red numbers indicate stations following Figure 5. Unit is minute for x-axis and cm for y-axis. **(b)** Results of STA/LTA scheme applying on the unit tsunamis. Note the sharp arrivals of tsunami waves make their picking robust. Unit is minute for x-axis and STA/LTA ratio for y-axis. The numbers to the right of each box is the picking arrival-time in minute.

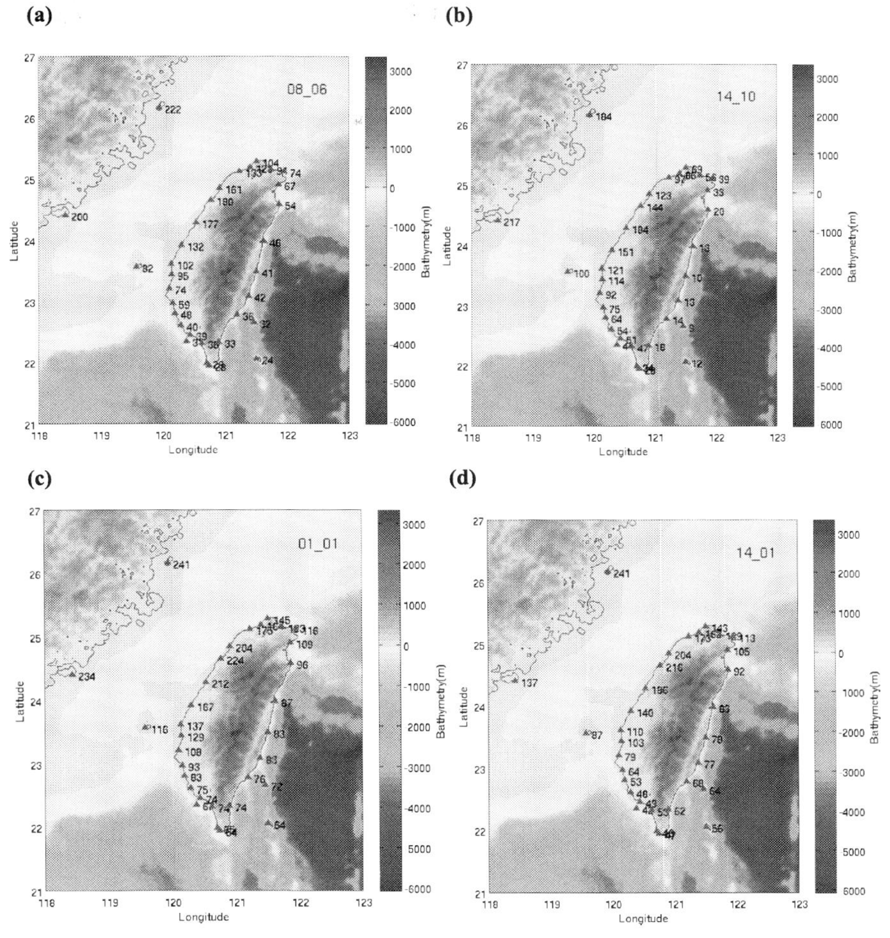

Figure 10. The arrival time maps for different unit sources in minutes. See Figure 4 for nomenclature of unit source. (a) For the 08_06 unit source. (b) For the 14_10 unit source (the top right corner). (c) For the 01_01 unit source (the bottom left corner). (d) For the 14_01 unit source (the top left corner).

The experiences learned by conducting the *W* inversion of past SCS earthquakes are valuable for future implementation of real-time warning and are addressed in the following. (1) The frequency band of *W* phase used for inversion as a function of earthquake size as suggested by Table 1 of Hayes *et al.* (2009) is robust and can be directly implemented. The only exception is that when the sizes of earthquake and, accordingly, the signal/noise ratios are relatively small (M_w≈6.0), under the circumstances, different frequency bands do matter and we need to manually try out the best scenario. Since earthquakes of such size rarely trigger significant tsunamis, we can just implement the empirical frequency band of Hayes *et al.* (2009) and focus on major events. (2) Having tried six scenarios in two groups, we find that grid search for epicenters of centroid location ([t_d+xy] location) does not necessarily decrease the discrepancies, compared to those of t_d location – probably due to the use of GCMT solutions as a reference. For the scenario of t_d location, the use of ZNE components (group II) only slightly reduces the discrepancies relative to those of Z component only, however, at the expense of processing time. Therefore, we can alternatively use group I t_d location as the scenario of real-time *W* phase inversion for a more efficient warning system. (3) The patterns of discrepancies exhibit similar trends for Mw and Kagan angles (Figures 7 and 8), presumably because both are derived from the same moment tensors – product of *W* phase inversion.

4.2. Unit tsunamis

We test with a scenario M_w9.0 earthquake in the Manila subduction zone to assess the applicability of predicting maximum amplitudes of tsunami waves using unit tsunami method. The vertical seafloor displacements caused by the scenario earthquake are calculated using Okada (1985)'s scheme [Figure 11(a)]. We first follow the conventional approach to simulate the propagation of the displacements, which takes time. Secondly, we derive the weighting factors of each unit source by averaging over the vertical displacements within each unit source [Figure 11(b)]. The tsunami waves are then synthesized by a linear combination of unit tsunamis – pulled from database – with corresponding weighting factors, which can be done instantaneously. Figure 12 shows resulting tsunami waves of the two approaches for comparison purpose. A maximum-amplitude map can also be created showing the magnitudes on the corresponding stations (Figure 13). By comparing (a) and (b) of Figures 12 and 13, we conclude that the unit tsunami method can produce waveforms and maximum amplitudes similar to those of conventional simulation method, however, only a much less time is needed.

Liu *et al.* (2009) have also proposed procedures to establish a tsunami early warning system in the SCS. The procedures also apply unit tsunami methods to pre-calculate tsunami waves of fault segments along the Manila subduction zone. The primary differences between the two systems are the ways in determining the weighting coefficients of unit tsunamis in synthetics - Liu *et al.* (2009)'s procedures invert the real-time observations of nearby ocean-bottom pressure sensors whereas we calculate from source parameters inverted by the *W* phase in this study. The two approaches actually complement each other in that the former is more accurate - direct observations - but rely on well-placed sensors and the latter has more observations – seismic stations – but indirect through source parameters.

(a)

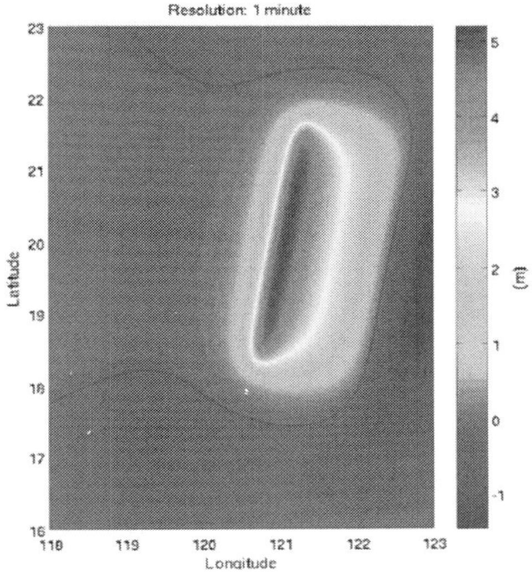

(b)

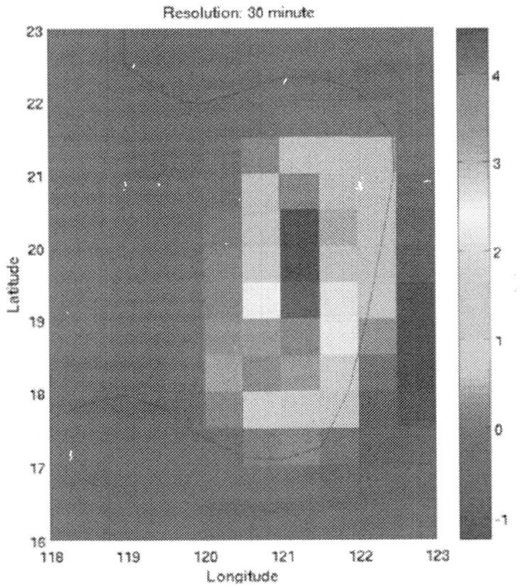

Figure 11. **(a)** The vertical seafloor displacements of the M_w9.0 scenario earthquake. **(b)** The weighting factors of each unit source corresponding to (a), determined by averaging displacements over the area of corresponding unit source. Scalar bar indicates the ratio of averaged displacements to that of 1 meter, the initial assigned displacement for unit source.

(a)

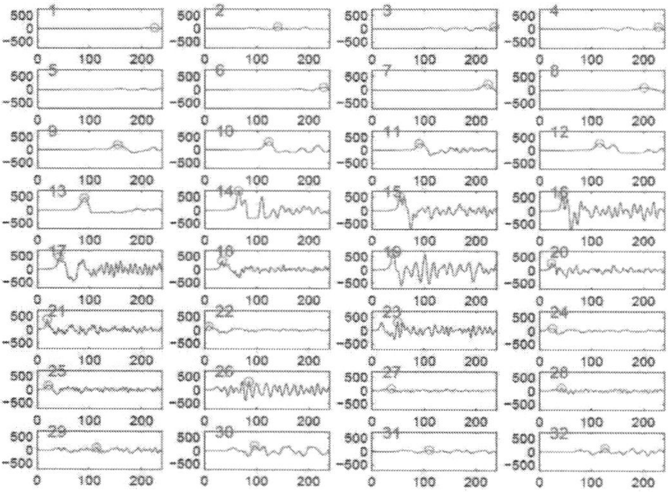

(b)

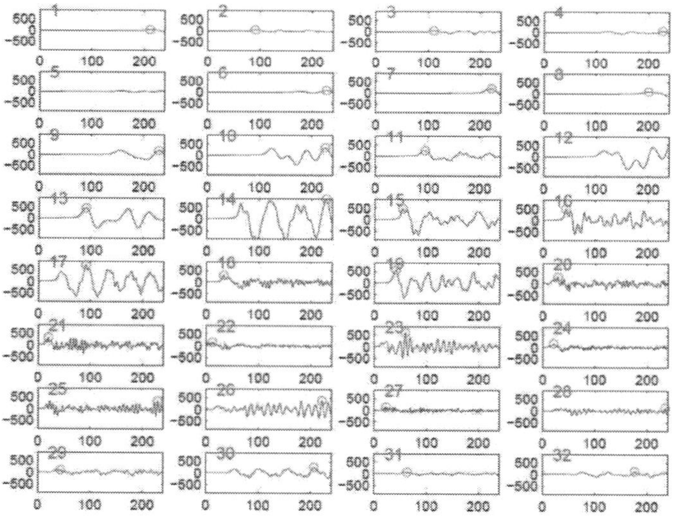

Figure 12. (a) Tsunami waves on the 32 virtual stations of the M_w9.0 scenario earthquake using conventional simulation of tsunami wave propagations. Red circles mark the position of maximum amplitude. **(b)** Same results but using unit tsunami methods. Note that the majority of stations bear similar waveforms. Units are the same as those of Figure 9(a).

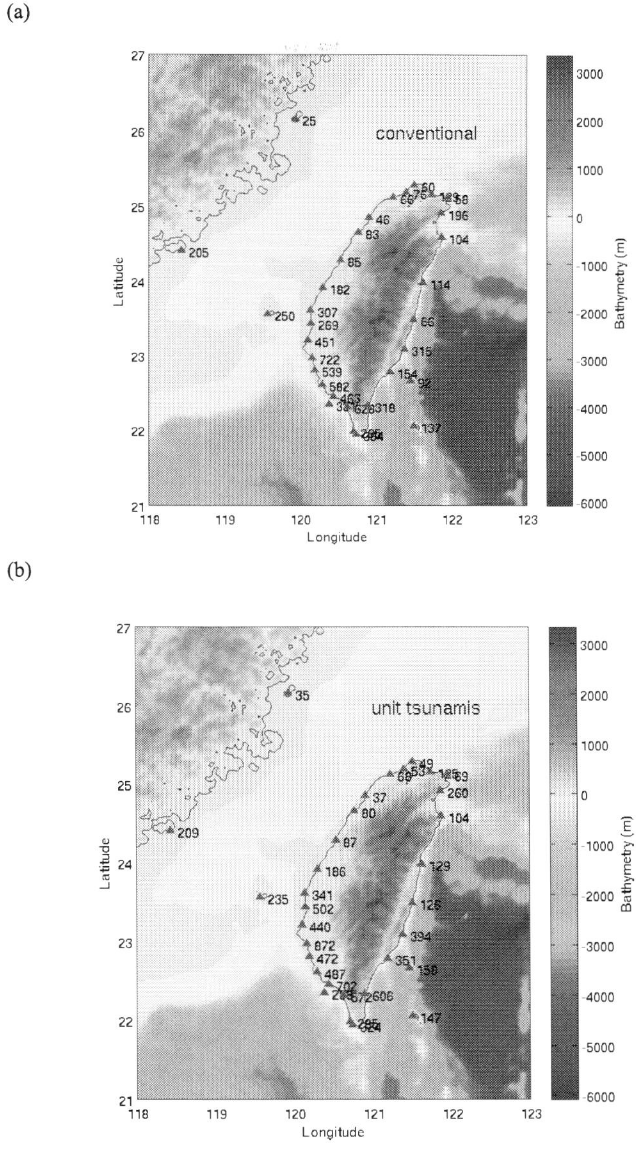

Figure 13. **(a)** The maximum amplitudes in centimeter at the 32 virtual stations of the tsunami waves triggered by the M_w9.0 scenario earthquake using conventional simulation of tsunami wave propagations. **(b)** The same results as (a) but using the unit tsunami methods. Note that the results are largely comparable between (a) and (b).

As a final remark, we point out that the unit tsunami methods are not compromised on regions of land or land-ocean boundaries. For pixels on land exclusively, the calculated unit tsunamis are zero for all stations and thus have no contributions to the synthetic tsunamis despite their non-zero weighting factors. Likewise, for pixels on coast lines, the calculated unit tsunamis only constitute contributions from the ocean parts, thus automatically proportional to the ocean-land ratio within the pixel. As a result, they will contribute proportionally to the synthetic tsunamis in the processes of linear combinations. As for regions of complicated bathymetries such as trenches, the vertical seafloor displacements in the regions tend to deviate from those estimated by Okada (1985)'s schemes, which assume a half-space elastic medium. However, both conventional approach and unit tsunami methods will suffer from the same deviations. The unit tsunami methods are equally applicable to warnings of local, regional, and global scales. However, strategies in building database of unit tsunamis should be adjusted accordingly; in global scale with greater potential source regions, we may use greater pixels to reduce the size of database, at the expense of reducing spatial resolutions of seafloor displacements; in local scale, the pixels cannot be too small in order that the conditions of long wave validate.

5. Conclusions

In this study, we propose to combine *W* phase inversion and unit tsunami methods to build a tsunami warning system in Taiwan for earthquakes in the SCS region. The *W* phase inversion allows us to rapidly determine moment tensors of large earthquakes for the calculations of vertical seafloor displacements. The applicability of *W* phase inversion for SCS earthquakes using BATS stations and its extension has been attested and expects to be improved, pending future completion of extended BATS. We have built a database of unit tsunamis for the source region of the Manila subduction zone and the prediction of arrival times is readily available once the epicenter of tsunamigenic earthquake is known. Upon assessment of a $M_w9.0$ scenario earthquake, the prediction of maximum amplitudes on various coasts in Taiwan using unit tsunami method is comparable with that of conventional simulation, but with a much less time needed.

Author details

Po-Fei Chen*
Department of Earth Sciences, National Central University, Taiwan

Wen-Tzong Liang
Institute of Earth Sciences, Academia Sinica, Taiwan

Bor-Yaw Lin
Seismological Center, Central Weather Bureau, Taiwan

---

* Corresponding Author

Acknowledgement

This work was supported by Central Weather Bureau with grant number MOTC-CWB-101-E-09 and National Science Council (NSC) in Taiwan with grant number 99-2116-M-008-040.

6. References

Allen, R. (1982). *Automatic phase pickers: their present use and future prospects.* Bull. Seismol. Soc. Am., Vol.72(6), pp.225-242.

Dziewonski, A. M., Chou, T.-A. & Woodhouse, J.H. (1981). *Determination of earthquake source parameters from waveform data for studies of global and regional seismicity.* J. geophys. Res., Vol.86, pp.2825-2852.

Ekström, G., Dziewonski, A.M., Maternovskaya, N.N. & Nettles, M. (2005). *Global seismicity of 2003: centroid-moment-tensor solutions for 1087 earthquakes.* Phys. Earth planet. Inter., Vol.148(1-2), pp.327-351.

Global CMT catalog; GCMT. http://www.globalcmt.org.

Hayes, G.P., Rivera, L. & Kanamori, H. (2009). *Source inversion of the W-phase: real-time implementation and extension to low magnitudes.* Seismological Research Letters., Vol.80(5), pp.817-822, doi: 10.1785/gssrl.80.5.817

Huang, B.-S., Le, T.S., Liu, C.-C., Toan, D.V., Huang, W.-G., Wu, Y.-M., Chen, Y.-G. & Chang, W.-Y. (2009). *Portable broadband seismic network in Vietnam for investigating tectonic deformation, the Earth's interior, and early-warning systems for earthquakes and tsunamis.* J. Asian Earth Sci., Vol.36(1), pp.110-118.

Kagan, Y.Y. (1991). *3-D rotation of double-couple earthquake sources,* Geophys. J. Int., Vol.106(3), pp.709-716.

Kanamori, H. & Rivera, L. (2008). *Source inversion of W phase: speeding up seismic tsunami warning.* Geophys. J. Int., Vol.175, pp.222-238 doi: 10.1111/j.1365-246X.2008.03887.x

Kao, H., Jian, P.R., Ma, K.F., Huang, B.S. & Liu, C.C. (1998). *Moment-tensor inversion for offshore earthquakes east of Taiwan and their implications to regional collision.* Geophys. Res. Lett., Vol.25, pp.3618-3622.

Lee, H.J., Cho, Y.S. & Woo, S.B. (2005). *Quick tsunami forecasting based database.* In: Satake, K. (ed.) Tsunamis: Case studies and recent developments. pp.231-240.

Liu, P.L.-F., Woo, S.-B. & Cho, Y.-S. (1998). *Computer Programs for Tsunami Propagation and Inundation.* Technical Report, Cornell University.

Liu, P.L. -F., Wang, X. & Salisbury, A.J. (2009). *Tsunami hazard and early warning system in South China Sea.* J. Asian Earth Sci., Vol.36(1), pp.2-12.

Mansinha, L. & Smylie, D. E. (1971). *The displacement fields of inclined faults.* Bull. Seismol. Soc. Am., Vol. 61, pp.1433-1440.

Okada, M. (1985). *Surface deformation due to shear and tensile faults in a half-space.* Bull. Seismol. Soc. Am., Vol.75(4), pp.1135-1154.

Ruff, L. & Kanamori, H. (1980). *Seismicity and the subduction process.* Phys. Earth Planet. Int., Vol.23, pp. 240-252.

Satake, K. (2002). *Tsunamis.* In: Lee, W.H.K., Kanamori, H., Jennings, P.C., Kisslinger, C. (eds.) International Handbook of Earthquake and Engineering Seismology. Vol.81A, p437-451.

Soloviev, S.L. & Go, Ch.N. (1984). *A catalogue of tsunamis on the western shore of the Pacific Ocean (173-1968).* Nauka Publishing House, Moscow, USSR, 310 pp., 1974. Can. Transl. Fish. Aquat. Sci. 5077.

Titov, V.V. (1997). *Numerical modeling of long wave runup.* PhD thesis. University of Southern California.

Section 3

Geological, Biological and Chemical Evidence – The Physical Aspect of a Tsunami

Chapter 9

Ancient Cliffs and Modern Fringing Reefs: Coupling Evidence for Tsunami Wave Action

Tetsuya Kogure and Yukinori Matsukura

Additional information is available at the end of the chapter

http://dx.doi.org/10.5772/50515

1. Introduction

The height of a tsunami as it surges into the coasts is altered by the presence of a coral reef. In particular, wave propagation over the reef crest induces wave breaking, and energy dissipation increases to 50-90% (Munk & Sargent, 1948; Kono & Tsukayama, 1980; Roberts & Suhayda, 1983). Shibayama et al. (2005) reported that the heights of tsunamis in Sri Lanka caused by the Sumatra earthquake in 2004 were lower at coastlines with coral reefs than without. This reduction in the height of the tsunami is due to friction between the tsunami and coral reefs seaward of the coasts. Nott (1997) stated in contrast that tsunamis pass through the gaps in coral reefs; the Great Barrier Reef does not act as an effective barrier against tsunamis, since the coastline near Cairns in Australia has been impacted by tsunamis. Coral reefs are recognized as a breakwater against invading tsunamis, but details of the behavior of tsunamis around coral reefs are unknown.

Such "modern reefs" would form coastal cliffs if the reefs uplifted. Vertical limestone cliffs formed by the uplifted coral reefs are seen in tropical-subtropical areas around the world (e.g. Tjia, 1985; Maekado, 1991). The limestone cliffs, i.e., "ancient reefs", are subject to wave erosion while the "modern reefs" could control wave energy. Notches carved by the wave erosion often induce cliff collapses (Maekado, 1991; Kogure et al., 2006; Kogure and Matsukura, 2010a). Kogure et al. (2006) presented a mechanical equation for calculating the critical notch depth for cliff collapse. In addition to these cases, waves also break down or smash geomorphological environment of rocky coasts. Notched cliffs can be collapsed by waves as well as gravity (e.g. Kogure and Matsukura, 2010b). Here, one question has been raised about the relationship between the tsunami action affected by "modern" fringing reefs and cliff collapses induced by the tsunami: are there any relationships between the developments of "modern" and "ancient" fringing reefs?

We can see an interesting interaction between "modern" and "ancient" fringing reefs through tsunami wave action in Kuro-shima, a small island in the Yaeyama Islands. The Yaeyama

Islands, in the south-western part of Japan, have been attacked by several large tsunamis in the last few thousand years (e.g. Nakata & Kawana, 1995). The largest tsunami on record is the 1771 Meiwa Tsunami caused by the Yaeyama Earthquake of April 24, 1771. The maximum run-up height of the tsunami was estimated by Nakata & Kawana (1995) to be more than 5 m on the east and south side of Kuro-shima. In Kuro-shima, a low-lying notched cliff of height 3–4 m, made of Ryukyu Limestone, currently has a flat top surface that is partly destroyed. Many blocks, apparently due to cliff collapses, are scattered on the reef flat. Based on the observed correspondence between geometries of these blocks and the scars on nearby cliffs, the blocks have been cleaved from the cliffs. In Kuro-shima, collapses of coastal cliffs have been induced not only by enlargement of the notch at the cliff base but also by the attack of extreme waves such as tsunamis or bores during a storm (Kogure & Matsukura, 2010a, 2010b).

The present study shows stability analysis models to evaluate the cliff collapses due to extreme waves and discusses the effects of wave heights on the cliff collapses. Finally, this paper presents the effect of coral reefs in reducing the height of extreme waves, especially tsunamis, by comparing the distribution of blocks produced by wave-induced collapses of coastal cliffs and the development of coral reefs around Kuro-shima.

2. Study area

Kuro-shima is a low-lying island made of uplifted coral limestone, located in the south-western area of the Ryukyu Islands (Fig. 1). Geomorphic, geologic and lithologic features are described below.

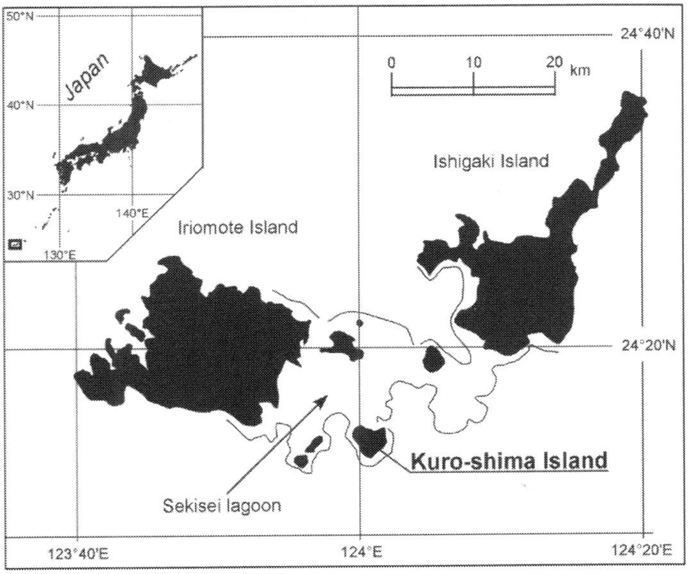

Figure 1. Location of Kuro-shima Island.

2.1. Geomorphological setting

The Pleistocene coral limestone is known as Ryukyu Limestone. The circumference of the island is about 10 km, and its highest point is about 10 m above mean sea level. Fringing reefs have developed around the island, except for the northern part; the coral reefs that are developing at the north of the island meet a large lagoon. Coastal areas in the south-east and south-west areas of the island face the Pacific Ocean. The maximum and minimum distance from the coast to the offshore reef edge is about 850 m in the south-east and 200 m in the south-west (Fig. 2). Waves break at the reef edge at low tide, and reach the coastal cliffs at high tide. A fault running NW–SE has developed in the north-eastern part of the island, and development of the coral reef is discontinuous along an extension of the fault (Fig. 2). A range of coastal cliffs, 6 km long and less than 5 m high (Fig. 3a), has developed in the area, except at the northern sandy beaches (Fig. 2). A reef flat has developed in front of the coastal cliff. The seaward side of the reef flat is a lagoonal environment, with depth of water 1–3 m (Fig. 3b).

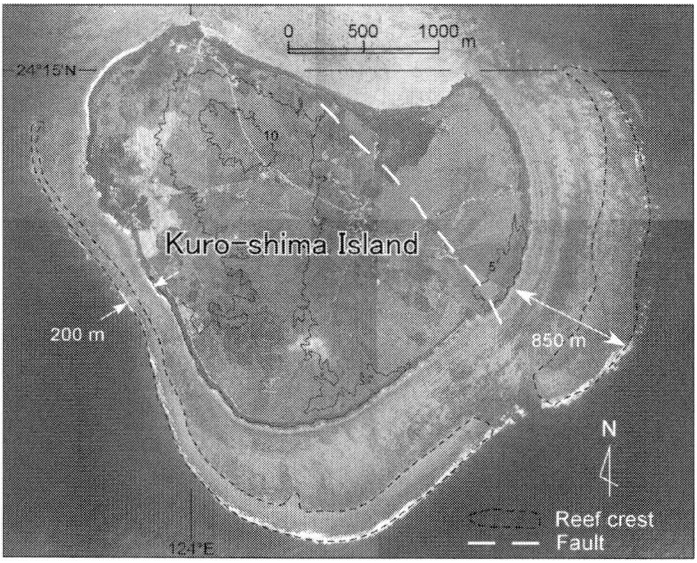

Figure 2. Photograph of Kuro-shima Island from the air. Fringing reefs are surrounded by the black dotted line. The fringing reefs are wider in the south-eastern part of the island. The fringing reefs are cut in the south-eastern part by a fault running NW-SE.

There is no visible vegetation on the top surface of the coastal cliff, certainly not near the shore. The cliff has a notch at its base; this notched cliff has a visor extending seaward (Fig. 3b). The depth of the notch, defined as the horizontal distance of the visor from the retreat point of the notch to the seaward end of the visor, is 3–4 m at most sites. The elevation of the retreat point of the notch is almost exactly equal to the maximum sea level (HSL) at spring tide, which is about 1 m higher than mean sea level (MSL); see Fig. 3b.

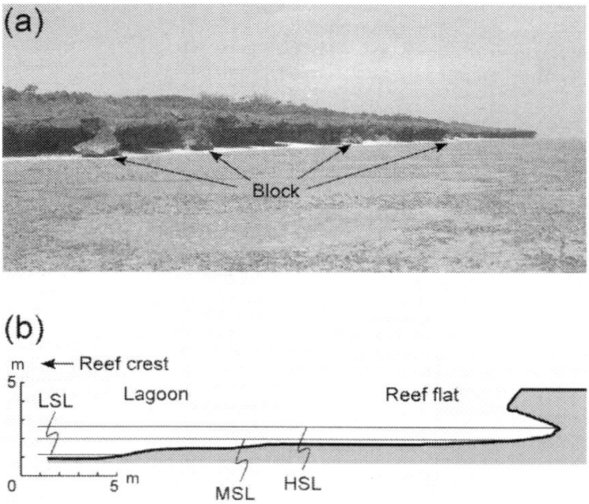

Figure 3. Coastal cliffs at Kuro-shima (Kogure & Matsukura, 2010a). (a) Overview of coastal cliffs and fallen blocks. The cliffs are less than 5 m high. The picture was taken at low tide. (b) Typical profile of a coastal cliff. There is no vertical exaggeration (x : y = 1 : 1).

Vertical joints with cracks having width between ten and several tens of centimetres have developed on coastal cliffs or reef flats. These joints appear to develop systematically rather than randomly, so that they are believed to have come about by geological processes. The depth of a joint is usually large enough to incise both cliffs and reef flats (Fig. 4a). These joints intersect coastal cliff faces at various angles, perpendicular, oblique (Fig. 4b) or parallel to the cliff. A cliff bounded by vertical joints therefore appears to be separated from the main cliff. The degree of development of a horizontal bedding plane on a cliff face varies from area to area; cliffs with and without bedding planes are distributed along the rocky coasts of this island with no apparent pattern.

2.2. Cliff collapses and blocks

Many angular blocks are scattered at cliff bases, and have apparently been produced by failures of notched cliffs (Fig. 3a). Along the 6 km coastal cliffs, more than 150 blocks were found having one side of dimension exceeding 1 m. The blocks are composed of Ryukyu Limestone identical to the cliffs behind the blocks. Vegetation or joints are rarely seen on the surfaces of the blocks. Some blocks can be clearly identified as having originated from the cliffs immediately behind them (Fig. 5a, 5b); the place of origin of other blocks cannot be identified because they appear to have moved. Many identifiable blocks are inclined seaward, and have a triangular or quadrangular flat surface which appears to coincide with the top surface of the cliff (Fig. 5a, 5b). Although the lower part of these blocks is inundated at high tide, the flat surface displays no development of a notch, as shown on the cliff. A few blocks have a breadth of 10 m in the direction following the shoreline; in most cases the breadth is approximately 5 m, as shown in Fig. 5a and 5b, and the maximum thickness of the blocks is about 3 m.

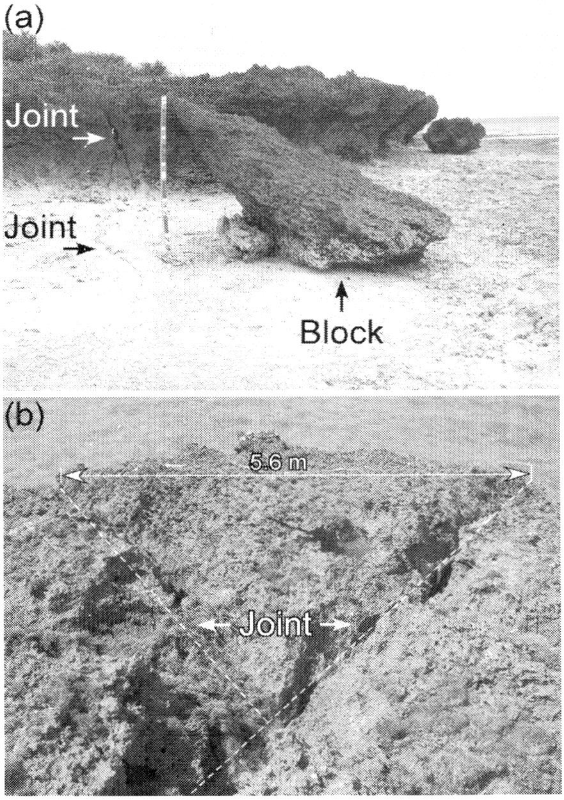

Figure 4. Cliffs with vertical joints (Kogure & Matsukura, 2010a). (a) A vertical joint visible on coastal cliffs and reef flats. The picture was taken at low tide. (b) Example of vertical joints running oblique to the cliff face.

Some cliffs behind a fallen block show a failure scar. The shape of the scar on the cliffs corresponds to that of the block, having a triangular or quadrangular flat surface (Fig. 5c and 5d). Many scars have similar features, including: (1) the cliff above the retreat point of a notch has detached from the main cliff (Fig. 5b); (2) the scar consists of a nearly horizontal surface and bounding vertical surfaces (Fig. 5d), and has no sliding or shearing striation; (3) the height of the horizontal surface is almost equal to that of the retreat point of the notch;

and (4) the vertical surfaces intersect the horizontal surface, with the vertical extension of the joints reaching the reef flat (Fig. 5b and 5d). In view of the resulting visible joints on the top surface of a cliff (Fig. 4b), Kogure & Matsukura (2010a) inferred that the vertical surfaces on the scar are joint surfaces, and that the horizontal surface is a failure surface at which cliff collapse has occurred. The cliff, which is separated from the main cliff by vertical joints, will collapse when the notch extends sufficiently far inward from the sea.

Figure 5. Cliff collapses around coastal cliffs (modified from Kogure & Matsukura, 2010a). Fallen blocks with (a) quadrangular and (b) triangular flat surface. The pictures were taken at low tide. (c) Bird's-eye view of a fallen block with a quadrangular failure surface. The failure surface is bounded by a single vertical joint running parallel to the cliff face and two joints perpendicular to the face. (d) Cliff collapse scar with a triangular failure surface. The failure surface is bounded by two vertical joints.

Many fallen blocks have a quadrangular or triangular surface. In many cases, the cliffs behind the blocks have a matching failure surface. Kogure & Matsukura (2010a) denoted collapses having a quadrangular or triangular failure surface as H_q-type or H_t-type collapse, respectively (Fig. 6). H_q-type collapse takes place at a cliff where three joints have intersected each other to form a quadrangular shape (Fig. 5a and 5c); two parallel joints develop perpendicular to a cliff face, and one joint runs almost parallel to the cliff face. Fig.

5b and 5d show a H_t-type collapse. Two vertical joints that were developing oblique to a cliff face intersected each other a few meters inland from the cliff face. The cliff bounded by the two vertical joints has collapsed at the horizontal surface (Fig. 5d). Cliff collapses in Kuro-shima therefore display two shapes of failure surface, quadrangular and triangular, according to the number of joints and the angle between the joints and the cliff face. Blocks of both H_q- and H_t-type are distributed widely around the coasts of Kuro-shima. The distributions of H_q- and H_t-type collapses appear to be controlled by the angle between the vertical joints and the cliff face (Kogure & Matsukura, 2011).

2.3. Tsunamis

The most significant tectonic processes in the area of the Ryukyu Islands are the rifting of the Okinawa Trough and subduction of the Philippine Sea Plate. As a result of these processes the Ryukyu Arc is characterized by intense seismic and tectonic activity, leading to raised and faulted Pleistocene and Holocene coral reef tracts. Seismic activity often induces tsunamis. Coastal cliffs in the Ryukyu Islands have suffered damage from large tsunamis several times in history (e.g. Imamura, 1938; Kato & Kimura, 1983; Nakata & Kawana, 1995).

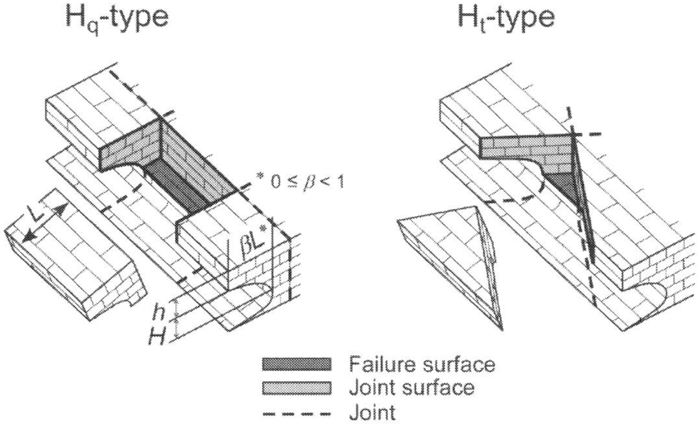

Figure 6. Schematic illustrations of H_q- and H_t-type collapse (Kogure & Matsukura, 2010b); β denotes the ratio of notch depth to the block length, L. An H_q-type block derives from a cliff with a quadrangular failure surface, and an H_t-type block originates from a cliff with a triangular failure surface.

The largest tsunami on record is the 1771 Meiwa Tsunami caused by the Yaeyama Earthquake (M 7.4) on April 24, 1771. The maximum run-up height of the tsunami was estimated by Nakata & Kawana (1995) to be more than 30 m on the south-eastern side of Ishigaki Island, and 5 m on the east and south side of Kuro-shima. Kato & Matsuo (1998) have indicated the possibility of coastal cliff collapses due to historical tsunamis, especially the 1771 Meiwa Tsunami, in Kuro-shima. Nakata & Kawana (1995) inferred the action of other large tsunamis prior to the Meiwa Tsunami, based on radiocarbon dating analysis. Many tsunami blocks that are composed of Holocene fossil corals are much older than the Meiwa period; in a few islands around Ishigaki Island, some are at much higher locations than the run-up height of the 1771 Meiwa Tsunami. According to Nakata & Kawana (1995), such large tsunamis recur on timescale of hundreds or thousands of years in the southern Ryukyu Islands. They asserted that the southern Ryukyu Islands were inundated by large tsunamis about 600, 1000 and 2000 years before the present (yr BP); the tsunami at 2000 yr BP was much higher than the Meiwa Tsunami.

The Ryukyu Islands are in a subtropical area, and in the summer and autumn they often experience typhoons. These give rise to bores that manifest as an extreme wave in the Ryukyu Islands. Severe damage to the coastal environment as a result of these bores has been reported from the late 1980s (e.g. Nakaza et al., 1988). Also in recent years, and especially in July 2007, extreme waves generated by a strong typhoon destroyed some social infrastructures in the southern part of Okinawa Island. Despite studies of the wave height and estimated wave pressure of extreme offshore waves during typhoons around the Ryukyu Islands, there is little information about these parameters for bores in coral reef lagoons which permits us to estimate the wave pressure. Below, we discuss the possibility of cliff collapses due to tsunamis.

3. Methods

3.1. Stability analysis models

According to Kogure & Matsukura (2010b), wave-induced collapse proceeds as follows: (1) a developing notch forms a visor-like cantilever beam; (2) tensile stress is generated, from its own weight, on the horizontal plane that is level with the retreat point of the notch; (3) collapse occurs due to stress from wave impact, resulting in an 'upset' mode of collapse. The cantilever beam model (Timoshenko & Gere, 1978, pp. 108-110) is suitable for estimating the critical notch depth. To obtain the critical failure stresses in H_q- and H_t-type collapse, equations were derived based on this model.

Fig. 7 shows the relation between wave height, H_w, and cliff height, H^*. When H and h are as shown in Fig. 6, and H' denotes the height from the reef flat to the retreat point of the cliff, H^* is equal to the sum of H', H and h. The relation can be divided into four cases (Kogure & Matsukura, 2010b): (1) $H_w \leq H'$ (Case 1), in which that wave pressure does not operate on the notched roof; (2) $H' < H_w \leq H' + H$ (Case 2), in which horizontal stress acts only on the notched roof; (3) $H' + H < H_w \leq H^*$ (Case 3), in which horizontal stress acts on the notched roof and part of the vertical cliff surface; and (4) $H^* < H_w$ (Case 4), in which horizontal stress acts on the notched roof and the vertical cliff surface. In Case 1, waves have no effect on collapse because no stress acts on the notched roof. Stability analysis models are derived below for Cases 2, 3 and 4.

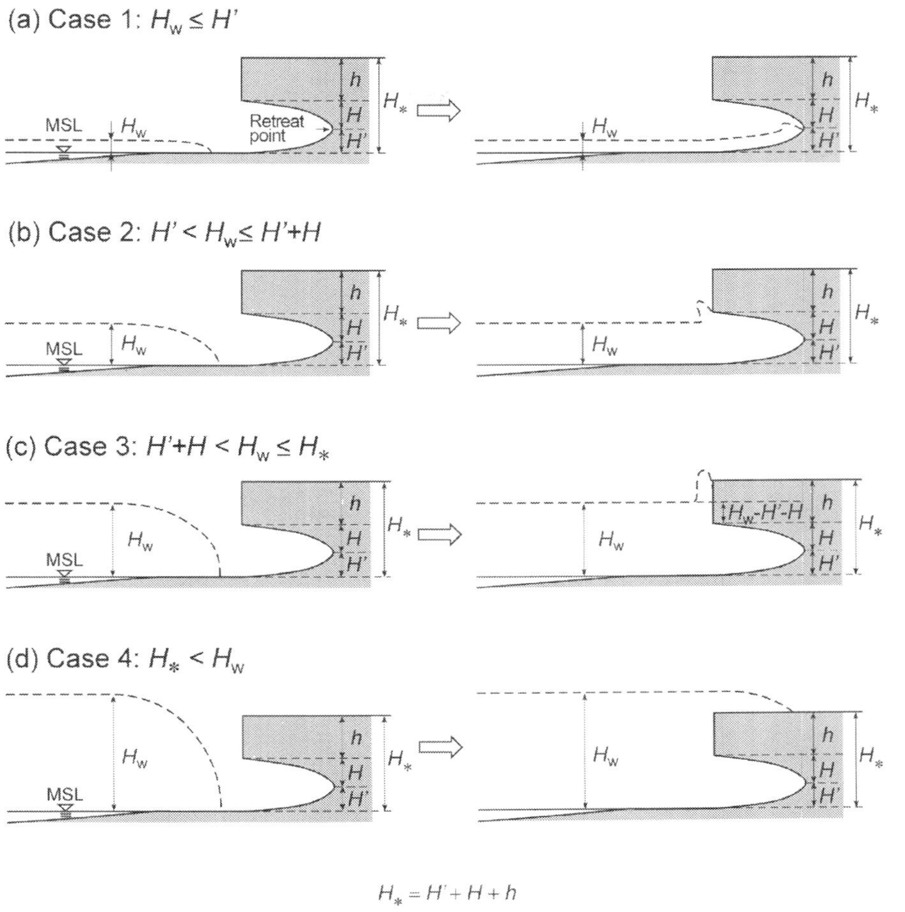

Figure 7. Relationships between wave height and cliff height used for calculating the stresses on a cliff (Kogure & Matsukura, 2010b). Symbols denote: H_w, wave height; h, cliff height corresponding to the vertical front face; H, cliff height corresponding to notched roof; H', height from reef flat to a retreat point of a notch; $H^* = H' + H + h$.

Three forces are involved in cliff instability: (1) downward stress due to gravity, σ_{tgmax}, (2) upward stress due to waves, σ_{twmax}, and (3) the strength of the cliff material, i.e., the tensile strength, S_t. The following equation gives the condition for wave-induced collapse:

$$\sigma_{twmax} \geq \sigma_{tgmax} + S_t \tag{1}$$

For a H_q-type cliff having a cliff height above the retreat point (referred to henceforth as the cliff height), $H + h$, the value of σ_{tgmax} operating on the retreat point of the notch (bold E in

Fig. 8a) is shown as the sum of gravity-induced stress, σ_A, and downward loading due to gravity, σ_B. (Both are shown in bold arrows in Fig. 8a.) This relation is

$$\sigma_{tgmax} = \sigma_A + \sigma_B \tag{2}$$

The values of σ_A and σ_B are:

$$\sigma_A = \frac{\beta^2 \rho g (H + 3h)}{(1-\beta)^2} \tag{3}$$

$$\sigma_B = \rho g (H + h) \tag{4}$$

where ρ is the rock density and g is the acceleration due to gravity. The value of σ_{tgmax} is given by substituting Eqs. (3) and (4) into Eq. (2), as:

$$\sigma_{tgmax} = \frac{\beta^2 \rho g (H + 3h)}{(1-\beta)^2} + \rho g (H + h) \tag{5}$$

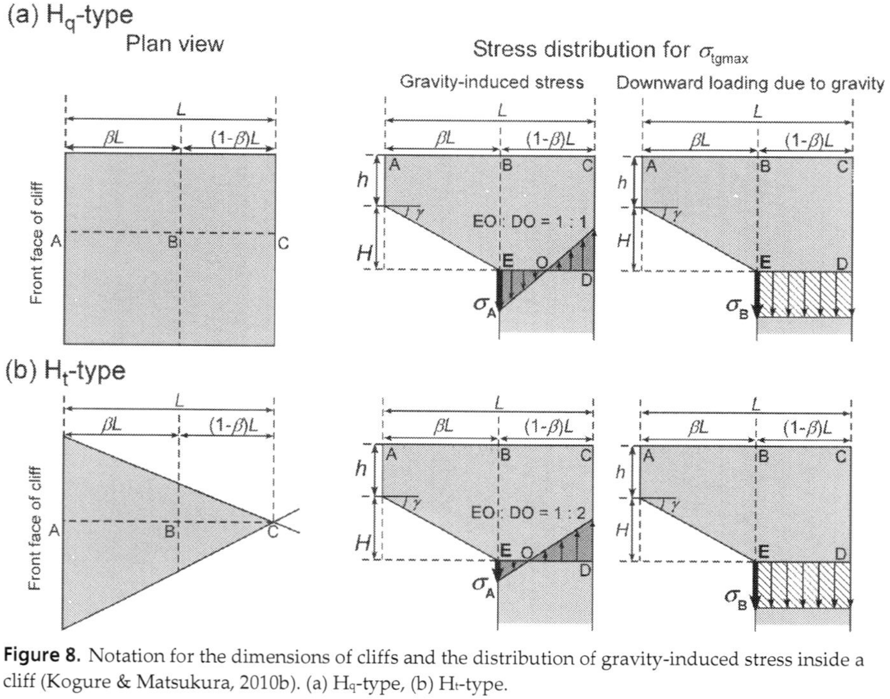

Figure 8. Notation for the dimensions of cliffs and the distribution of gravity-induced stress inside a cliff (Kogure & Matsukura, 2010b). (a) H_q-type, (b) H_t-type.

Wave pressure acts on a notched roof when waves fill up a notch; a cliff experiences both vertical and horizontal components of wave pressure through the whole of the notched roof (Fig. 9). Cliffs are also subject to buoyancy according to the underwater volumes of the cliffs. For a H_q-type cliff, in Case 2 (Fig. 7b), having cliff height $H + h$, with distance L between the cliff face and a vertical joint parallel to the cliff face, notch depth βL ($0 \le \beta < 1$) and angle γ, the maximum value of σ_{twmax} at E in Fig. 9a is given by:

$$\sigma_{twmax} = \frac{3\beta^2 P\cos\gamma}{(1-\beta)^2} + \frac{3H^2 P\sin\gamma}{(1-\beta)^2 L^2} + \frac{\beta^2 \rho_{sw} g H}{(1-\beta)^2} \tag{6}$$

where P is the wave pressure and ρ_{sw} is the density of sea water. In Cases 3 and 4, the value of σ_{twmax} is given respectively by:

$$\sigma_{twmax} = \frac{3\beta^2 P\cos\gamma}{(1-\beta)^2} + \frac{3H^2 P\sin\gamma}{(1-\beta)^2 L^2} + \frac{3P(H_w - H' - H)^2}{(1-\beta)^2 L^2} + \frac{4\beta^2 \rho_{sw} g(2H_w - H' - H)(H+3h)}{(1-\beta)^2 (H+2h)} \tag{7}$$

$$\sigma_{twmax} = \frac{3H^2 P\sin\gamma}{(1-\beta)^2 L^2} + \frac{3hP(2H+h)}{(1-\beta)^2 L^2} + \frac{3\beta^2 P\cos\gamma}{(1-\beta)^2} + \frac{\beta^2 \rho_{sw} g(H+3h)}{(1-\beta)^2} \tag{8}$$

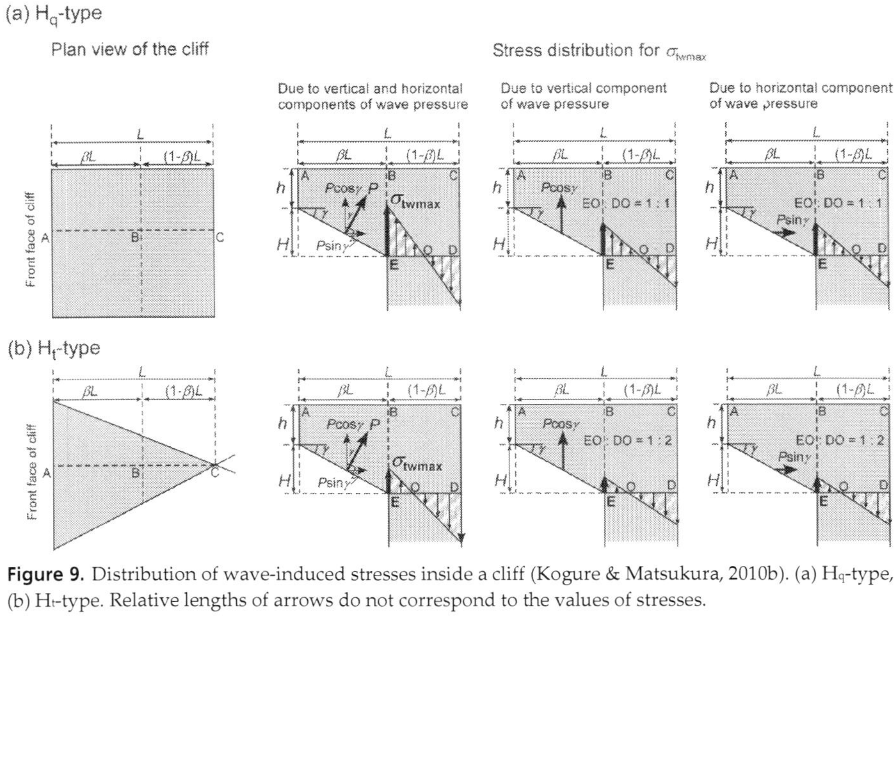

Figure 9. Distribution of wave-induced stresses inside a cliff (Kogure & Matsukura, 2010b). (a) H_q-type, (b) H_t-type. Relative lengths of arrows do not correspond to the values of stresses.

For a H_t-type cliff (Figs. 8b and 9b), the values of σ_{tgmax} and σ_{twmax} can be obtained as above for a H_q-type cliff. The equations obtained for a H_t-type cliff are set out in Table 1, numbered from Eq. (9) to (12).

H_q-type	H_t-type
σ_{tgmax}	
Equation (5): $\frac{\beta^2 \rho g(H+3h)}{(1-\beta)^2}+\rho g(H+h)$	Equation (9): $\frac{\beta^2 \rho g}{(1-\beta)^3}\{2(H+3h)-\beta(H+2h)\}+\rho g(H+h)$
σ_{twmax}	
Case 2	
Equation (6): $\frac{3\beta^2 P\cos\gamma}{(1-\beta)^2}+\frac{3H^2 P\sin\gamma}{(1-\beta)^2 L^2}+\frac{\beta^2 \rho_{sw} gH}{(1-\beta)^2}$	Equation (10): $\frac{2\beta^2(3-\beta)P\cos\gamma}{(1-\beta)^3}+\frac{2(3-\beta)H^2 P\sin\gamma}{(1-\beta)^3 L^2}+\frac{\beta^2(2-\beta)\rho_{sw} gH}{(1-\beta)^3}$
Case 3	
Equation (7): $\frac{3\beta^2 P\cos\gamma}{(1-\beta)^2}+\frac{3H^2 P\sin\gamma}{(1-\beta)^2 L^2}+\frac{3P(H_w-H'-H)^2}{(1-\beta)^2 L^2}+\frac{4\beta^2 \rho_{sw} g(2H_w-2H'-H)(H+3h)}{(1-\beta)^2(H+2h)}$	Equation (11): $\frac{2\beta^2(3-\beta)P\cos\gamma}{(1-\beta)^3}+\frac{2(3-\beta)H^2 P\sin\gamma}{(1-\beta)^3 L^2}+\frac{6P(H_w-H'-H)^2}{(1-\beta)^3 L^2}+\frac{\beta^2 \rho_{sw} g}{(1-\beta)^3}\{(2-\beta)H+2(3-\beta)(H_w-H'-H)\}$
Case 4	
Equation (8): $\frac{2(3-\beta)H^2 P\sin\gamma}{(1-\beta)^3 L^2}+\frac{6hP(2H+h)}{(1-\beta)^3 L^2}+\frac{2\beta^2(3-\beta)P\cos\gamma}{(1-\beta)^3}+\frac{\beta^2 \rho_{sw} g}{(1-\beta)^3}\{(2-\beta)H+2(3-\beta)h\}$	Equation (12): $\frac{2(3-\beta)H^2 P\sin\gamma}{(1-\beta)^3 L^2}+\frac{6hP(2H+h)}{(1-\beta)^3 L^2}+\frac{2\beta^2(3-\beta)P\cos\gamma}{(1-\beta)^3}+\frac{\beta^2 \rho_{sw} g}{(1-\beta)^3}\{(2-\beta)H+2(3-\beta)h\}$

Table 1. Equations for estimating the value of σ_{tgmax} and σ_{twmax} for stability analysis (Kogure & Matsukura, 2010b).

3.2. Parameters for stability analysis

The stability analysis requires information on the dimensions of cliffs and blocks, H', H, h, L, βL and γ, the physical and mechanical properties S_t and ρ of the cliff material, i.e. Ryukyu limestone, and the wave pressure, P. These parameters are characterized as follows.

3.2.1. *Dimensions of blocks and cliffs and rock properties of Ryukyu limestone*

Of the blocks scattered around the coastal cliff, blocks of clearly identifiable type with identifiable place of origin on the cliff face were chosen for the present analysis. In addition to the blocks, the dimensions of two cliffs identified as H_t-type, including that in Fig. 4b, were measured by a laser finder and a measuring stick. Table 2 shows the data for the heights (H and h) and lengths (L) of the blocks and cliffs. The values of H, h, L and βL are 0.9-1.5 m, 0–1.7 m, 2.0–7.0 m, and 0.7–4.8 m for H_q-type blocks, 1.3–1.5 m, 0–1.7 m, 2.4–6.1 m and 0.9–3.6 m for H_t-type blocks, and 1.4 m, 0 m, 3.4–4.7 m and 2.0–2.7 m for H_t-type cliffs.

Kogure & Matsukura (2010b) measured the values of γ and H'. The value of γ is 20° for every block, and the value of H' is the same for every cliff in Kuro-shima, i.e. 1.5 m. The value of S_t for Ryukyu limestone can be estimated using the scaling equation proposed by Kogure et al. (2006); S_t is given in terms of the length of the horizontal failure surface, $(1-\beta)L$, as:

$$S_t = 1000\left[5.6\{100(1-\beta)L\}^{-0.6} + 0.6\right] \tag{13}$$

where the units of S_t and L are kPa and meters. The physical properties of Ryukyu limestone in Kuro-shima have been given by Kogure & Matsukura (2010a), and $\rho = 2.35$ Mg/m^3.

No.	Height			Length	Notch depth
	H (m)	h (m)	$H+h$ (m)	L (m)	βL (m)
H_q-1	1.5	1.1	2.6	7.0	4.8
H_q-2	1.5	0.5	2.0	3.4	2.3
H_q-3	1.5	1.5	3.0	5.0	3.6
H_q-4	1.5	0.4	1.9	2.8	2.2
H_q-5	1.5	1.5	3.0	3.0	2.4
H_q-6	1.5	1.2	2.7	3.8	3.2
H_q-7	1.0	0	1.0	3.6	2.0
H_q-8	1.5	0.3	1.8	4.6	2.2
H_q-9	1.2	0	1.2	2.3	1.9
H_q-10	1.5	0.7	2.2	3.9	2.0
H_q-11	1.5	1.0	2.5	3.2	1.4
H_q-12	1.5	0.1	1.6	5.4	3.5
H_q-13	1.0	0	1.0	3.5	0.7
H_q-14	1.5	0.8	2.3	6.8	3.6
H_q-15	1.5	0.2	1.7	2.5	2.1

H_q-16	1.5	1.1	2.6	3.6	2.7
H_q-17	1.5	0.9	2.4	4.2	2.7
H_q-18	1.5	0.3	1.8	5.0	3.9
H_q-19	1.5	0.9	2.4	3.2	2.6
H_q-20	1.5	0.6	2.1	2.5	1.9
H_q-21	1.5	0.9	2.4	3.6	3.0
H_q-22	1.5	1.1	2.6	2.5	1.9
H_q-23	1.5	0.8	2.3	3.9	2.5
H_q-24	1.5	1.0	2.5	3.7	3.1
H_q-25	1.4	0	1.4	2.0	1.6
H_q-26	1.4	0	1.4	2.6	1.5
H_q-27	0.9	0	0.9	3.7	3.1
H_q-28	1.0	0	1.0	4.0	2.0
H_q-29	1.2	0	1.2	2.5	1.4
H_q-30	1.5	0.5	2.0	4.4	3.6
H_q-31	1.5	1.7	3.2	3.4	2.6
H_q-32	1.3	0	1.3	2.0	1.3
H_q-33	1.5	0.4	1.9	3.9	2.9
H_q-34	1.5	0.3	1.8	4.1	2.0
H_q-35	1.3	0	1.3	2.2	1.6
H_t-1	1.5	0.8	2.3	3.8	2.1
H_t-2	1.5	0.5	2.0	2.6	0.9
H_t-3	1.5	0	1.5	3.3	1.9
H_t-4	1.4	0	1.4	2.4	1.4
H_t-5	1.5	0.8	2.3	5.0	3.2
H_t-6	1.5	0.8	2.3	4.7	2.2
H_t-7	1.5	0.7	2.2	3.1	1.7
H_t-8	1.5	1.7	3.2	6.1	3.6
H_t-9	1.5	0.9	2.4	5.1	3.1
H_t-10	1.5	0.5	2.0	4.3	2.6
H_t-11	1.5	0.9	2.4	2.7	1.5
H_t-12	1.5	0.6	2.1	2.8	1.8
H_t-13	1.5	0.5	2.0	4.5	2.8
H_t-14	1.3	0	1.3	3.3	2.1
H_t-15	1.5	0.2	1.7	3.3	1.9
H_t-16	1.5	0.8	2.3	3.4	1.8
H_t-17	1.3	0	1.3	3.4	1.0
H_t-18	1.5	0.6	2.1	3.7	1.9
H_t-19	1.4	0	1.4	2.8	1.5
H_t-C1*	1.4	0	1.5	3.4	2.0
H_t-C2*	1.4	0	1.5	4.7	2.7

*Cliff

Table 2. Results of measurements of blocks (Kogure & Matsukura, 2010b) and cliffs.

3.2.2. *Estimation of wave pressure*

The dynamic pressure of waves at a cliff, *P*, used in Eqs.6–8 and 10–12 (Table 1), is given as a simple product of parameters such as ρ_{sw}, *g*, and wave height, H_w (e.g. Hom-maand Horikawa, 1964) as:

$$P = k\rho_{sw} g H_w \quad (14)$$

where *k* is termed the coefficient of wave pressure. Eq.14 provides the maximum dynamic pressure exerted on a vertical seawall in the horizontal direction, derived from experimental data. Use of Eq. 14 in this study should take into account the shape of the target object (vertical seawall and notched cliff) and the direction of wave pressure (horizontal and obliquely upward). In spite of detailed research into coastal engineering, no equations have been derived for the wave pressure in the present situation. Extreme waves compacted into a notch are believed to generate the same pressure in all directions, because the water conforms to Pascal's principle during such extreme events. If a significant notch developed on the cliff, it would experience high pressure on its face. We therefore use Eq. 14 to calculate the wave pressure under the assumption that large waves filling up the entire notch exert the same pressure on the notched face perpendicularly and in the horizontal direction.

Some studies have been made of the value of *k* for tsunamis. Ikeno et al. (1998) considered the *k*-value of a tsunami in which the wave crest is divided into two or more parts in propagation at a shallow beach, and the crests collapse just before reaching a coast having cliffs and breakwaters. This type of tsunami causes severe damage to the coastal environment. Ikeno et al. (1998) proposed a *k*-value of 3.5, and Kogure & Matsukura (2010b) used this value. We do the same.

4. Results and discussion

4.1. Calculation of critical wave height

To determine the effect of wave impact on the cliffs, we perform a stability analysis for the blocks and the cliffs listed in Table 2. The critical condition for wave-induced collapse is equality of the left- and right-hand side of Eq. 1. The right-hand side is given by Eqs. 5 and 13 for H_q-type, and by Eqs. 9 and 13 for H_t-type:

$$\sigma_{tgmax} + S_t = \frac{\beta^2 \rho g (H + 3h)}{(1-\beta)^2} + \rho g (H + h) + 1000\left[5.6\{100(1-\beta)L\}^{-0.6} + 0.6\right] \quad (15)$$

$$\sigma_{tgmax} + S_t = \frac{\beta^2 \rho g}{(1-\beta)^3}\{2(H+3h) - \beta(H+2h)\} + \rho g(H+h) + \\ +1000\left[5.6\{100(1-\beta)L\}^{-0.6} + 0.6\right] \quad (16)$$

To estimate the wave pressure in Eq. 14, Kogure & Matsukura (2010b) took $k = 3.5$ and $\rho_{sw} = 1.02$ Mg/m^3; this is the density of sea water at 26.3 °C, which was the annual sea surface temperature observed in 1998 by Ishigaki-jima Local Meteorological Observatory. Also, $g = 9.81$ m/s^2.

No.	H^* (m)	H_{wc} (m)	Case	No.	H^* (m)	H_{wc} (m)	Case
H_q-1	4.1	2.8	2	H_q-29	2.7	4.9	4
H_q-2	3.5	2.4	2	H_q-30	3.5	1.0	1
H_q-3	4.5	2.6	2	H_q-31	4.7	2.2	2
H_q-4	3.4	1.1	1	H_q-32	2.8	2.4	2
H_q-5	4.5	1.7	2	H_q-33	3.4	1.5	1
H_q-6	4.2	1.3	1	H_q-34	3.3	6.8	4
H_q-7	2.5	5.4	4	H_q-35	2.8	1.4	1
H_q-8	3.3	7.6	4	H_t-1	3.8	2.1	2
H_q-9	2.7	0.6	1	H_t-2	3.5	3.6	4
H_q-10	3.7	4.6	4	H_t-3	3.0	1.3	1
H_q-11	4.0	4.4	4	H_t-4	2.9	1.2	1
H_q-12	3.1	2.7	2	H_t-5	3.8	1.4	1
H_q-13	2.5	76.0	4	H_t-6	3.8	3.4	3
H_q-14	3.8	5.5	4	H_t-7	3.7	2.0	2
H_q-15	3.2	0.7	1	H_t-8	4.7	2.4	2
H_q-16	4.1	1.9	2	H_t-9	3.9	1.7	2
H_q-17	3.9	2.8	2	H_t-10	3.5	1.5	1
H_q-18	3.3	1.1	1	H_t-11	3.9	2.0	2
H_q-19	3.9	1.3	1	H_t-12	3.6	1.1	1
H_q-20	3.6	1.4	1	H_t-13	3.5	1.3	1
H_q-21	3.9	1.2	1	H_t-14	2.8	0.9	1
H_q-22	4.1	1.7	2	H_t-15	3.2	1.5	1
H_q-23	3.8	2.6	2	H_t-16	3.8	2.4	2
H_q-24	4.0	1.2	1	H_t-17	2.8	12.0	4
H_q-25	2.9	0.8	1	H_t-18	3.6	2.6	2
H_q-26	2.9	4.2	4	H_t-19	2.9	1.8	2
H_q-27	2.4	0.5	1	H_t-C1*	4.2	2.0	2
H_q-28	2.5	8.1	4	H_t-C2*	4.1	2.2	2

*Cliff

Table 3. Critical wave heights for collapses.

Kogure & Matsukura (2010b) calculated the critical wave height H_{wc} at which wave-induced collapse occurs, as follows: (1) the value of $\sigma_{tgmax} + S_t$ was calculated by substituting the data for each block and cliff into Eq. 15 for H_q-type, or Eq. 16 for H_t-type; (2) Eq. 14 giving the wave pressure was substituted into Eqs. 6 or 7 or 8 (H_q-type), or Eqs. 10 or 11 or 12 (H_t-type), to calculate the value of σ_{twmax}; (3) once the data for each block and cliff except H_w had been substituted into those equations, H_{wc} was determined as the critical wave height, so that the value of σ_{twmax} matched $\sigma_{tgmax} + S_t$. The value of H_{wc} shows the minimum wave height required to induce collapse for each block-and-cliff combination. Table 3 shows the calculated values of H_{wc}, with reference to the relationship between wave height and cliff height (Fig. 7). The resulting values of H_{wc} are 0.5–76.0 m. Kogure & Matsukura (2010b) argued that H_{wc} = 76.0 m for H_q-13 is not a realistic value, and the analysis therefore excludes H_q-13.

4.2. Distribution of wave-provided blocks

To distinguish between blocks produced by gravitational or wave-induced collapse, the blocks and the cliffs were classified into Cases 1–4 in Fig. 7 according to the relationship between H^* and H_{wc} (Table 3). Blocks classified as Case 1 appear to have been produced by gravitational collapse, because the waves cannot reach a notched roof in this case (Fig. 7a). Blocks classified into Cases 2–4 may involve collapse due to waves (Fig. 7b, 7c and 7d). Especially, blocks classified as Case 4 derive from wave-induced collapse. The cliffs H_t-C1 and C2 were classified as Case 2.

The blocks broken off by tsunamis were plotted on an air photograph of Kuro-shima, together with blocks fallen due to gravity (Fig. 10). The tsunami-induced blocks are concentrated in the south-eastern and south-western coasts, whereas the gravity-caused blocks are distributed evenly along the coastlines.

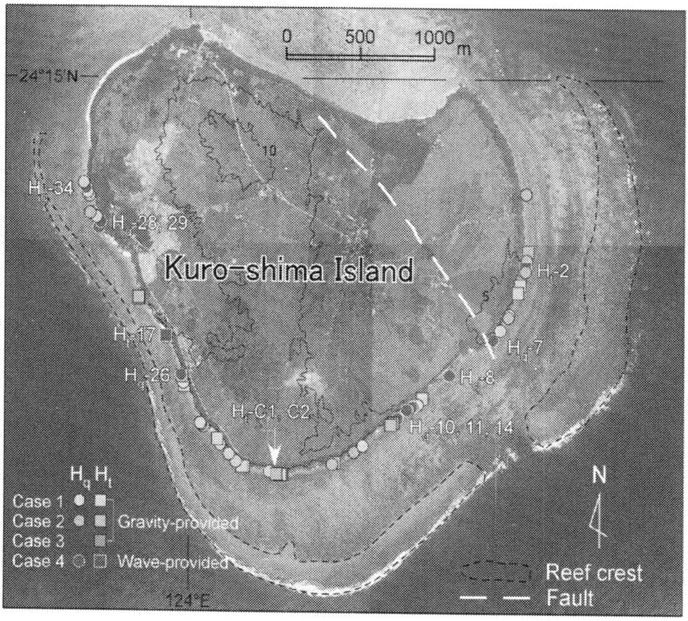

Figure 10. Comparison of distributions between blocks and fringing reefs around Kuro-shima Island. The names of blocks classified as Case 4 are shown on this figure in addition to those of H_t-C1 and C2.

The distance from the coast to the offshore reef edge is about 800 m in the south-eastern area, the farthest offshore in Kuro-shima and indicating that the fringing reefs are highly developed. These fringing reefs are cut by a fault running NW-SE which makes them discontinuous. Tsunamis must therefore have invaded the south-eastern coastal area through this gap, giving rise to the cliff collapses. In the south-western area, the distance between the coastline and reef edge is about 200 m, the smallest value in Kuro-shima. Additionally, the reef crest is poorly developed and its width is the narrowest in the island. Reduction of the height of tsunamis by these fringing reefs must have been inadequate to prevent the collapses.

4.3. Effects of fringing reefs in reducing the height of tsunamis

Fig. 10 indicates that there are two factors affecting the height of tsunamis invading these coasts. One factor is the distance between the coastlines and reef edge; the other is continuity of the fringing reefs and their development.

Comparison of the distribution of the blocks between the southern and south-western areas clearly shows the effect of distance from the coastlines to the reef edge. For all blocks classified as Case 4 (H_q-7,8,10,11,14,26,28,29,34,H_t-2 and 17), the average value of H_{wc} is 6.6 m. It follows that tsunamis having height approximately 6.6 m must have attacked the south-eastern and south-western coasts, causing collapse of the cliffs. In contrast, the average value of H_{wc} is 2.1 m for the cliffs (H_t-C1 and C2) on the south coast where the distance between the coastline and reef edge is about 700 m (see the white arrow in Fig. 10). This implies that H_t-C1 and C2 have never experienced tsunami(s) higher than 2.1 m, and have not yet fallen even though the heights of tsunamis are almost the same everywhere before the reef crest around Kuro-shima. We infer that the heights of tsunamis reaching the coasts decrease with increasing distance between the coastlines and the edge of the offshore reefs. The height of the reef crest may also affect the tsunami height, but we do not consider this further here.

Fringing reefs do not necessarily act as an effective barrier against invading tsunamis if fringing reefs develop discontinuously. Some wave-induced blocks are seen at south-eastern coasts although the distance between the coastline and offshore reef edge is the greatest in Kuro-shima. This appears to be due to the cutting of the fringing reefs by the fault running in the NW-SE direction. Surging tsunamis from the Pacific ocean could be concentrated at the gap. This phenomenon is often referred to as the "channelling effect". Overall, reefs will always absorb some of the wave energy, but, by channelling the water masses, the concentrated impact could be greater on coastal stretches with nearby reefs (Cochard et al., 2008). Our observation is consistent with records by Fernando et al. (2005) and Marris (2005), who reported the effects of channelling on the heights of tsunamis in the 2004 tsunami in Sri Lanka. Also, the height of tsunamis would be amplified, not reduced, during propagation through narrow paths or patchy reefs in the Great Barrier Reef (Nott, 1997).

5. Conclusion

This study showed stability analysis models to evaluate the cliff collapses due to extreme waves. We determined the distribution of the blocks caused by cliff collapses due to tsunamis, by means of stability analysis. Comparison of the distributions of blocks caused by tsunamis and by gravitational processes shows that two factors influence the height of tsunamis that reach the coasts. One factor is the distance between the coastline and reef edge. The heights of tsunamis reaching coasts decrease with increasing distance between the coastline and offshore reef edge. The other factor is continuity in the fringing reefs and their development. The height of tsunamis invading through the coral reef must have been constant or amplified by deep and narrow gaps, known as the channelling effect. Tsunamis which passed through the gap were able to reach the coasts without any loss of height and induce collapses of coastal cliffs. Therefore, broad and continuous "modern" fringing reefs may act as an effective barrier against tsunamis to collapse coastal cliffs, i.e., "ancient" fringing reefs.

Author details

Tetsuya Kogure
Research Institute of Innovative Technology for the Earth, Japan

Yukinori Matsukura
University of Tsukuba, Japan

6. References

Cochard, R., Ranamukhaarachchi, S.L., Shivakoti, G.P., Shipin, O.V., Edwards, P.J. & Seeland, K.T. 2008. The 2004 tsunami in Aceh and Southern Thailand : A review on coastal ecosystems, wave hazards and vulnerability. *Perspectives in Plant Ecology, Evolution and Systematics*, Vol. 10, pp. 3-40.

Fernando, H.J.S., Mendis, S.G., McCulley, J.L. & Perera, K. 2005. Coral poaching worsens tsunami destruction in Sri Lanka. *EOS Transactions*, Vol. 86, pp. 301-304.

Hom-ma, M.&Horikawa, K. 1964. Wave force against sea wall, *Proceedings of 9th Conference on Coastal Engineering*, Reston, 1964.

Ikeno, M., Matsuyama, M.&Tanaka, H. 1998. Shoaling and soliton fission of tsunami on a shelf and wave pressure for tsunami-resistant design of breakwater by large wave flume-experiments, *Proceedings of 45th Japanese Coastal Engineering Conference*, Akita, November, 1998 (in Japanese).

Imamura A. 1938. On the earthquake zone of the Ryukyus and the large Meiwa Tsunami. *Zishin*, Vol. 10, pp. 431–450 (in Japanese).

Kato Y. & Kimura M. 1983. Age and origin of so-called "Tsunami-ishi", Ishigaki Island, Okinawa Prefecture. *The Geological Society of Japan*, Vol. 89, pp. 471–474 (in Japanese).

Kato, Y. & Matsuo, K. 1998. Estimation of tsunami behavior based on identification of original tsunami boulders: an example in Kuroshima Island. *Kaiyo Monthly*, Vol. 15, pp. 183–187 (in Japanese).

Kogure, T., Aoki, H., Maekado, A., Hirose, T. & Matsukura, Y. 2006. Effect of the development of notches and tension cracks on instability of limestone coastal cliffs in the Ryukyus, Japan. *Geomorphology,* Vol. 80, pp. 236–244. DOI. 10.1016/j.geomorph.2006.02.012

Kogure, T. & Matsukura, Y. 2010a. Critical notch depths for failure of coastal limestone cliffs: Case study at Kuro-shima Island, Okinawa, Japan. *Earth Surface Processes and Landforms,* Vol. 35, No. 9, pp. 1044-1056.

Kogure, T. & Matsukura, Y. 2010b. Instability of coral limestone cliffs due to extreme waves. *Earth Surface Processes and Landforms,* Vol. 35, No. 11, pp. 1357-1367.

Kogure, T. & Matsukura, Y. 2011. Effect of vertical joints on the planar shape of fallen blocks in coastal cliff collapses at Kuro-shima, Okinawa, Japan. *Transactions, Japanese Geomorphological Union,* Vol. 32, pp. 15-28.

Kono, T. & Tsukayama, S. 1980. Wave transformation on reef and some consideration on its application to field. *Coastal Engineering in Japan,* Vol. 23, pp. 45-57.

Maekado, A. 1991. Recession of coastal cliff made of Ryukyu Limestone: Arasaki coast, southern end of Okinawa Island, Japan. *Bulletin of the Okinawa Geographical Society,* Vol. 3, pp. 63–70.

Marris, E. 2005. Tsunami damage was enhanced by coral theft. *Nature,* Vol. 406, pp. 1071.

Munk, W. & Sargent, M.C. 1948. Adjustment of Bikini Atoll to ocean waves. *Transactions of American Geophysical Union,* Vol. 29, pp. 855-860.

Nakata, T. & Kawana, T. 1995. Historical and Prehistorical large tsunamis in the southern Ryukyus, Japan. In: *Tsunami: Progress in Prediction, Disaster Prevention and Warning,* Tsuchiya, Y., & Shuto, N. (Eds.), pp. 211-222, Kluwer Academic Publishers, ISBN: 978-0-7923-3483-5, Netherlands.

Nakaza, E., Tsukayama, M., Hino, M. & Ohosiro, T. 1988. A study on waveforces induced by wave groups, *Proceedings of 35th Japanese Coastal Engineering Conference,* Ehime, November, 1988 (in Japanese).

Nott, J. 1997. Extremely high-energy wave deposits inside the Great Barrier Reef, Australia: determining the cause – tsunami or tropical cyclone. *Marine Geology,* Vol. 141, pp. 193-207.

Roberts, H.H. & Suhayda, J.N. 1983. Wave-current interactions on a shallow reef (Nicaragua, Central America). *Coral Reefs,* Vol. 1, pp. 209-214.

Shibayama, T., Okayasu, A., Wijayaratna, N., Sasaki, J., Suzuki, T. & Jayaratne, R. 2005. The 2004 Sumatra Earthquake Tsunami, tsunami field survey in southern part of Sri Lanka, *Proceedings of 52th Japanese Coastal Engineering Conference,* Aomori, November, 2005 (in Japanese).

Timoshenko, S.P. & Gere, J.M. 1978. *Mechanics of Materials,* Van Nostrand Reinhold, New York.

Tjia, H.D. 1985. Notching by abrasion on a limestone coast. *Zeitschrift für Geomorphologie N. F.,* Vol. 29, pp. 367-372.

Chapter 10

Traces of Coral Bearing Deposits on Lanai, Hawaii, and Implications for Their Origin (Island Uplift vs. Giant Tsunami)

Barbara Keating Helsley and Charles E. Helsley

Additional information is available at the end of the chapter

http://dx.doi.org/10.5772/50875

1. Introduction

Harold Stearns carried out geologic studies on the island of Lanai during 1936. The strand line deposits he described were interpreted as a series of high stands of sea level (Stearns 1938). But, he realized that this was not consistent with Sea Level Variation (SLV) records elsewhere in the world. Stearns (1985) interpreted the strandlines as being due to a combination of **SL**V and **I**sland u**P**lift (here referred to as SLIP). Moore and Moore (1984, 1988) proposed that these same deposits were debris thrown up on the island by giant waves generated by slope failure on the SW flank of the island of Hawaii (hereafter referred to as the Giant Wave Hypothesis, GWH). 1.1 Validity of Giant Wave Hypothesis vs. SLIP

Keating and Helsley (2002) revisited the three ancient Lanai shorelines reported by Stearns (1978) in Kaluakapo Crater (SE Lanai) with the objective of testing the validity of the GWH and SLIP (Sea Level oscillations and Island Uplift) scenarios. In this paper we report the results of field studies west of the previously studied Manele Bay and Kaluakapo Crater areas. These studies as well as geologic excursions around Lanai allowed the authors to: 1) observe the outcrops reported in Harold Stearns field notes and publications, 2) study the features assigned a GWH origin, 3) extend Stearns original observations, and 4) compare the coral bearing deposits. Evidence for SLIP is reported and geologic inconsistencies with the GWH are discussed.

1.1. Setting

Corals in boulder beach deposits as well as SLV notches were found preserved on the arid southern flank of Lanai where maximum precipitation (in the mountains) is less than 25 cm/yr. The field areas are located on the leeward side of the island of Lanai, which in itself is situated in the lee of the island of Maui (Fig. 1). Thus the modern environment is desert, with vegetation dominated by mesquite (locally called kiawe, *Prosopis pallida*, introduced in 1828) and other drought tolerant vegetation. Furthermore, due to long-term drought conditions and fires, vegetation was minimal providing excellent opportunities to view outcrops. Similar deposits are not present on the wet, windward sides of the islands, where erosion rapidly removes these deposits.

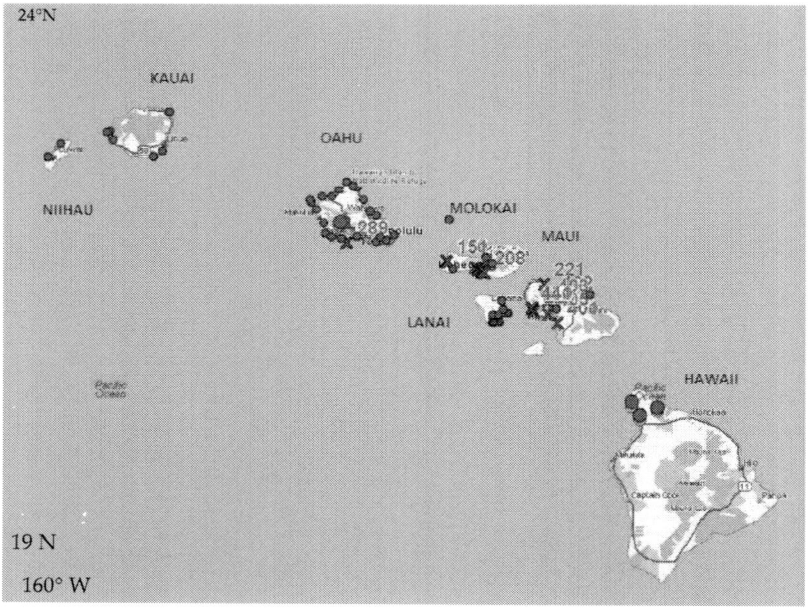

Figure 1. Map of the Hawaiian Islands showing sites of sedimentary carbonate rock outcrops (red dots) and drill core sites with sedimentary units (x-symbols). These units include: dunes, reefs, and marine conglomerates. The marine conglomerates have also been called gravels associated with the GWH.

1.2. Methodology

The geologic field area is situated west of the Manele Bay Resort golf course. Traverses on foot were made, beginning at the top of the slope at an elevation of nearly 800 ft, down the dry gullies, to elevations of 30 masl (meters above sea level). The coral and basalt deposits were found at various elevations within the dry gullies, and recorded as GPS landmarks, and were concentrated on the leeward eastern side of the gullies (side least exposed to waves at the time of formation. The exploratory traverses followed the gullies down slope (Fig. 2), and then were extended across the interfluves between gullies, GPS mapping of the coral and basalt bearing deposits show that the pattern of deposits extends across the island flank, in bath-tub ring fashion (Fig. 3) without any deposits on interfluves and rarely on the slopes between the 'bath-tub rings'.

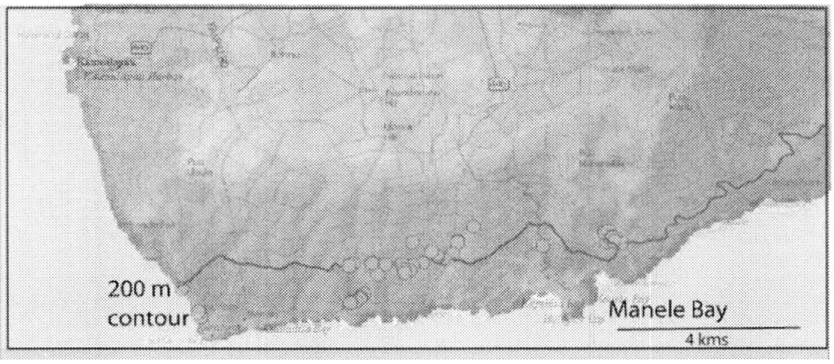

Figure 2. Map showing the sampling sites (mapped by dots), situated on the southern flank of the island of Lanai associated with this study, and Poopoo and Anapuka gullies in particular. The cluster of sites on the extreme right (East) are situated at Kaluakapo Crater and extreme left (West) at Kaunolu Point. The elevations of fossil sampling sites were corrected with barometric measurements. Up to 25 fossils were collected at each site. At most sites, the marine fossils were abundant enough that the number collected needed to be restricted, due to the difficulty in transportation to the nearest access point.

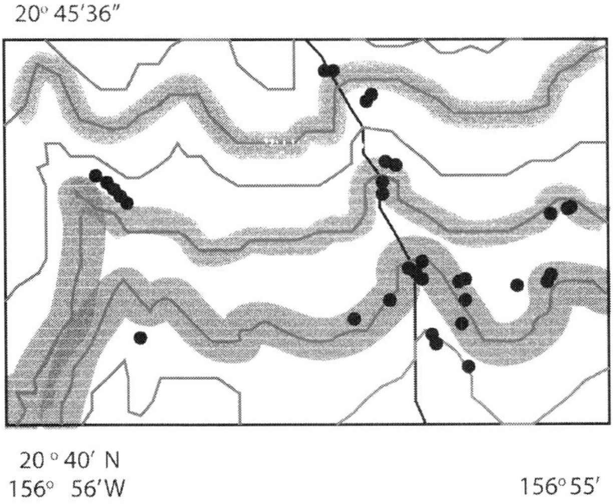

Figure 3. Detail of the center portion of Fig. 2, showing the location of sample sites, marked with black dots. Contour lines have been highlighted in gray to show the parallel "bathtub-ring" distribution of the deposits. Note that fossils were not found on the interfluves (surface between gullies) or on the slopes between the 'bath-tub rings'.

1.3. Prior related studies

During the 1980's Harold Stearns (then retired) contacted several scientists, encouraging them to evaluate the GWH. Several publications have resulted from those discussions, including: Jones (1993, 1995) Grigg and Jones (1997), and later Felton, *et al.* (2000), Keating and Helsley (2002), Felton, *et al.* (2006), and Crook and Felton (2008). Felton *et al.* (2006) published a detailed facies analysis, describing a series of 14 beds separated by eight discontinuities. Three of these discontinuities were characterized by truncated paleosols; indicative the deposits accumulated over long periods of time, and thus were from several geological events separated in time. Keating and Helsley (2002) showed that the ancient shorelines described by Stearns at 190 and 170 m in Kaluakapo crater showed undisturbed marine deposits at 190 m, abundant marine carbonate deposits at 170 m elevation, and a total absence of fossil bearing deposits between 200 and 365 m.

Age dating studies by Grigg and Jones (Grigg 1997) and Rubin *et al.* (2000) provided evidence for more than one geologic event on Lanai. Sedimentary studies of SLV which bear on this study were carried out on Oahu (Fletcher and Sherman, 1995; Fletcher and Jones, 1996; Hearty, 2002; Hearty, 2011; and McMurtry *et al.*, 2011).

Numerical modeling of the GWH tsunami were carried out by Johnson and Mader, (1994), Jones and Mader (1995), and these models suggest a wave height of less than 100 m. Keating *et al.* (2011) published a compilation of a Tsunami Deposit Database that summarized the range of tsunami run-up values reported elsewhere in the literature. The maximum run up level reported was 70 m, and the compilation indicated that a 70 m run up is a rare and extreme event.

1.4. Field observations on Lanai

During this study, traverses were concentrated in the SW portion of Lanai across a series of dry gullies west of the Manele Bay Resort Golf Course. The Poopoo and Anapuka areas display an outcrop pattern of fossil-bearing deposits preserved in dry gullies, and in small notches eroded into basaltic rock and filled with lithified coral and basaltic detritus with rare boulders of beach rock, that display lateral continuity. These parallel notches, benches, and deposits display lateral continuity that is apparent in the map view of sampling sites shown in Fig. 3. Examples of the outcrops are shown in Figs. 4 through 6.

Figure 4. (Right) Photograph of a notch within lava rock containing shallow marine sediments and corals. This site is designated as GPS Landmark 241 at 140m above sea level in Poopoo Gullies. (Left) Photographs of a notch in the hillside filled with coral-bearing deposits. This site is designated GPS Landmark 172 at an elevation of 165m on the Poopoo Gullies.

Figure 5. (Top left) Image of the dry gully that corresponds with the hill side drawn in the lower left corner in the landscape sketch. Fig. 5b. Center, cartoon drawing showing the distribution of geologic features. Fig. 5c. The floor of the embayment is shown in the photograph at the lower left. Fig. 5d. The image at the lower right shows an ancient boulder beach, situated on the left side of the embayment at the center of the drawing (Fig. 5c, center). Had there been a giant tsunami at this site, the boulder beach would have been disrupted. Its presence is a very strong indication that the Poopoo and Anapuka area was not affected by giant waves. Fig. 5d. Upper right, image of a small wave cut cliff at the toe of the slope on the right side of the embayment.

In the main Anapuka drainage, there are a series of parallel notches with coral fillings, between elevations of 35 and 240 m (Fig.4a & b). These deposits are similar in rock types and lithologies found at the present time along modern Hawaiian rocky shorelines including beach rock (cemented coral sand), rounded boulders, and coral and basalt conglomerates with coral fragments.

In the Anapuka area a semi-flat area was found, showing evidence of small wave-cut cliffs, Fig. 5b, boulder-clad shorelines (Fig. 5c), and outcrops of corals with one coral having grown upon another (Fig. 5d). Large pieces of lithified beach rock (up to 1 m across) are present in places along the margin of the platform. This area formed in a small sheltered embayment (see configuration Fig. 5c) that obviously favored the growth of abundant shallow marine organisms.

Figure 6. Photograph showing the position of fossil-bearing deposits on the margin of a dry gully. The fossil-bearing outcrop extends from the horizon marked by the horizontal bar and the feet of the observer, standing at the left edge of the picture. Notice the step in topography in the background.

Caliches (gravel, rocks, soil and alluvium cemented with soluble salts) are common throughout the arid south and leeward side of Lanai. They occur as: 1) fracture fillings within and between lava flows, 2) irregular masses and sheets between lava flows, 3) cementation within the marine conglomerates, and 4) casts of plant roots. The caliches are very common in lava flows roughly 10 m above the 190-m level platform at Poopoo and Ananpuka and elsewhere on the southern coast of Lanai. Caliches also occur at natural seeps, and at watering stations where there is, or has been, long-term water seepage. Ancient sea stacks were observed at several locations on the slopes of Lanai, e.g. near Kaluakapo Crater (illustrated in Keating and Helsley, 2002) and near the Naha Road/trail (east of Kaluakapo Crater). On the old Naha trail, a broad terrace at roughly 180 m on the southeastern flank of Lanai contains laminated ash deposits. Elsewhere on the central and west flank of Lanai outcrops of lava contain vein-filling caliche are common (non-fossiliferous carbonates) at roughly the 180 m elevation. In the Poopoo and Anapuka area, there is a transition at roughly 180 m, from deeply weathered red basaltic boulders (above) to smaller, more-rounded black volcanic boulders (lacking weathered rims) lower on the slopes.

1.5. Fossils

Symbols mark the location of sampling sites containing coral in Fig. 2. In general, the coral-bearing rocks are well cemented conglomerates and were collected using a sledgehammer and chisel. Samples collected from individual sites contained fossil specimens (mostly coral and mollusk). In some small sites, essentially all the mega-fossils exposed on the outcrop were collected. On the Anapuka plain, corals were far too numerous to be collected in bulk and transported in large numbers. Instead, a comprehensive collection was made of representative samples for laboratory study. Drs. Allison Kay, Julie Brock and Richard Grigg made fossil identifications. Dr. Richard Grigg made the selection of fossils that display the least alteration for radiometric age dating. The coral *Porites lobata* was the most abundant species found, followed by *Porites compressa*. The bioturbation was identified as moderate and typical of shallow Hawaiian coastal waters (A. Kay, P. Comm.). Some outcrops contain numerous shells and corals, that are whole, unbroken and abrasion-free, cemented in a matrix of sand; others are dominated by broken clasts. Clasts of volcanic material are found in most outcrops. At one locality there is a clear vertical progression from course fossil bearing rubble at the base to fine-grained fossiliferous gravel upward in the unit. At the 190 m elevation in the Anapuka drainage, an aeolian sand unit, similar to other back beach sand deposits in Hawaii, overlies the coral and basalt bearing units, and further up slope, there are weathered red basalts with abundant caliche filled fractures.

The major occurrences of fossils in the Poopoo and Anapuka dry gullies are at elevations of 180, 168-171, 156, 150 m (600, 560-570, 520, and 500 ft) above sea level. Stearns (1978) suggested wave-cut gently sloping platforms lie at altitudes of 45 and 168 m (150 and 560 ft) and between 97.5 and 112.5 m (325 and 375 ft). These platforms appear to be associated with in-situ deposits we identified at 45 m, 168-171 m, and the bench deposits at 120-135 m.

1.6. Ages

Abundant fossils are present in the strandline deposits we have studied and they display a general increase in weathering, and surface darkening with elevation. The coral *Favia* is no longer present in Hawaiian waters, but it was found in 7 sites between 147.3 and 181.5 m elevation (identified by Grigg). The abundance of the fossils no longer living in Hawaiian waters increases with elevation. This interval roughly corresponds with radiometric dating sample LanS (Rubin *et al.*, 2000), collected from an elevation of 155m, which yielded an age of 134 kya This observation is consistent with an overall pattern of increasing fossil age with elevation. Moore and Moore report U-series ages from the island of Hawaii, dated at 110 ka +/- 10). Moore and Moore (1988) reported 3 U-series ages from Lanai that yielded ages of 108 ka +/-5, 101 ka +/- 4 from Kawaiu Gulch and 134 ka +/-7 from Kaluakapo Crater. (These are equivalent to the Waimanalu sea level stand on Oahu).

Additional radiometric ages for Lanai were published by Rubin *et al.* (2000; U-series) and Grigg and Jones (1997; ESR, Electron Spin Resonance). ESR, U-series, TL (thermoluminence) and OSL (optical dating), age determinations have produced comparable results to those of Yoshida and Brumby (1999), Carew and Mylroie (1995), Tanaka *et al.* (1997) where ages are generally correlated to specific faunal assemblages. The original studies of the type Manele Bay deposits were concentrated on the lower 50 m (Felton *et al.*, 2000, 2006) of the Manele Bay outcrops. Most of the studies are from corals in the Manele Bay Resort complex at elevations less than 50 m. The studies of Poopoo and Anapuka gullies concentrate on the exposed deposits at higher elevations from 45 m to 200 m.

Plots of sample age and sample site occurrences versus elevation are shown in Fig. 7. The outcrops between 50-200 m elevations represent rocks equivalent to older Marine Isotope Stages (MIS stages 7 and 9 and perhaps 11). Numerous fossils with minimum alteration were collected for dating (by Ken Rubin). And, according to Rubin, the new collection should provide much more suitable candidates for dating, than those previously utilized.

Young ages of 2-3 thousand years ago (kya) occur near sea level (0.5 m). Ages of roughly 130 kya are observed up to roughly 50 m with one at 155 m (Szabo, 1984 reported in Moore and Moore, 1988). A cluster of ages between 200-250 kya occurs at elevations up to 80 m. One age of 350 kya has been reported at 190 m within Kaluakapo Crater.

Using the Marine Isotope Record as a proxy for a relative sea level curve for Lanai (Fig. 8) and using the oldest ages reported by Rubin *et al.* (2000), Moore and Moore (1984), and Grigg and Jones (1997) for controls; ages around 120 kya (MIS 5.5) occur up to 23 m elevation; 211-230 kya (MIS 7.1, 7.3) occur between 28-35 m; one age of 250 kya is reported at 58 m (MIS 7.5); a single age of 350 kya is reported from 171 m (MIS 9.3). A theorized uplift scenario for Lanai is illustrated using the proxy marine oxygen isotope records from Shackleton (2000) and others (Fig. 9). A comparison of radiometric ages versus elevation, and the uplift SLV curve are shown in Fig 9. These comparisons are similar to those Toscano and Lundberg (1999) and Tanaka et al, (1999) for Florida, Barbados and Haiti. A list of observed notches filled with corals, depositional benches, and platforms with ages are given in Table 1.

In conclusion, the complicated internal stratigraphy documented by Felton *et al.* (2000) and Felton et al. (2006) is much easier to explain if the interbedded soils are interpreted as erosional surfaces within marine carbonates and clastic deposits accumulated during low sea level stands during glacial periods.

Samples collected in submersible dives by Grigg (2002), yielded an age of 8 ka for submerged coral at -61m (-200 ft). Fig. 7a shows a pattern of increasing age of fossiliferous deposits with increasing elevation. Grigg and Jones (1997) report that a trend of increasing age of coral beach deposits with elevation (on Oahu and Molokai) have been interpreted as support for island uplift.

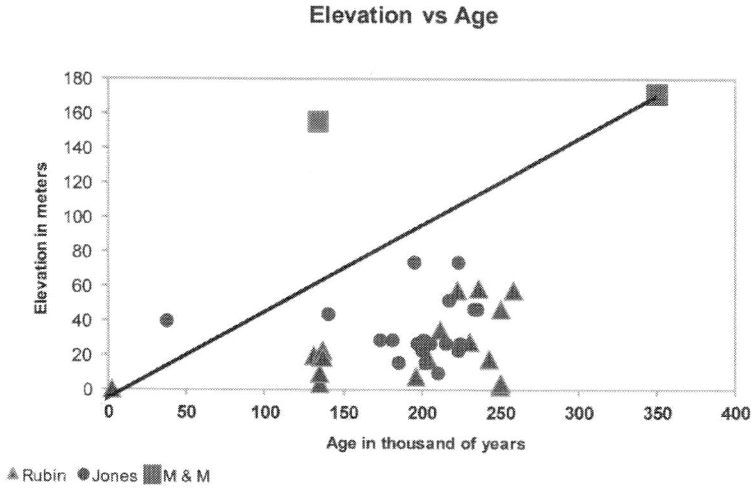

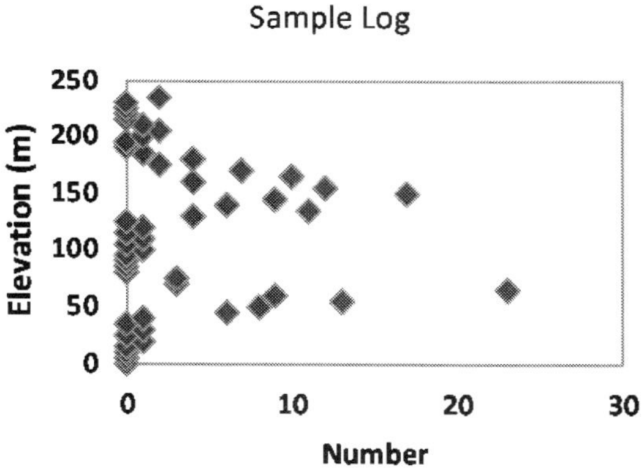

Figure 7. Fig. 7a (top) Plot of age versus elevation for coral samples from the southern slope of Lanai. The symbols represent the published radiometric ages: U-Series ages from Rubin et al (2000), ESR ages from Grigg & Jones (1997), and two other ages reported by Moore and Moore (1988). The line showing the uplift rate from today to the oldest ages date is a solid line.

Fig. 7b. A plot of sample occurrences versus elevations is shown above. The plot shows that a great number of the samples collected for this study came from elevations higher than those previously published. The number refers to the number of sampling bags that were collected: each bag contained from one to 25 rocks. Zero values indicate that no samples were found between the elevations occurring in a 'bath-tub' ring distribution.

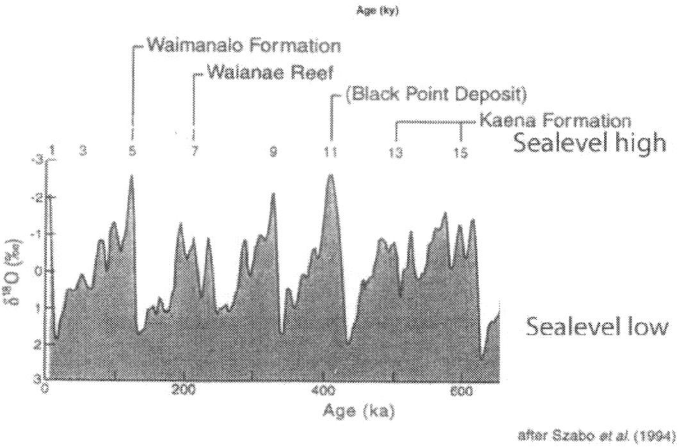

Figure 8. Sea Level Variations for Hawaii after the record by Szabo *et al.* (1994). The labels at the top list the equivalent rock units on the island of Oahu.

Elevation	Age (thousand yrs)	Isotopic Stage
225 m		(*Likely road construction material)
186-192 m		MIS11
180 m		
168-171 m	350 ka	MIS9.3
156 m		MIS 9.1
150 m		MIS 9.1
120-135		MIS 8.5
97.5-112.5m		MIS 8.3
78 m		MIS 8.3
45 m		MIS
35 m		Low stand 7/8
30 m	211-230 Ka	MIS 7.1, 7.3
23 m	120 ka	MIS 5.5
14 m		MIS 6.5
5 m		MIS 3

Table 1. Elevation (m), Age (ka), Correlated MIS

2. Mechanism for island uplift

The sinking of the active islands due to the unbalanced load applied to the crust by growing volcanoes has been well documented and sinking rates of one to three mm/y have been documented by many authors studying coral reef deposits and erosional surfaces near the big island of Hawaii. The consequences of this loading have been examined by Watts and ten Brink (1989), Wessel and Keating (1994) Smith and Wessel (2000), and others.

The volcanic edifice adds a load to the crust that causes the formation of a moat around the volcanic load and at greater distances the seafloor is uplifted in a broad low arch. This phenomenon can be seen in the bathymetry around the 'Big Island' as well as all along the Hawaiian Island chain where a portion of the sea floor has been raised a few hundred meters as two parallel ridges on either side of the chain. What is not generally considered is the uplift also affects the pre-existing islands immediately downstream of the hot spot. Here, a balance is struck between the continued slow sinking of the volcanic edifice due to the massive load on the crust with the flow of sub crustal material away from the latest volcanic load that forms the arch around that load. In the case of Lanai, it is at the appropriate distance for the inflow of the displaced sub crustal asthenosphere to form uplift in excess of the residual sinking of the island edifice due to its residual crustal loading. The SLIP hypothesis suggests that this balance of forces became positive, i.e. in favor of uplift, about 450 kya at the time that the volcanic edifice beneath the Kohala volcano ceased to be active. The peak uplift rate is probably past, since the center of volcanic edifice building has shifted to the SE in the past few hundred thousand years and Lanai is now at a distance equivalent to the outer edges of the arch to either side of the Big Island. It is our contention that the elevated strand lines we have observed are the result of coastal erosion and concurrent offshore and strand line deposition at times in the past when the island uplift rate and the rate of eustatic sea level rise due to changes in ice volume were approximately equal (Fig. 9). Comparably, low stand deposits did form at times when the uplift rate and the beginnings of sea level rise are again comparable. But in the low stand case, corals can continue to grow and sediments can continue to be accumulated beneath the sea through the sea level rise period and even into the sea level fall portion of the cycle. The deposits in the Manele Bay area (those between current sea level and 70 masl are part of the deposits that formed during previous low stands for they contain interbedded marine and non-marine clastic layers including soils and irregular erosion surfaces characteristic of karst surfaces as well as offshore reef material.

For simplicity, we have assumed an island uplift that begins about 450 kya and that remained constant until today. A more realistic model would be to assume the uplift rate rose from zero uplift to a maximum one as the island passed over the arch and then to a lower uplift rate after the peak had passed. But we have too few age constraints for the older and more elevated strand lines at present so a simple constant uplift rate has been used. The model, regardless of exact uplift history, is constrained by the strand line deposits we have observed at high elevation (190 m) on the slopes of Lanai (Fig. 9). Modelling results imply uplift rates for the past 350 kya plus years must average about 0.5 mm per year. We have added this cumulative time dependent uplift to the proxy sea level data (Waelbroeck 2002) to provide a model uplift to estimate where sea level might have been relative to the observed strandline and other carbonate deposits on Lanai. The high elevation strand lines are probably associated with the 400 kya maximum based on a single age of 350 kya for a sample collected at 170 meters. We also need to point out once again that the carbonate deposits near the present coast line with ages between 100 and 220 kya are most likely low stand accumulation deposits.

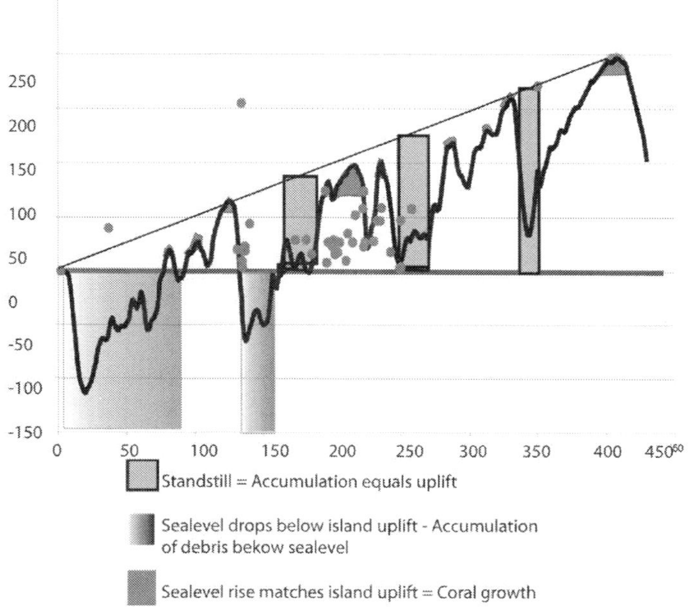

Figure 9. Plot of the MIS Proxy of sea level (Waelbroeck *et al.*, 2002) vs. elevation (on vertical axis) with the origin of the plot at modern sea level (0 on horizontal axis). The published ages are shown in red dots. The ages from Manele are from publications by Grigg and Jones (1997) and Rubin *et al.* (2000). The plot is constrained from modern sea level 0 to the maximum age of 350 kya at 200 m elevation. The slope of the line above the curve is interpreted as the rate of island uplift. The green shaded vertical bars mark periods of standstill when the accumulation of sediments matches the rate of uplift, at these times corals accumulate. The two gray shaded areas represent times when low stands of sea level allow accumulations of debris eroded from the island to accumulate below sea level. The violet colour represents times of maximum sea level when rate of coral growth equals the rate of uplift. High resolution sea floor mapping in the shallow water between the islands of Lanai and Molokai reveals submerged lagoons and terraces on the submerged flanks Lanai and Molokai and shows many submerged lagoons and terraces are present on the submerged flanks of Lanai (see the University of Hawaii SOEST Website www.soest.hawaii.edu/HMRG/Multibeam/3d.php) and Grigg *et al.*, 2002).

Many others have recognized the need for sea level variation (SLV) to be combined with island uplift to explain the uplifted reef deposits in the Hawaiian Islands -Stearns (1978) Stearns (1970) 0.3 mm/y, Bruckner and Radtke (1989); Jones (1993b); Muhs and Szabo (1994) 0.036-0.050 m/kya; Szabo *et al.* (1994); Grigg and Jones (1997) 0.33 +/- 0.03 mm/y; Grossman *et al.* (1998) 0.05 mm/y; Hearty et al. (2002) 0.03mm/y; (Hearty et al. 2011). There is a wealth of evidence that indicates the fossiliferous deposits above sea level in the Hawaiian Islands likely formed by a combination of normal geologic processes, rather than extreme events.

Elevation (1978)	**Locations (1978)** Other Islands	**Locations (1978)** Lanai
360 m (1200 ft)	Maybe Oahu, W. Maui, Molokai	Mahana Shoreline, Lanai
190 m (625 ft)		Kaluakapo Shoreline, Lanai
168 m (560 ft)	Oahu, Molokai	Manele Shoreline, Lanai
75m (250 ft)	Olowalu, Maui and Oahu	
29 m (95 ft)	Kaena, Oahu	
21 m (70 ft)	Laie, Oahu	
16.5 m (55 ft)	Kahuku, Oahu	
12 m (40 ft)	Waialae Shoreline, Oahu, Kohala, HI	
7.6 m (22 and 27 ft)	Waimanolo, Oahu PCA Lualualei Valley, Oahu	
3.6 m (12 ft.)	Kailua Shoreline, Oahu	
0.6 (2 ft.)	Leahi, Oahu	

Table 2. List of terraces recognized by Harold Stearns (1978).

3. Is the mega-tsunami concept valid?

Succinctly stated, the giant wave hypothesis (GWH) suggests a submarine landslide southeast of Lanai triggered three "giant waves" that rushed toward Lanai with initial velocities of 149 m/s, at intervals of only one and a half minutes. The first wave reached 190 m elevation on Lanai and eroded the soils and churned up boulders. The second wave reached the 375 m elevation, and picked up the gravels in suspension and stripped the terrain. The third wave reached 190 m high on the island slope taking boulders in suspension, then accelerated down slope, stripping soil and molding the boulders into mound-shaped bed forms. The GWH presents an exciting but problematic scenario of events.

As part of an effort to date drowned reefs around Hawaii (Moore 1987; Moore and Campbell, 1987; Ludwig, Szabo, Moore and Simmons, 1991; Campbell, 1986; Moore and

Moore, 1984; Moore, 1987) used both tide gauge records and radiometrically dated drowned reef corals to argue that the Hawaiian islands have sunk too quickly for there to be marine deposits remaining above sea level. This belief that all marine platforms in Hawaii are quickly drowned, led them to suggest the coral-bearing gravels on Lanai had been thrown above sea level by tsunami waves. But these surmises fly in the face of nearly a hundred years of published data that has documented in-situ marine carbonate deposits high above modern sea level. More recent studies report evidence inconsistent with the GWH (Jones, 1993b; Grigg and Jones, 1997; Rubin, *et al.* 2000; Felton, Crook and Keating, 2000; Keating and Helsley, 2002 and Felton *et al.* 2006, and this paper).

Basic Tenets of the Giant Wave Hypothesis:

1. Tide gauge records and dated drowned reefs indicate island subsidence.
2. Shell fragments have been reported at 326 m and 183 m above sea level.
3. Soils were stripped from the island slopes.
4. Deposits consist of "locally continuous gravel beds."
5. Gravel beds consist of 95% basalt, 5% coral and beach rocks.
6. Boulders form a continuous formation that thins landward to vestiges at 326 m.
7. Corals are not in growth position.
8. A cemented layer at the base thought to be "encrusted".
9. The gravels are clast-supported deposits with thickness and size of clasts decreasing inland.
10. The gravels originally blanketed the region as a single wedge-shaped deposit.
11. The deposits correspond to ages around 110 kya.
12. The deposits were ripped up from a coral reef.
13. The lower layers of gravel contain coral clasts that represent upsurge. The upper layers contain angular clasts representing tsunami wave retreat.
14. Deposits are the result of giant waves triggered by a giant landslide.

Table 3.

Our field observations and other evidence from the literature suggest that these tenets are not supported by field evidence. The reasoning behind this statement is provided in the point by point discussion that follows.

4. Problems with the GWH

4.1. Subsidence

A basic tenet of the GWH is the contention that the island of Lanai was sinking too quickly for marine deposits to be preserved near sea level. Yet, elevated strand line related, sedimentary deposits are found on each of the main Hawaiian Islands, and described in the classic geologic texts describing the Hawaiian Islands (Macdonald and Abbot, 1970; MacDonald, Abbot and Peterson, 1983; Stearns, 1966, Stearns, 1985) as well as our own publications. The presence of numerous marine deposits throughout the Hawaiian island chain refutes this basic tenet.

4.2. Marine fossils at 326 m

The GWH attributes observations of coral and shell fragments at high elevations on Lanai to Harold Stearns but ignores Stearns's evidence for ancient shorelines preserved in the GWH inundation zone. Instead the GWH theorizes that tsunami wave run up was responsible for fossils at 326 masl. Jim Moore (P. Comm., 1997) reports that he was unable to locate the Stearns' fossil outcrop. Subsequently numerous scientists (including: Grigg, Jones, Walter, Gavenda, Felton and Crook) concentrated a great effort looking for the reported 326 m limestone outcrop without success. Keating and Helsley (2002) used Stearns's field maps and notebooks to identify the Stearns' outcrops. But, these geologists failed to find marine limestones at the 326 m elevation. Instead, the authors observed sub-aerially deposited caliche vein-fillings (Fig. 10b). Caliches are cements that fill cracks in existing rocks associated with the wetting and drying of rocks within an arid environments (Fig. 10a &b). The caliche results from meteoric waters leaching minerals from the rocks and the precipitation of these minerals at rock surfaces as the waters evaporate. This deposition is subaerial not submarine (Thomas, 1994). With rainfall on Lanai less than 25 cm/yr (Ziegler, 1986) caliche cements form extensively along the southern arid coasts. Samples were collected of vein-fillings between 326 and 200 m at Kaluakapo Crater and the samples were examined for fossils by micropaleontologist J. Resig who found no evidence of biological structures in any of the samples. Thus, despite extended research (involving on the order of ten man-days) no evidence exists to verify the presence of any tsunami-related marine deposits at 326 m on Lanai.

There are root casts in these caliche deposits at Stearns Swale that may have been thought to be fossils, but it is more likely that a bag of Stearns' samples became mislabeled and thus a misidentification of material has occurred. On the same day Stearns visited a locality at approximately 150 m elevation to the east of the 326 m site that is known to be quite fossiliferous. His notes indicate that he made extensive fossil collections that included corals, pelecypods, and gastropods. In our view, it is likely that some of these samples became mixed or mislabeled with the caliche samples collected at 326 m. A subsequent study that bears on this topic will be raised in the discussion (Crook and Felton, 2008).

4.3. Soil stripping and scouring

A basic tenet of the GW Hypothesis is that soils were stripped from the island, "at an altitude of 365 m, is an irregular boundary below which the thick red soil typical of the uplands has been removed, presumably by wave erosion associated with deposition of the Hulopoe Gravel."

Figure 10. (Top) Caliches are common throughout southern Lanai. Caliche deposits from the Stearn's Swale site at 326 masl, were proposed as the upper limit of tsunami deposits in the GWH. Fig. 10b. The photograph at bottom was taken on the margin of a dry gully near the shore in the Poopoo-Anapuka area.

Soil scientists (Foote, Hill, Nakamura, and Stephens, 1972) in fact described the southern slopes of Lanai as very stony ground. During repeated visits to Lanai, both red and black soils were found within the hypothesized giant tsunami inundation zone (Keating, 1997). With the assistance of Robert Gavenda (U. S. Department of Agriculture) test pits were dug on interfluves along the south coast of Lanai. Soils of up to 1.2 m thickness were observed (Fig. 11a-e). A road cut on the interfluv (the area between dry gullies) adjacent to the type section shows 50 cm of soil. Natural outcrops reveal roughly 50 cm of black soil exposed at the Kaunolu Archeological site on southwest Lanai. In tens of archaeological test pits dug throughout the Manele/Hulopoe Bay area, 70 cm of soil were found lying directly on basaltic bedrock (Kaschko, 1991). These same black, montmorillonite-rich soils were also found in excavations for foundations at the Manele Bay Resort (Cacalda, 2000). Other significant deposits of the highly expansive soils are present throughout the proposed tsunami inundation area. Soil scientists report that black soils are commonly found in coastal zone deposits around the Hawaiian Islands a result of coastal emergence from the sea. These soils expand when wet and then shrink when dried, leading to serious toppled walls and damaged concrete foundations. The expansion and shrinking of the soils works rocks upward from the underlying substrate until the rocks are exposed at the surface. Subsurface coring shows no loose rocks remain in the soils but instead, the loose rocks occur at the surface as an one-clast thick layer, thus classified as very stony ground. The unusual rock studded surface was misinterpreted as tsunami deposits.

There is an abundance of fine-grained material within the study area. Fig. 12 shows an example of the suspension of fine grain soil washed off the island after a winter storm. Also, ash deposits have been found in the proposed inundation zone along the Naha road (Fig. 13a). It is surprising that the ash would not be eroded away by "Giant waves."

Scour

A compilation of the characteristics of tsunami deposits was published as a data base (TDDB) by Keating *et al.* (2008). The compilation shows that during the drain back phase of tsunami waves, tsunami waves do strip coastal sediments (particularly sands) and they scour the pre-existing drainages. This observation is inconsistent with the GWH. On Lanai the Hulopoe and Kapihua gullies were filled rather than scoured.

4.4. Locally continuous gravel beds

The GWH describes locally continuous gravel beds. We observed numerous locally continuous gravel beds that are commonly associated with strandline carbonate deposits. We did not observe a continuous gravel sheet.

4.5. Lithostratigraphy

The original GWH publication (1984) described a lithologic section along the major gulch which drains into the northernmost extent of Kapihua Bay, directly west of Hulopoe Bay, about 200 m from the shore. The series of beds were originally described as follows: a bed 5

Figure 11. Soils are common throughout the GWH type field. Fig. 11A (Top left). Photograph of an example of a fossil soil in Kapihua Gully showing soil overlain by carbonate fragment gravel. These soils at the base of the gravels are approximately 0.3 m thick. Fig. 11b (Top right) Geologists investigate bed forms situated off the west side of the Manele Bay Road and above the Manele Bay Resort. There, lava boulders were stacked together to form small basins that were filled with the black-brown soil. This is the traditional method of growing sweet potatoes by native Hawaiians and is most likely a cultural artefact rather than a 'bedform'. Fig. 11c (Lower and middle left). Digging through the rock-strewn surface of the GWH type area reveals a thick soil layer beneath it. The lower left photograph shows a hole in the expansive soils (roughly 0.3 m deep) extending down to basaltic rocks. The rocks have been worked upward by the shrinking and swelling of the soils under dry and wet weather conditions. The surface of the expansive soils is now covered by the loose boulders derived from the lava flows below, and the soils are free of loose expelled rocks. Fig. 11d (Lower right) A geologist points out the thickness of the soils in the GWH type area, approximately 0.5 m thick.

Figure 12. Rain showers from the day prior, washed abundant soils off shore at Hulopoe gully, adjacent to the Manele Bay Resort Hotel. Clearly, the soils forming the light colour plume in the ocean off the Manele Bay area were not stripped by extreme tsunami waves. They are instead present and are being eroded by normal geological processes.

m thick overlying nonweathered basalt consisting of 2 layers: a lower layer of subrounded to rounded clasts of basalt (about 95%) and limestone (about 5%) representing the upsurge of the wave and an upper layer of sub-angular to angular clasts of basalt ranging from 20 cm to 1.5 m assigned as debris from the drain back of the same wave." Four years later, these original descriptions were revised with the description of the Hulopoe Gravel section calling for three beds (a lower bed of 2m, an intermediate bed of 4 m, and an upper bed of 2 m), with the limestone clasts mainly confined to the lower third of the three beds. The publication states, "in the limestone-free upper part of the bed, which is distinctly bimodal in size distribution, basalt boulders are enclosed in a silty pebbly matrix that in places contains abundant marine debris..."

Felton *et al.* (2000) published detailed lithological descriptions of the section. Thirteen beds were documented at this location showing the limestone-to-basalt clast ratio varying significantly from bed to bed (from roughly 5-40% limestone). The initial lithological description of the Lanai section is poorly representative of the southern Lanai rock unit. These rock units are considered to represent a sequence of marine gravels and non-marine soils by Felton et al. (2006) and that paper contains detailed descriptions of the rocks and environments of deposition. Most likely, these deposits represent subaerial low stand deposits.

4.6. Boulders form a continuous formation that thins landward to vestiges at 326 m elevation

Our observations do not support a continuous and thinning boulder bed. Patches of well-rounded beach boulders are present to an elevation of 190 m (see Fig. 5c). Locally derived quasi-rounded boulders are present throughout the slopes on lanai, irrespective of elevation. Other than the boulder beds associated with strand line carbonates, the boulders rarely occur in concentrations. We conclude that the isolated boulders are ubiquitous and are boulders formed by chemical weathering of the basaltic substrate (see Fig 6).

4.7. No corals in growth position

The GWH publications indicate that the Hulopoe Gravels unit contains no corals in growth position. Indeed, most corals do not occur in growth position, but A. Kay (P. Comm., 1998) has found assemblages of mollusks that display in-situ preservation and indicate a water depth of 80 m. If the Hulopoe gravels were deposited along a highly energetic rock-bound coast like the modern one, even large boulders would be moved by storm waves, so corals growing on these boulders would not be found in growth positions. But biological assemblages, particularly micro-mollusks living within sheltered patches between boulders, are preserved in growth position, and an isolated site reported here showed single corals growing one on the top of another, suggesting that remnants of a short-lived embayment remain.

4.8. Encrusted base

The GWH states that a cemented layer at the base of the Hulopoe gravel is an "encrusted" surface (Fig. 13b). We have observed this layer and identify it as caliche, a subaerial deposit (formed after the gravels were deposited).

Figure 13. (Left) Fine grain ash deposits are present within the inundation area, on the Naha Road, west of Kaluakapu Crater. Fig. 13b. The base of the gravel unit in Kapihua gully is marked by caliche.

4.9. Clast-supported deposits

The GWH publications describe clast-supported gravels in the Hulopoe gully. The argument was made that clast-supported boulders were water-laid, not transported in a submarine debris flow (which would produce matrix-supported debris), the implication being that the powerful tsunami waves had removed the finer grain material leaving only large boulders, free of fine-grained material. But the deposits exposed in gullies only 100 m and 400 m west of the type section do have sand and clay size components clearly present. Modern boulder deposits in coves along the modern shore lack often lack sand filing yet are clearly storm derived deposits. This is not a diagnostic feature.

Sorting and grain size

The GWH contends that the original thickness of the gravels and size of clasts decrease systematically with distance and elevation above sea level. But extended field observations upslope (approximately 60 m upslope of the Hulopoe section and up to 200 m elsewhere), show that boulders up to 1m can be found in many places. There is no size trend with either elevation or distance from shore. All of our observations suggest that the modern shore line assemblage is characteristic of these upland boulder deposits. Observations in the Poopoo and Anapuka drainage gullies also confirm that large boulders are found upslope (particularly around 100 m) that do not conform to the size/elevation distribution described by the GWH. Boulder pavements are associated with each of our 'strandlines' and are present at the modern shore. Also, we have frequently observed boulders of up to 1 m diameter, but no mega-boulders (over 3 m) were found. Had a giant tsunami taken place, it would be expected that mega-boulders would occur somewhere along the southern coast of Lanai.

4.10. Wedge-shaped unit

The GWH indicates that the coral-bearing gravels originally blanketed the region as a continuous wedge-shaped deposit. Instead, horizontal fossil-bearing notches are exposed in gullies in SW Lanai (this publication) that occur in a "bathtub ring" fashion. These notches and coral-rich deposits extend for kilometers along slope with the interfluves between the gullies barren of fossils. The deposits have the appearance of "high stand deposits" described elsewhere in the geologic literature and thus are contradictory to a proposed wedge-shaped gravel unit.

Inside Kaluakapo Crater (east of Manele Bay), Keating and Helsley (2002) found: no in situ fossils between 190-326 m, the absence of weathered rounded boulders (i.e., lacking weathering rinds) between 190-365 m, the preservation of an essentially undisturbed boulder and coral platform deposit at 190 m, the presence of marine deposits confined to limited stratigraphic intervals with a systematic internal stratigraphy at 170 m, the presence of both fine-grained and coarse material in distinct stratigraphic relationship in Kaluakapo Crater, and the presence of fragile fossils in deposits at 170m, and sea stacks preserved All of these geologic observations conflict with the notion of a wedge-shaped tsunami debris unit.

4.11. Ages

Uranium-series dating methods were used for dating rocks ages on Lanai reported by Moore and Moore (1984), Rubin *et al.* (1995), and Bryan, et al. (1997). Reported age clusters around 100 kya and 200 kya. (The rocks from Molokai had older ages, clustering around 200 kya and 350 kya, according to Jones and Grigg, 1997).

In 1988, the GWH theorized that the coral and basalt units were formed by a tsunami generated by the submarine collapse of the margin of Hawaii. Later, with an established age of roughly 105 kya from the coral clasts, the source of the tsunami was suggested to be the Alika landslide on the west coast of Hawaii, believed to have a corresponding age. During a 1997 Geological Society of America field trip to the outcrops, Jim Moore (1997, P. Comm.) concluded the radiometric ages indicated there had been more than one tsunami event on Lanai.

Numerous investigators have published radiometric-dating results for Hawaiian coastal deposits. The radiometric dating includes: 13 different techniques described in roughly 40 publications. The published ages for Lanai are incorporated in the plot in Fig. 7a.

Published radiometric analyses for other Hawaiian Islands have largely been concentrated on corals in-situ (i.e., growth position), with fewer analyses of marine conglomerates or sand units. But where the comparison of dates from the same site derived from in-situ coral versus dates of coral clasts extracted from coral and basalt conglomerates, reveals that no significant age difference exists between the two rock types (Muhs and Szabo, 1994). The deposits (Fig. 14) are lithologically and stratigraphically very similar to modern shoreline deposits (e.g. Queen's Beach on Oahu) and they are generally considered to be uplifted reef and coastal material derived from normal coastal processes (Keating, Whelan et al. 2004). On Oahu, marine conglomerates occur along the shore of Queens Beach, Makai Pier, and Waianae where sand and shell fragments fill the crevices between cracks in volcanic rocks.

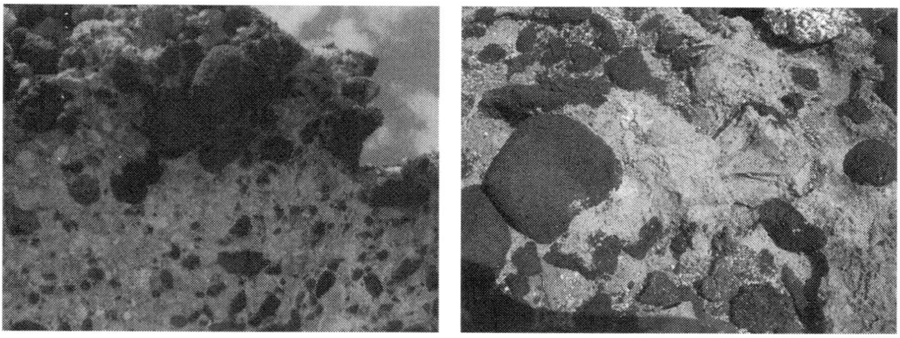

Figure 14. The GWH refers to the rocks as "gravels", but they are also referred to as marine conglomerates. The outcrop at left is from the island of Lanai, while the marine conglomerates at right are from the modern shoreline at Queen's Beach, Makapu'u, Oahu.

Like Lanai, age dating of sedimentary deposits at Waimanalo, S.E. Oahu, demonstrates that erosional notches were cut by wave activity at 7- 9 m (22 and 27 ft) above modern sea level at 120 kya. Sherman *et al.* (1993) report two distinct high stands on the Ewa Plain (SW Oahu) associated with the Waimanalo Formation, with ages correlated to Marine Isotope sub stage 5e. Studies reported by Stearns (1985); Muhs and Szabo (1994); Ku et al. (1974) and Szabo et al. (1994) prove that numerous outcrops of Waimanalo Limestone around Oahu have a similar age range. Radiometric dates of sediments from Kapapu Island, E. Oahu were reported by Grossman *et al.* (1989) and dates from an intertidal notch from Mohulua Island (Grossman and Fletcher, 1998) were in agreement with the Stearns (1935; 1978) high stand prior to 3889-3665 cal ya. These ages help constrain Oahu's long-term average uplift rate (0.03-0.07 mm/yr) based on Pleistocene age shorelines (Hearty, 2002). The coral bearing deposits at about 2 m elevation on Hawaii are reported to be a similar age to the Waimanalo Formation on Oahu (McMurtry, Fryer et al. 2004b) (While these latter authors suggest these deposits are associated with a submarine landslide generated tsunami (GWH), these rock ages are equivalents of the Waimanalo Formation exposed on other islands, consistent with the SLIP.)

Problems with radiometric dating

Hearty (2011) commented on the age of rocks in a publication (McMurtry, Campbell et al. 2011) and Reply (McMurtry, Campbell et al. 2011) and concludes, "their age data are flawed and lack supportive field and proxy evidence." Furthermore he writes, "such allochtonous cobbles have been emplaced by younger transgressions or tsunami any time after coral growth, but screening protocals (e.g. Mortlock *et al.,* 2005) should exclude Ko Olina and Lualualei. . ." See the publication of Sherman *et al.* (1999) that described marine and meteoric diagenesis from Oahu emergent sediment and the work of Oliver Chadwick and colleagues for pertinent studies on the island of Hawaii

Hearty concludes, "unsupportable interpretations such as theirs should not become embedded in the literature." Hearty writes, "ages as incontrovertibly unreliable due to excessive recrystallization and detrital Th greater than 1-2 ppb.... Lacking reliable ages and ESL's [Eustatic Sea Level], it is not possible to determine accurate uplift rates...."

Basing geologic models on single sample radiometric dates or age correlated with distant sites need also to be reconsidered. The original (1984) estimate of the age of the deposits within the GWH and the revised ages Moore and Moore (1988) are problematic. One age is from the island of Hawaii, with no supportive field evidence between the two islands. Other dates use less than ideal numbers of samples.

4.12. Source material

The GWH states that tsunami waves ripped up corals from an offshore reef as tsunami waves reached shallow depths.

Outcrops in gullies draining into Manele Bay and Hulopoe Bay contain individual specimens of corals, not aggregations of corals common on modern and ancient reefs, e.g., corals exposed in emergent outcrops on the Ewa plain of Oahu Island, Hawaii. Additionally, the maximum size of the coral clasts is small (a few to 20 cm across) compared to the basalt boulders. If a reef were present offshore, a tsunami could have moved coral clasts comparable to the size of the basalt boulders. Since it did not, it is concluded that the carbonate source material was composed of isolated corals of restricted size growing on boulders, like those observed in the modern shore and shallow offshore environment, as opposed to a developed reef complex.

4.13. Wave couplets

The GWH suggested that bedding couplets rich with coral clasts occur in their lower layers (representing upsurge of the wave) while the upper layers contain angular clasts (representing wave retreat). Detailed lithological descriptions are now available Felton, *et al.* (2000) and Felton *et al.* (2006) that show that rather than couplets, 14 beds were found along with 8 unconformities between beds.

Paleosols are identified at several boundaries within the gully-filling deposits proving a subaerial history for parts of the deposits. Root clasts are found as well as insect remains according to Resig (1999). At least one bed of the GWH locality is alluvial, i.e., a stream deposited unit of subaerial origin (Felton et al., 2006). These observations are inconsistent with the GWH. Had a giant tsunami taken place, it is extremely doubtful that insect remains would be preserved, since they are light enough to be blown away in the wind or washed away by rain.

4.14. Bed forms

The GWH describes branching dune-like gravel ridges (called bed forms) generally 1 m high. The deposits are situated in a location that was long occupied by native Hawaiians. In their preliminary survey of the area, archaeologist Steve Athens identified 182 archaeological structures. Ancient Hawaiians made extensive use of basalt boulders for construction in the Manele/Hulopoe Bay area including: platforms, grinding stones, game stones, hammer stones, hearths, rock walls, canoe sheds, cairns marking upland trails, cairns such as fishing shrines, holding pens, housing (oval terraces, rectangular shelters, curved wall enclosures), rock shelters, temporary fishing shelters, burial structures, boundary walls (Athens, 1991). The structures described as GW-derived bed forms (Fig. 15a-e) appear to be the remains of ancient anthropogenic structures. The photograph in GWH publication (1984, p. 1313, Fig. 4) even looks like an archaeological site illustrated by Emory (1924, Fig. 4).

Athens (1991) remarks on a description of the south and west coast of Lanai made by Captain King in 1785, a member of Captain Cook's expedition, who wrote, "the country to the south is high and craggy; but the other parts of the island had a better aspect and appeared to [be] well inhabited. We were told it produced very few plantains, and breadfruit trees; but that it abounds in roots such as yams, sweet potatoes and tarrow [taro]." Furthermore, Stearns (1940) wrote, "around the old native village sites especially at

Kaunolu and Manele are traces of rock terracing representing considerable industry. These terraces are reported to have been sweet potato gardens." Since the structures were built on expansive soils, they have toppled.

Figure 15. (Upper Left) the GWH describes bed forms in the Manele Bay Resort property. However, there are archaeological sites spread throughout the area that can be easily misinterpreted as bed forms. Athens (1991) described 182 structures built by native Hawaiians. Fig. 15b. (Upper right) The image shows a Hawaiian structure, while the rock walls have toppled, the rectangular shape of the foundation is still apparent. Fig. 15c. (Middle left) A drawing made by Emory (1924) of some of the numerous archaeological sites that dot the southern slope of Lanai (compare this drawing to Fig. 12b). Fig. 15d. (Lower left) The hillside above the Manele Bay Resort is marked by rock wall structures. They consist of rock walls laid in c-shaped configuration with the expansive black soils filling the structures to provide raised beds for growing sweet potatoes. Similar structures are in use in Hawaii currently. When the Captain Cook Expedition reached the Hawaiian island chain, the Lanai sweet potatoes were obtained for the expedition. Fig. 15e. (Lower right) Geologists examine the rock walls and soils still present in the growing beds.

4.15. Additional considerations

The GWH compares the postulated catastrophic tsunami flooding on Lanai to the catastrophic draining that took place when dammed glacial melt water of Lake Missoula (W. USA) broke through a topographic barrier and eroded basaltic terrain to produce the "scablands" terrain of western Oregon and Washington (Baker, 1973). (These scabland deposits were examined by the authors on several occasions.) The "scablands" surfaces, display extensive scarring and breakage of the boulders and cobbles and bedrock. In marked contrast, the Lanai deposits lack the scarring obvious in the scablands. The coral clasts instead display abrasion and breakage comparable to that of the modern boulder beach and are generally associated within the gravel or boulder deposit, that is, they are beach deposit, not giant wave deposits.

Furthermore, basalt boulders found below 200m m generally have a well rounded nature typical of beach boulder deposits. Basalt exposed on small cliffs generally have well-preserved irregular a'a surface textures as well as delicate surface structures (Fig. 16). If the boulders had been transported by tsunami waves, these delicate surface textures would likely have been destroyed or at least heavily scraped and scarred. A cone of red cinder sits at roughly 120 m elevation at the west side of the Manele Bay section. Felton et al. (2000) report that red cinders were found in a bed in the nearby gully. Given the extreme height, turbulent, and erosive nature of the proposed "giant waves", the red cinders would be expected to be distributed throughout the presumably tsunami-derived gravels of southern Lanai, yet they are not. The observation that this rock type is restricted indicates that its deposition is due to local erosional processes that segregate rocks rather than a giant catastrophic event that would widely distribute distinctive cinders.

Figure 16. A photograph of a highly vesicular basalt boulder shows delicate surface of lava preserved. The boulders in the GWH type area are not marked by heavy scratching causes by abrasion in a mass flow. They have not been heavily abraded similar to debris left by the flooding of glacial Lake Missoula.

Alignment of boulders has also been cited as evidence for tsunami. Nanayama *et al.* (1998) examined the deposits of the 1993 tsunami (with maximum wave run up of 8.6 m) on the Hirahama coast at Taisei town, southwest Hokkaido, Japan. They found that the imbrication of gravel was restricted to a small area of only a few meters indicating an outward (seaward flow) and beach drainage (the long axis of gravels inclined seaward). Dominey-Howes (1999) reported similar observations from the Mediterranean. There is no obvious imbrication observed in Lanai deposits (see .Fig. 17).

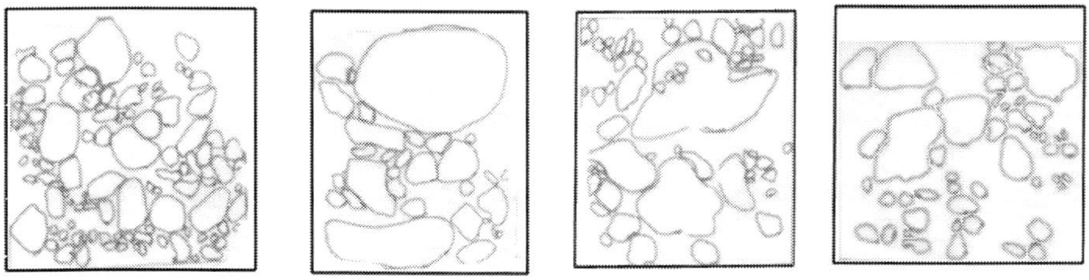

Figure 17. The figure above consists of a set of tracings of outcrop photographs, made to examine the alignment of clasts. The photographs were taken perpendicular to the outcrop so that the upslope direction is to the left and down slope direction is to the right, in the gully adjacent to Hulopoe Gully. In general there is little evidence of alignment. The images are roughly 1.5 m, in dimension. The grain size varies from boulders to pebbles in a caliche-encrusted matrix. The clasts range from well-rounded to angular shapes, and the largest clasts are roughly 0.3 m in diameter.

4.16. Erosion versus deposition

Shepard *et al.* (1950) documented the destructive effects of the 1946 tsunami in Hawaii. The photographs in that report show areas of maximum flooding marked by floating debris (including wooden buildings), large areas swept clean of debris other than sand deposited as a sheet, and eroded and fragmented corals and shells found along drainages and in poorly drained, low-elevation coastal areas. Sheppard *et al.* (1950) remark that some areas were so thoroughly swept clean of debris by the tsunami waves that they could be used to land an airplane.

The pattern seen in the historic Hawaiian tsunami is one of erosion with sporadic deposition of limited nature and breakage of corals and shells. This pattern of tsunami erosion along drainages is often reported in the literature (see, Keating et al, 2008; Keating *et al.*, 2011). This pattern is entirely different from the pattern suggested by the GWH. On Lanai, rather than seeing evidence of erosion sweeping the drainages clean, rocks are deposited in the drainages. If the southern slope of Lanai was the site of giant tsunami waves, it seems logical that the waves on the moderately steep slope would have preferentially scoured out existing drainages, rather than using them as "sediment traps" for the rock mobilized by wave action, as proposed by the GW hypothesis.

The southern coast of Lanai seldom receives rain, but when it does, it is often in the form of a winter storms with several inches or feet of rain fall within several hours to several days. The surface run-off from these storms can move large rocks significant distances downhill and erode soils. This normal process of reworking of rocks from higher to lower elevations has taken place throughout the history of the island and should be expected to produce a scattered pattern of ages but retain a pattern of increasing age with elevation. Also, the SLV during the last 400 kya has provided repeated opportunities for rock deposition (and erosion) at any given elevation within the Poopoo and Anapuka gullies.

High elevation fossils are also a problem. Crook and Felton (2008) report on the microscopic study of samples from Stearns Swale, a site situated at 326 m on the outside of Kaluakapo crater. They report "we found only veins of carbonate concentrated in cracks in lava outcrops....'" Despite numerous visits to the site by numerous scientists, no fossils have been found. And, all samples collected by Keating and Helsley from the site were studied by micro-paleontologist J. Resig (specialist in foraminifera) who found no biological structures in any of the roughly 30 samples collected.

Crook and Felton examined a sample collected by Harold Stearns during 1936 that was stored in the U.S. National Museum, in Washington, D.C. Stearns in 1936 (Stearns 1936) recorded in his notes, "Rode down into fault country between Manele and Kawaio [Naha side of Kaluakapu Crater] (by horseback). While riding thru here I notice [sic] some fossil limestone fragments but that they had been packed up here after riding past them I realized how weathered they were and that no one would pack [carry] them up so I went back and found plenty of loose fragments and two places where fossiliferous limestone fills cracks in the bedrock...."

In 1936, Stearns visited two sites on the same day. Since no fossils have been found at the swale by subsequent investigators, we question whether the sample examined by Crook and Felton (2008) was mislabeled when collected. This question arises because Helsley observed rocks of a similar nature to those in the Stearns collection at the locality that Stearns visited later that day. Further field work on the Naha side of the crater will be required to resolve this question.

Stearns wrote, "Henry says the old Hawaiians used this white (montmorillonite?) in the gulch east of the high coral locality for chalk." The only white material in this area of the swale is crack filling caliche, some having thicknesses of more than 3 cm. These caliches have root casts and other 'organic' looking structures but under the microscope, all such structures are inorganic in origin (as reported by Resig). This raises another concern; there is no gulch east of the Stearns Swale. The swale is located on the outside of Kaluakapo Crater, on the west flank of the crater. There is a small gully inside the crater but no white montmorillonite deposit. There is a significant deposit of white material in the gully to the east near the outcrops that Helsley thinks are the source of the fossils in the Stearns Collection at the Smithsonian Museum.

Crook and Felton (2008) also refer to disturbance of the Stearns site due to a military installation. Our understanding is that the high point served as an observation point, rather than any full-blown military installation. The disturbance seems to be limited to the construction of a jeep road into the swale area that terminates short of the observation post. This construction does not seem to impact areas beyond the jeep road but it is possible that the road destroyed the small outcrop that Stearns described. Also, the publication cites that Stearns mentioned that limestone would not have been carried up to the site. Keating raised the question to Manele Bay Resort game manager Gary Onuno. He reported that limestone was used in ceremonials when young Hawaiian boys came of age and that it was carried from the coastal area to the uplands, and it would not have been a miniscule amount. One such route from Manele bay to the interior passes by the swale that Stearns defines as the source of his samples.

Giant Wave Hypothesis Basic Tenets	**Our Observations**
Island Subsidence	Island Uplift and Sea Level Variation
Shells at 365 m and Shells at 183 masl	Multiple strandlines below 190 m. No fossils at 326 m. Marine Terrace at 190 masl
Soils Stripped	Highly expansive vertisols present having a thickness of 50 cm to 1 m.
Locally continuous gravel beds	Coral bearing strandlines and notches And gully filling gravels near sea level
95% basalt, 5% corals	Complex stratigraphy with four discontinuity bounded transgressive-regressive cycles with soils and insects marking subaerial intervals
Continuous deposits to 326 m	No carbonates above 190 m. Carbonates not continuous – tends to be elevation limited strandlines
Corals not in growth position, repositioned by tsunami inundation and retreat	Growth position corals are not expected in rocky shoreline littoral deposits – some evidence for in-growth position in biofacies and growth on substrate.
Encrusted layer	Caliche (post depositional)
Clast-supported with thickness and clast size decreasing inland.	Rocky shoreline deposits with wave cut notches in bathtub-ring configuration.
Single wedge-shaped deposit	Multiple strandlines representing both high and low stands
Ages around 110,000 ybp, then revised to include another event around 220,000 ybp	Multiple age strandlines with older material at higher elevations, also fossils of corals no longer living in Hawaiian waters occur at high elevations.
Source: coral reef	Shallow water transgressive strandlines and regression hiatuses and low stand deposits
Bimodal lithofacies representing tsunami upsurge and tsunami wave retreat	Complex stratigraphy representing multiple depositional events of high stand and low stand origin.
Disturbance resulting from giant waves	Multiple lines of evidence represent strandline origins

Table 4. Comparison of the basic tenets of the Giant Wave Hypothesis with the geologic features described in this publication.

5. Conclusion

Our observations on coral-bearing deposits on the south flank of Lanai are totally supportive of the hypothesis that these deposits represent uplifted strandlines formed during island uplift and glacial eustatic sea level variations documented elsewhere in the world. The formation of these uplifted coral-bearing deposits does not require a giant wave origin. Moreover, had a Giant Wave(s), of the sort proposed by others, occurred, the loose fragmented coral, sand, and boulder deposits characteristic of some of these strandlines would have been disturbed or destroyed and some of the transported material would have been deposited at higher elevations. Yet, no evidence of coral bearing deposits was found above 190 masl.

Erosion and boulder stripping is a normal process and all the geomorphic features described by the Giant Wave Hypothesis can be attributed to normal geologic processes and occasional but significant rainfall events, such as winter storms or hurricanes. The large number of geologic publications describing exposed carbonate units in the Hawaiian islands, the abundance of published radiometric dates, the consensus of scientists that these rocks represent deposition by normal coastal processes and the display of a pattern of increasing rock ages with elevation contradict a mega-tsunami origin for deposits on Lanai. An extreme catastrophic event is not necessary to explain these marine deposits. While the GW hypothesis is an exciting notion, the concept is not supported by field evidence on Lanai.

Author details

Barbara Keating Helsley and Charles.E. Helsley
University of Hawaii, SOEST, Honolulu, HI 96822, USA

Acknowledgement

Harold Stearns, Doak Cox, and Anne Felton were instrumental in intriguing Keating and Helsley, into examining the nature of sedimentary outcrops in Hawaii and eventually led to field studies on Oahu, Lanai, Molokai, Kauai, and Maui. The fieldwork has been greatly enhanced by the participation of Chuck Helsley and Anne Felton. These studies have benefited from conversations with many geologists involved in the study of the Hawaiian Islands. We thank our colleagues at the University of Hawaii (particularly Richard Grigg, Ken Rubin, Alison Key (deceased) Bob Gauldie (deceased) and Julie Bailey-Brock), USGS, and others for discussions that have improved this manuscript. The compilation of radiometric age dates was completed with the help of Matt Wanink and Franciska Whelan. This investigation and publication was supported by the University of Hawaii Vice Chancellor for Research, Sea Grant Project Development Funds, and personal funds.

6. References

Athens, J. S. (1991). *Archaeological Investigations at the Canoe Shed Complex of Hulopoe Bay, Lanai Site 85-40-98-85*; Report to the Lanai Co, Inc. Honolulu, International Archaeological Research Institute.

Baker, V. R. (1973). *Paleohydrology and sedimentology of lake Missoula flooding in eastern Washington*. Boulder, Co., Geol. Soc. Am.

Bruckner and Radtke, 1989, Fossile strande und korallenbanke auf Oahu, Hawaii. *Essener Geogr. Arb.* 17, 291-308.

Bryan, W. B., Ludwig, K. R., Moore, J. G. (1997). "U series ages of clasts in Pleistocene marine conglomerate, Lanai, Hawaii (abstract)." *Geol. Soc. America, Cordilleran section Meeting, Kailua-Kona, Hawaii,* Abstracts with Program.

Cacalda (2000). "Pers. Comm."

Campbell, J. F., (1986). Subsidence Rates for the Southeastern Hawaiian Islands Determined from Submerged Terraces. *Geo-marine Lett.,* 6 (1986), 139-46.

Carew, J. L., and Mylroie, J. E. (1995). "Quaternary Tectonic Stability of the Bahamian Archipelago: Evidence from Fossil Coral Reefs and Flank Margin Caves." *Quat. Sc. Rev.* 14(2): 145-153.

Crook, K. A. W. and Felton, E. A. (2008) Sedimentology of rocky shorelines 5: The marine samples at +326 m from 'Stearns swale' (Lanai, Hawaii) and their paleo-environmental and sedimentary process implications. *Sed. Geology.* 206, 33-41.

Emory, K. P. (1924). "The island of Lanai, a survey of native culture." *B. P. Bishop Museum Bull.* 12: 1-129.

Felton, E. A., Crook, K. A. W. and Keating, B. H. (2000) The Hulopoe Gravel, Lanai, Hawaii: New Sedimentological Data and Their Bearing on the Giant Wave (Mega-Tsunami) Emplacement Hypothesis. *Pure and Appl. Geophys.,* 157, 1257-1284.

Felton, E.A., Crook, K.A.W., Keating, B. H. and Kay, E.A. (2006). Sedimentology of Rocky Shorelines 4. Coarse Gravel Lithofacies, Molluscan Biofacies, and the Stratigraphic and Eustatic Records in the Type Area of the Pleistocene Hulopoe Gravel, Lanai, Hawaii. *Sed. Geology,* 184, 1-76.

Fletcher, C. H., III, and Jones, A. T. (1996). Sea-Level Highstand Recorded in Holocene Shoreline Deposits on Oahu, Hawaii. *J. Sed. Research A Sedimentary Petrology and Processes,* 66, no. 3 (1996), 632-41.

Fletcher, C. H., III, and Sherman, C. E. (1995). Submerged Shorelines on O'ahu, Hawaii: Archive of Episodic Transgression During the Deglaciation? *J. Coastal Research* Climate, Sea Levels and sedimentation. Special Issue 17 (1995), 141-52.

Foote, D. E., Hill, E. L., Nakamura, S., and Stephens, F. (1972). *Soil Survey of the Islands of Kauai, Oahu, Maui, Molokai, and Lanai, State of Hawaii.,* USDA- Soil Conservation Service.

Grigg, R. W. and Jones, A. T. (1997) Uplift caused bu lithospheric flexure in the Hawaiian Archipelago, as revealed by elevated coral deposits, *Mar. Geology*, 141, 11-25.

Grigg, E. W., Grossman, E. F., Earle, Gittings, Lott, McDonough, (2002) Drowned reefs and antecedent Karst topography, Au'au Channel, S. E. Hawaiian Islands, *Coral Reefs*, 21, 73-82.

Grossman, E. F., Fletcher, C. H., III, and Richmond, B. M. (1998). The Holocene Sea-Level Highstand in the Equatorial Pacific Analysis of the Insular Paleosea-Level Database. *Coral Reefs,* 17 (1998), 309-27.

Grossman, E. F., and Fletcher, C. H., (1988) Sea level higher than present 3,500 yrs ago on the northern main Hawaiian Islands. *Geology,* 26: 363-366.

Hearty, P. (2002). The K'eana Highstand of O'ahu, Hawai'i: Further Evidence of Antarctic Ice Collapse during the Middle Pleistocene, *Pacific Science,* 56 (2002), 65-81.

Hearty, P. J. (2011). Uplift of Oahu, Hawaii, during the past 500 k.y. as recorded by elevated reef deposits: Comment. *Geology,* 39, 3, e234-e235.

Hearty, Paul J., Kaufman, D. S., Olson, L. Storrs and Helen, F. J. (2000). Stratigraphy and Whole Rock Amino Acid Geochronology of Key Holocene and Last Interglacial Carbonate Deposits in the Hawaiian Islands. *Pacific Science,* 54, no. 4 (2000), 423-42.

Johnson, C. & Mader, C., 1994, Modeling the 105 Ka Lanai Tsunmi, *Sc. Tsun. Haz.,* 11, 33-38.

Jones, A. T. (1995). Geochronology of Drowned Hawaiian Coral Reefs. *Sed. Geology,* 99, no. 3/4 (1995), 233-42.

Jones, A. T. (1993a). Review of the Chronology of Marine Terraces in the Hawaiian Archipelago. *Quaternary Science Reviews* 12 (1993), 811-23.

Jones, A. T. (1993b). Elevated Fossil Coral Deposits in the Hawaiian Islands: A Measure of Island Uplift in the Quaternary, *Ph.D. Dissertation,* University of Hawaii, 274 pp.

Jones, A. T. and Mader, C. (1995) Modeling of tsunami propagation directed at wave erosion on southeastern Australia coast 105,000 years ago., *Sc. Tsun. Hazards,* 13 (1) 45-52.

Kaschko, M. W. (1991). *Archaeological Test Excavations and Site Mapping for the Manele Multi-family Residental Developement Area, Island of Lanai, Hawaii.* Report for M & E Pacific, Inc., pp 1-58. Honolulu, International Archaeological Research Institute, Inc.

Keating, B.. H., (1997) Are the Coastal gravels on Lanai tsunami deposits? *Abstracts with Program, Geological Soc. Am.,* Cordillerian Section, Kona, 1997.

Keating, B. H. and Helsley, C. E. (2002). The Ancient Shorelines of Lanai, Hawaii, Revisited. *Sed. Geology,* 150 (2002), 3-15.

Keating, B. H., Helsley, C. E., Wanink, M., and Walker, D. (2011) Tsunami Deposit Research: Fidelity of the Tsunami Record, Ephermal Nature, Tsunami Deposit Characteristics, Remobilization of Sediment by Later Waves, and Boulder Movements, In *The Tsunami Threat, Research and Technology* (ed. Nils-Axel Morner), Intech Open, pp. 389-422

Keating, B. H., F. Whelan, M. Wanink, and Businger, S. (2004) Queen's Beach and Sandy Beaches, S. E. Oahu, Comparison of Storm Deposits Versus Historic Tsunami Deposits. *Science of Tsunami Haz.,* 22, 1, 23-39.

Keating, B. H., F. Whelan, et al. (2004). "Tsunami Deposits at Queen's Beach, Oahu, Hawaii - Initial Resukts and Wave Modeling." Sc. Tsunami Hazards 22(1): 33-43.

Ku, T. L., M. A. Kimmel, M. A., Easton, W. E. H. and O'Neil, T. J. (1974). Eustatic Sea Level 120,000 Years Ago on Oahu, Hawaii. *Science,* 183, 959.

Ludwig, K. R., Szabo, B. J., Moore, J. G., and Simmons, K. R. (1991). Crustal Subsidence Off Hawaii for the Last 450 Ka Determined by 234U/238U Ages of Drowned Coral Reefs. *Geology,* 19 (1991), 171-74.

Macdonald, G. A. a. A., A. T. (1970). *Volcanoes in the Sea.* Honolulu, University of Hawaii Press.

Macdonald, G. A., A. T. Abbot, and Peterson, F. L. (1983). *Volcanoes in the Sea,* Second Edition, University of Hawaii press, Honolulu, HI

McMurtry, G., Fryer, G. J., Tappin, D. R., Wilkinson, I. P., Williams, M., Fietzke, J., Garbe-Schoenberg, D., and Watts, P. (2004) Megatsunami deposit on Kohala volcano, Hawaii, from flank collapse of Mauna Loa, *Geology*, 32, 9, 741-744.

McMurtry, G. M., Watts, P., Fryer, G. J., Smith, J. R. and Imamura, F. (2004) Giant landslides, mega-tsunami, and Paleo-sea level in the Hawaiian Islands. *Marine Geology*, 203, 219-233.

McMurtry, G. M., J. F. Campbell, et al. (2011). "Uplift of Oahu, Hawaii, during the past 500 k.y. as recorded by elevated reef deposits: Reply." Geology: e236-e237. .

McMurtry, G. M., G. J. Fryer, et al. (2004b). "Megatsunami deposits on Kohala volcano, Hawaii, from flank collapse of Mauna Loa." Geology 226: 741-744. .

Moore, G. W. and Moore, J. G. (1988). Large Scale Bedforms in Boulder Gravel Produced by Giant Waves in Hawaii. *Geological Society of America Special Paper*, 229 (1988), 101-10.

Moore, J. G. (1987). Subsidence of the Hawaiian Ridge, *Volcanism in Hawaii, U. S. Geol. Soc. Prof. Pap.*, 1350 (1987), 85-100.

Moore, J. G. and Campbell, F. (1987). Age of Tilted Reefs, Hawaii. *J. Geophys. Res.*, 92, no. B3 (1987), 2641-2636.

Moore, J. G., and Moore, G. W. (1984). Deposit from a Giant Wave on the Island of Lanai, Hawaii. *Science*, 226 (1984) 1312-15.

Moore, J. G., Bryan, W. B., and Ludrig, K. R. (1994). "Chaotic deposition by a giant wave, Molokai, Hawaii." *Geol. Soc. Am. 106*: 962-967.

Muhs, D. R., and Szabo, B. J. (1994). New Uranium-Series Ages of the Waimanalo Limestone, Oahu, Hawaii: Implications for Sea Level During the Last Interglacial Period. *Marine Geology*, 118, no. 3-4 (1994), 315-26.

Resig, J. (1999). "Pers. comm.".

Rubin, K. H., Sherman, C. E., and Fletcher, C. H., III (1995). "Ages of emerged coral deposits in Kaihua Gulch. Lanai, Hawaiian Islands and speculation about their environment of deposition (abstract)." *Geol. Soc. Am. Cordilleran Sections Meeting Abstracts* (76): F307.

Rubin , K.H., Fletcher, C. H., Sherman, C. (2000) Fossiliferous Lanai deposits formed by multiple events rather than a single giant tsunami. *Nature*, 408, 675-681.

Sherman C. E., Glenn, C. R., Jones, A. T., Burnett, and W. C. Schwarcz, H. P. (1993) New evidence for two highstands of the sea during the last interglacial, oxygen isotope substage 5e. *Geology*, 21, 1079-1082.

Sherman , C., Fletcher, C. H. and Rubin, K. (1999) Marine and meteoric diagenesis of Pleistocene carbonates from a nearshore submarine terraces, Oahu, Hawaii, *J. Sed. Res.*, 69, 1083-1097.

Shackleton, N. J. (2000) The 100,000 year ice-age cycle identified and found to lag temperature, Carbon Dioxide and Orbital Eccentricity, *Science*, 289, 1897-1901.

Smith, J. R., and Wessel, P. (2000). Isostatic Consequences of Giant Landslides on the Hawaiian Ridge. *Pure and Applied Geophys.*, 157, no. 6/7/8 (2000), 1097-114.

Stearns, H. T. (1970). Ages of Dunes on Oahu, Hawaii. *B. P. Bishop Museum Occ. Papers*, 24, no. 4 (1970), 49-72.

Stearns, H. T. (1975). PCA 25-Ft Stand of the Sea on Oahu, Hawaii. *Geol. Soc. Am. Bull.*, 86 (1975), 1279-80.

Stearns, H. T. (1985) *Geology of the State of Hawaii,* Pacific Books, Palo Alto, Ca, 266 pp.

Stearns, H. T. (1935). Pleistocene Shorelines on the Islands of Oahu and Maui, Hawaii. *Geol. Soc. Am. Bull.,* 46, no. 10 (1935), 1927-56.

Stearns, Harold T. (1939). Geologic Map and Guide of the Island of Oahu, Hawaii. *Division of Hydrology, Territory of Hawaii Bulletin,* 2 (1939), 75 pp.

Stearns, H.T. (1938). Ancient shorelines on the island of Lanai, Hawaii. *Geological Society of America, Bulletin* 49, 615– 628.

Stearns, H. T. (1940). "Geology and ground-water resources of the islands of Lanai and Kahoolawe, Hawaii [including a Geolgoic Map of Lanai]." *Hawaii Division of Hydrography Bulletin* 6: 1-177.

Stearns, H. T. (1966). Geology of the State of Hawaii. Palo Alto, Calif., Pacific Books.

Stearns, H. T. (1978). Quaternary Shorelines in the Hawaiian Islands. *Bernice P. Bishop Mus. Bull.,* 237 (1978) 57.

Stearns, H. T. (1936). Srearns Field Book. U.S. Geol. Survey Field Office Library Honolulu, HI. 17-2: 104-131 (Copy of original)

Tanaka, K., Hataya, R., Spooner, N. A., Questiaus, D. G., Saito, Y., Hashimoto, T. (1997). "Dating of Marine Terrace Sediments by ESR, TL, OSL Methods and their Applicability." *Quat. Sc. Rev.* 16(3-5): 257-264.

Tanaka, K., Toscano, M. A. and J. Lundberg (1999). "Submerged Late Pleistocene reefs on the tectonically-stable S. E. Florida margin: high-precision geochronology, stratigraphy, resolution of Substage 5a sea-level elevation and orbital forcing." *Quat. Sc. Rev.* 18(6): 753-767.

Szabo, B. J., K. R. Ludwig, D. R. Muhs, and K. R. Simmons. "Thorium-230 Ages of Corals and Duration of the Last Interglacial Sea-Level High Stand on Oahu, Hawaii." *Science,* 266 (1994): 93-96.

Thomas, M. F. (1994). *Geomorphology in the Tropics.* New York, John Wiley & Sons.Waelbroeck, C., Labeyrie, L., Michel, E., Duplessy, J. C., McManus, J. F., Lambeck, K., Balbon, E., Labracherie, M. (2002). "Sea-level and deep water temperature changes from benthis foraminifera isotope records." *Quat. Sc. Rev.* 21: 295-305.

Watts, A. B., and ten Brink, U. S. (1989). Coastal Structure, Flexure, and Subsidence History of the Hawaiian Islands. *J. Geophys. Res.,* 94 (1989), 10,473-10500.

Waelbroeck, C., Labeyrie, L., Michel, E., Duplessy, J. C., McManus, J. F., Lambeck, K., Balbon, E., Labracherie, M. (2002). "Sea-level and deep water temperature changes from benthis foraminifera isotope records." Quat. Sc. Rev. 21: 295-305.

Wessel, P., and Keating, B. H. (1994). Temporal Variations of Flexural Deformation in Hawaii, *J. Geophys. Res.,* 99 (1994), 2747-2756.

NEW PERSPECTIVES IN FLUID DYNAMICS

Edited by **Chaoqun Liu**

New Perspectives in Fluid Dynamics

http://dx.doi.org/10.5772/59432
Edited by Chaoqun Liu

Contributors

Jorge Flores, José Simão Antunes Do Carmo, Daniel Marinho, Pedro Forte, Tiago Barbosa, Medzid Muhasilovic, Chaoqun Liu

CBS, Edition **2017**

Published by InTech

Janeza Trdine 9, 51000 Rijeka, Croatia

Publishing Process Manager Iva Lipovic
Technical Editor InTech DTP team
Cover InTech Design team

Additional hard copies can be obtained from orders@intechopen.com

New Perspectives in Fluid Dynamics, Edited by Chaoqun Liu
p. cm.
ISBN 978-953-51-2228-9

INTECH

open science | open minds

Contents

Preface

Although fluid dynamics, a classical discipline, has been studied for many centuries, there are many questions still that remain unanswered, especially on turbulent flow. With significant advances in computational technology and decrease in hardware costs, a rapid evolution in computational fluid dynamics (CFD) techniques and models has been noticed. Many industrial engineering sectors are also making a concerted effort in the development and advancement of fluid dynamics. With the increasing demand for research, I believe the future for the advancement of fluid dynamics could not be brighter. Thus, the purpose of this book is to invite the leading researchers engaged in various research areas, relating to fluid dynamics, to come together and present their recent achievements and invoke further discussions. This book will mainly focus on the individual expert's experiences and discoveries. Although we realize that a single chapter with selected topics cannot provide exhaustive representations, it should give readers great insight into experts' accomplishments.

This book contains five chapters. The new turbulence theory and the advances and wide applications of CFD are described in each of the chapters as follows.

In the first chapter, "Self-Contradictions of Current Turbulence Theory and Liu's New Turbulence Generation Theory", Dr. Liu and Dr. Chern give a critical review of the classical and current turbulence theories and point out many self-contradictions existing in the classical and current turbulence theories. In this chapter, the authors provide more than 10 topics about the misconceptions created of the classical and current turbulence theories, especially baffling with two totally different concepts, vortex and vorticity tubes. Turbulence is one of the most important but unresolved problems in the modern fluid dynamics. Mathematically, one counter example can overthrow a theory without the need of a second example. However, classical and current turbulence theories are not only filled with one but many self-contradictions. Therefore, we really do not have the right turbulence theory. The transition community believes that turbulence is generated by "vortex breakdown,", but the turbulence community believes that there are coherent structures in the fully developed turbulence after vortex breakdown. In other words, studying the structure of a house after its collapse is meaningless. There are many other such contradictions. This chapter introduces a new turbulence generation theory given by Liu. According to Liu, (i) flow transition is not a process of vortex breakdown but turbulence vortex structure buildup; (ii) the nature of turbulence generation is that fluids cannot tolerate high shear, and shear must be transferred to the rotation and form a very fast rotation core; (iii) "shear layer instability" is the "mother of turbulence"; (iv) small-scale turbulence is generated by multi-level shear layers that are generated by multi-level sweeps, ejections, and negative and positive spikes; and (v) a large vortex provides energy to smaller vortices through fast rotation, which causes strong sweeps. According to Liu, the nature of the flow transition is mainly due to vorticity roll-up from the wall. The flow transition is vorticity redistribution and increment from near wall to whole boundary layer. In addition, flow transition is a process of non-rotational vorticity transferring to rotational vorticity.

In the second chapter, "Modeling of Wave Propagation from Arbitrary Depths to Shallow Waters: A Review", José Simão Antunes do Carmo points out that numerical models are a useful instrument for studying the complex superposition of wave–wave and wave–current interactions in coastal and estuarine regions and investigating the interaction of waves with complex bathymetries or structures built in nearshore areas. Moreover, since their applications are significantly less expensive and more flexible than the construction of physical models, they form a convenient tool to support design. The ability of the standard Boussinesq and Serre or Green and Naghdi equations to reproduce these nonlinear processes is well known. However, these models are restricted to shallow water conditions, and the addition of other terms of dispersive origin has been considered since the 1990s, particularly for approximations of the Boussinesq type. To allow applications covering a greater range of area, other than shallow waters, (where the water depth is at rest and so is the wavelength), a new set of Serre-type equations, with additional terms of dispersive origin, is developed and tested with the available data and with a numerical solution of a Boussinesq-type model, also improved with dispersive characteristics. Explicit and implicit methods of finite difference are implemented to solve both approximations of Boussinesq and Serre- types with improved dispersive characteristics. A finite element method is also implemented to solve an extension of a Boussinesq-type model that takes into account the wave–current interactions. Application examples to solve real-world problems are shown and discussed. The performances of 1HD models are compared with experimental data of very demanding applications, namely (i) a highly nonlinear solitary wave propagating up a slope and reflecting from a vertical wall and (ii) a periodic wave propagating in intermediate-depth waters upstream a trapezoidal bar, followed by very shallow waters over the bar, and again in intermediate-depth water conditions downstream.

In the third chapter, "CFD-Based Investigation of Wind-Strokes over Highway Bridge Section", Dr. Medzid Muhasilovic (M.Sc.), Kenan Imsirpasic (M.Sc), Prof. Dr. Karel Ciahotny, and Prof. Dr. Brano Sirok point out that there is an almost everlasting debate on the possibilities of investigation tools for their applications in a prospective fashion, while solving the engineering tasks, and against the fact that there is use of these engineering tools, while correcting the existing technological problems. A unique chance to compare these tools of research in traffic intentions, while setting the modern road communications through southern Bosnia and Herzegovina, with the natural occurrences in the atmosphere (such as a strong north wind in this geographic region), offers the highway section of Pocitelj-Zvirovici. In such cases (and before the actual construction of this highway bridge), "for the sake" of prospective engineering, the CFD mechanism (the "tool kit" for performing computational fluid dynamics) was applied to solve this atmospheric problem. Both in steady-state explorations (while applying the k-ε turbulence treatment) and in the time-dependent CFD-based mode, the authors explored the wind strokes at 10 m/s, 20 m/s, 30 m/s, and 40 m/s, an expected but certainly unwanted strong gaseous flows over the bridge, detecting the traffic safety edge points. Results obtained from the explorations of the CFD tool are explained and discussed in this chapter.

In the fourth chapter, "Computational Fluid Dynamics Achievements Applied to Optimal Crop Production in a Greenhouse", Jorge Flores-Velázquez, Abraham Rojano, Adriana Rojas-Rishor, and Waldo Ojeda Bustamante point out that computational fluid dynamics (CFD) has been successfully used in protected agriculture to simulate greenhouse weather as a physical process. The variables involved are velocity, wind direction related to absolute or relative humidity, temperature, as well as deficit vapor pressure and carbon dioxide, among others. The research evolution is changing from the traditional validation of new designs and management to the testing of efficient production with less environmental pollution. This study points out the assessment based on the physical principles of conservation of mass, momentum, and energy. Constitutive relationships like Darcy-Forchheimer porosity model in the momentum equation as well as the geometrical and physical properties of the materials involved are needed to fulfill particular issues of temperature, wind, and humidity. As a result, this study is enhanced by the effect of solar radiation in processes such as crop transpiration with dynamical meshes and condensation. Most of the processes that occur in a greenhouse are related to mass transfer of events (relative humidity and transpiration), momentum (velocity and species), and energy or exchange of heat (convection and radiation), and consequently the computational fluid dynamics is a robust tool for greenhouses and animal farm behavior simulations. The inclusion of culture and auxiliary climatic control systems is essential to the production process from an agricultural point of view. The results of this study indicate that once the models are experimentally validated, the use of these systems allows to know the spatial characteristics of the variables involved. From a mechanical point of view, the use of anti-insect screens and crop presence cause a drop of 80% in speed. A heating system for hot pipes under the atmospheric conditions of central Mexico is only necessary for specific periods before sunrise. A change in the position of the greenhouse windows in zigzag shape increases air movement and homogenized conditions inside, even if the same ventilation ratio is maintained, but a variation in the porosity of the mesh has side effects in the air renewal ratio. The prospect of computational fluid dynamics applied to the protected agriculture will focus on building a dynamic model with many variables to make an efficient use of resources and reduce the use of fossil fuels in crop production.

In the last chapter, "Technologic Appliance and Performance Concerns in Wheelchair Racing: Helping Paralympic Athletes to Excel,", Pedro Forte, Tiago M. Barbosa, and Daniel A. Marinho show that numerical simulations have provided useful evidence on helping several sportsmen to excel. This methodology enables a deeper understanding of the influence of equipment and sports techniques on sports performance. In wheelchair racing, technology was used without considering Olympic sports (once some of the Paralympics used the same technology of their Olympic counterparts) and induced unique changes in prosthetic and wheelchair devices. Eventually, technology has become an essential part of Paralympic sports, wheelchair racing being one of the most popular events. Numerical simulations can help us gather evidence on the effects of drag force, acting upon the athlete-chair system. Different types of wheelchairs are designed for racing (track and road races) and net and invasion sports. One of the various strategies to enhance performance is to minimize the aerodynamic drag of the frame, tires, helmet, sports outfit, and body posture. Numerical simulations can be used to predict fluid dynamics. The aim of this chapter is to review the state of the art on numerical simulations in wheelchair racing and suggest further studies. The chapter includes sections covering (i) the main determinants in wheelchair racing, (ii) the effect of aerodynamic force on performance in wheelchair racing, (iii) analytical models, experimental testing, and numerical simulations in wheelchair racing, (iv) numerical simulations on equipment and techniques. The wheelchair racing sports sciences lack research;

however, there is some information that could positively contribute to the performance increase in the athletes. While managing the mathematical models based on physics laws, coaches could identify the mechanical performance obstacles and try to minimize them. The drag reduction by the rolling resistance and air resistance access also contributes to the performance increase in the athletes. A better propulsion could also be obtained by reducing the total drag associated to a mass reduction and other possible aerodynamic positions. The purpose of the aerodynamic drag reduction in computational simulations and/or computer fluid dynamics should be an improvement in the winning time. No data pertaining to CFD tests in wheelchair racing were found. The stroke technique should also be focused on despite of the fact that there are indications on the number of hours of training that should be taken for an injury occurrence. It is stated that high angular velocities near the shoulder and the elbow, generated by the strokes, induce stress, increasing the risk for joint injuries. There are also some indications about the contact zones, force applied, and contact break at the rear wheels. Physiological variables should also be considered in the wheelchair racing athletes, which certainly lacks information regarding this sport. Thus, it is acceptable to say that a precise control of the total drag and an efficient use of the stroke technique allied to high levels of strength derived from the strength and physical training conditions will positively contribute to a better performance in the athletes.

I hope this book will be useful to scientists and engineers who are interested in fundamental fluid dynamics and its applications in engineering.

In conclusion, I would like to thank all the authors for their incredible contributions to this book and their patience in assisting us. Furthermore, I would like to acknowledge and thank the referees for their tiresome effort in making this book come to fruition. Last but not the least, I would like to thank my family, including Weilan Jin (my wife), Haiyan Liu (my daughter), and Haifeng Liu (my son), for their unconditional support.

Chaoqun Liu
Distinguished Scholar Professor
University of Texas at Arlington
Arlington, USA

Chapter 1

Self-Contradictions of Current Turbulence Theory and Liu's New Turbulence Generation Theory

Chaoqun Liu and Shuhyi Chen

Additional information is available at the end of the chapter

http://dx.doi.org/10.5772/61836

Abstract

Turbulence is one of the most important but unresolved problems in modern fluid dynamics. Mathematically, one counterexample can overthrow a theory and we really do not need a second example. However, classical and current turbulence theories are filled with not only one but many self-contradictions. Therefore, we really do not have the right turbulence theory. The transition community believes that turbulence is generated by "vortex breakdown," but the turbulence community believes there are coherent structures in fully developed turbulence after vortex breakdown. It is equivalent to say we study the structure of a house after the house collapses and breaks down. This is really ridiculous. There are many others. According to Liu, (1) flow transition is not a process of vortex breakdown but turbulence vortex structure buildup; (2) the nature of turbulence generation is that fluids cannot tolerate high shear and shear must transfer to rotation and form a very fast rotation core; (3) "shear layer instability" is the "mother of turbulence"; (4) turbulence small scales are generated by multiple-level shear layers that are generated by multiple level sweeps, ejections, negative and positive spikes; (5) large vortex provides energy to smaller vortices through fast rotation, which causes strong sweeps. According to Liu, the nature of the flow transition is mainly caused by vorticity rollup from the wall. Flow transition is vorticity redistribution and increment from near wall to whole boundary layer. In addition, flow transition is a process of non-rotational vorticity transferring to rotational vorticity.

Keywords: Vorticity, vortex, rotation, vortex buildup, shear layer, boundary layer transition

1. Introduction

Turbulence is still an unsolved scientific problem, which is not only important to science but also to industrial applications in aerospace engineering, mechanical engineering, energy engineering, bioengineering, and many others. Turbulence remains the most important unsolved problem of classical physics. Clearly, understanding of turbulence will help scientists and engineers cope with the broad range of turbulent flows. Nobel Prize winner Richard Feynman considered turbulence as "One of the most important unsolved problems of classical physics" [34]. Nobel Prize winner Werner Heisenberg said, "When I meet God, I am going to ask him two questions: why relativity? And why turbulence? I really believe he will have an answer for the first" [29]. These comments and addresses clearly show turbulence remains a top secret in nature and awaits more research activities.

Mathematically, one counterexample can overthrow a theory and we really do not need a second example. However, classical and current turbulence theories are filled with many self-contradictions. Therefore, we really do not have the right turbulence theory. The transition community believes that turbulence is generated by "vortex breakdown," but the turbulence community believes there are coherent structures in fully developed turbulence after vortex breakdown. It is equivalent to say we study the structure of a house after the house collapses and breaks down. This is really ridiculous. According to Liu, flow transition is not a process of vortex breakdown but turbulence vortex structure buildup.

In the current turbulence textbooks, there are many places filled with self-contradictions. We first say vortex never breaks down according to Helmholtz vorticity flux conservation law, and later we say turbulence is generated by "vortex breakdown." We first say vortex can only end on the wall surface and later say vortex can detach from the wall surface. How can a vortex attach to a wall, and then become detached, break down, and reconnect? These will leave vortex leg inside the flow field and directly violate vorticity flux conservation law.

Some people argue that "we do not have exact definition for vortex." If we really have no definition for "vortex," we then have no serious scientific research at all for turbulence study. In fact, the definition of a "vortex" is clear, which is a fluid rotation core but not vortex tube. Our textbooks say lambda vortex becomes hairpin vortex through "self-deform," but there should not be any "self-deform" in the world. "Deformation" is a motion and any motion must be driven by force. Some literatures say multiple vortex rings are auto-generated. However, the world should not have anything to be auto-generated. Everything in this world should be generated under certain mechanism.

Many people accepted the concept of "turbulence bursting" and "turbulence intermittency." However, this is a misunderstanding. Turbulence is generated by very organized activities, step by step. There is no possibility that the turbulence could suddenly appear and then suddenly disappear. It would be a question that if God manipulates the fluid flow as a result of that turbulence could suddenly appear and then suddenly disappear? This is a misunderstanding by some people who do not fully understand turbulence.

Richardson [31] believed there is an eddy cascade, but no one was able to find such a vortex cascade, even today when the 3-D PIV and laser equipment is quite advanced. Kolmogorov [11] believed the larger vortex gives energy to smaller vortices through "vortex breakdown," but there is no "vortex breakdown" in this world. Especially, no matter how one defines "vortex," turbulence has no way of being generated by "vortex breakdown."

Linear mode suppression technology was studied for decades for flow transition control, and tens of million dollars have been spent in US and Europe, but who succeeded? When the perturbation is larger than 2%, there are no linear modes. If flow transition was caused by linear modes and must experience the process of self-deform from lambda vortex to hairpin vortex, how could we explain "bypass transition" and "free stream turbulence"? Anyway, the classical and current turbulence theories are full of self-contradictions.

There are many others. This paper tries to revisit the classical and current turbulence theory, find some self-contradictions, and briefly introduces Liu's new turbulence theory.

After 25 years of intensive study by Dr. Chaoqun Liu and his students at the University of Texas at Arlington, Dr. Liu presented a new theory on turbulence generation and sustenance, which is consistent and has no self-contradictions. Compared with the current theories on turbulence, which is mainly empirical and derived on the basis of dimensional analysis, Liu's theory is based on accurate scientific computing and experiments. Therefore, the new theory is more trustworthy. The core of Liu's theory maintains that the nature of turbulence is such that shear must transfer to fast rotation cores in a flow field and that turbulence is not generated by vortex breakdown, but multiple level shear layer instability, which is generated by sweeps and ejections. "Shear layer instability is the mother of turbulence," and "it is the nature of turbulence that shear must transfer to fast rotation." Turbulence is not generated by "vortex breakdown" but "**vortex buildup**." This new theory may bring a revolution to not only the basic fluid mechanics, but also to practical engineering applications including turbulence modeling.

Wu and Moin [36] reported a new DNS for late flow transition on a flat plate. They obtained fully developed turbulent flow with structure of a forest of hair-ping vortices by flow transition at zero pressure gradients. However, they did not give the mechanism of the late flow transition. Actually, similar work for the whole process of K- and H-type transition has been reported by Liu et al. [12-14] 14 years ago, and Rist et al. [32] 7 years ago. The newer results have higher resolution, but all reported vortex structures are similar.

In order to arrive at a deeper understanding of the physics of turbulence generation and sustenance, we recently conducted a high-order direct numerical simulation (DNS) with 1920×241×128 gird points and about 600,000 time steps to study the mechanism of the late stages of flow transition in a boundary layer at a free stream Mach number 0.5 [2-6, 12-23, 24-28, 37-38]. The DNS results have been well-validated by UTA and NASA Langley researchers [10]. A visualization method combining λ_2 iso-surface [9] and vortex lines were used to describe the flow field.

According to the current flow transition theory, the flow transition process has been described as follows: (1) receptivity, (2) linear instability, (3) nonlinear growth and interaction, and (4) breakdown to turbulence. However, the authors believe that turbulence is not caused by "vortex breakdown" but "vortex buildup" and linear modes only play a role in triggering vorticity rollup, but not directly causing the flow transition. Therefore, the authors believe that the transition process should be described as follows: (1) perturbation and growth (which may include linear modes or other disturbances); (2) large vortex formation including vorticity rollup and shear layer instability; (3) multiple-level vortex structure buildup including sweeps, ejections, and small length scale generation; and (4) symmetry loss and being chaotic to turbulence [23].

First, contrary to current transition theory, there is no "vortex breakdown" but there is a "turbulence vortex structure buildup," which is just the opposite. "Vortex breakdown" is theoretically incorrect and is never observed by any experiment or DNS. At present, most flow transition papers just use one term, "vortex breakdown," to describe the last stage of flow transition. If "vortex breakdown," which never exists, means flow transition from the laminar state to turbulence state, the authors believe that we will need more than one hundred research papers to describe such a process, not just the one term, "vortex breakdown." According to Dr. Cai's high-resolution experiment (Figure 1, personal communication) with the highest resolution of 1 μm, while most of our experiments only have ***mm***-scale resolution, large vortices interaction could produce countless small vortices with the scale in the order of 1 μm but did not find any large vortex breakdown, even no large vortex deformation. The large vortices are still alive, which contradicts to Richardson's large vortex short turnover time, expected by ***l/u*** [7]. Cai et al conclude that the classical and current theory that small length vortices are produced by large vortex breakdown and that the energy is passed from large vortex to smaller vortex through "vortex breakdown" has no way to be correct. The other impressive qualitative agreement is that the vortex rings rotate fast with a rotation speed of around 10,000 circles per second in a jet flow, while our DNS shows that the rotation speed is around 8,000 circles per second in a boundary layer. According to the DNS observations by Liu, no small scales survive if the large vortex disappears, since the small scales are generated and supported by the large vortex structure.

Mathematically, one counterexample can overthrow a theory and we really do not need a second example. The classical and current flow transition theories are filled with many misunderstandings and self-contradictions. Thus, we do not have the right turbulence theory. The classical and current flow transition and turbulence theories must be revisited.

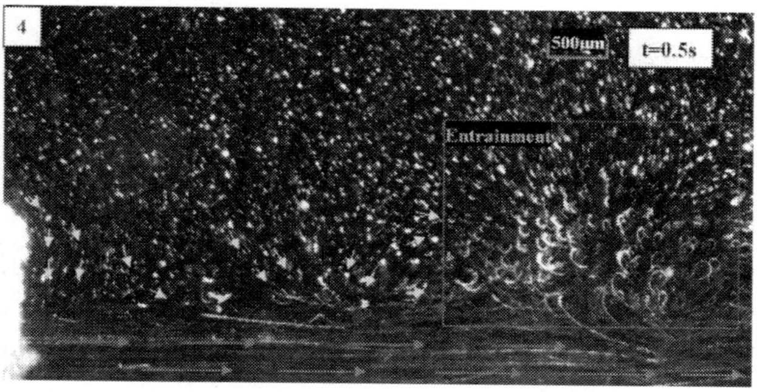

Figure 1. Vortices generated by water jet (Cai's experimental observation with highest resolution of 1 μm ; personal communication)

http://dx.doi.org/10.5772/61836

2. Self-contradictions of classical and current turbulence theories

As shown below, the classical and current turbulence theories have many serious self-contradictions:

1. **Vortex breakdown**: The transition community commonly agrees that after receptivity, linear instability and nonlinear instability stages, flow will "break down" to turbulence, which means the turbulent flow does not have structure at all due to the "breakdown." However, the turbulence community believes there are coherent structures in fully developed turbulence after the flow breakdown. It does not make much sense to study the house structure after the house collapses and breaks down into hundreds pieces of debris. If turbulence has coherent structure, the flow structure must never break down during the transition. This question must be answered for any serious scientific research on turbulence and it cannot be skipped or ignored. Liu et al. [23] believe there is no breakdown in any sense and both transitional flow and turbulent flow have the same mechanism for turbulence generation and sustenance and certain structure that can be accepted by Navier-Stokes equations. The fully developed turbulent flow has more small structures in a more chaotic manner. Liu et al. [23] also believe that when the transition community discusses the flow transition process and the turbulence community discusses the turbulence "coherent structure," they really discuss the same thing, which is "turbulence generation and sustenance" and should have the same mechanism. According to the authors, flow transition is not a process of vortex "breakdown," but just the opposite which is a process of turbulence vortex structure "**buildup.**" There may be some arguments that "breakdown" does not mean one vortex breaks to hundreds of pieces. However, no one was able to observe even that one breaks into two pieces, like that shown in many textbooks for Richardson's eddy cascade and Kolmgorov's vortex breakdown (e.g., [7]) In any case, one cannot believe the English word "breakdown" means "buildup," or house "breakdown" means we are building the house. Flow transition is a process of vortex "buildup." Some people may argue that we do not have the exact definition for vortex. Actually, if we define vortex as a vortex tube, vortex tube cannot break down. If we define vortex as a rotation core, the core is very stable and cannot break down either, like a tornado. However, if we have no definition for vortex, we then have no serious scientific research on turbulence. The definition of vortex is apparently a rotation dominant flow with a rotation core which has less dissipation. In practice, vortex breakdown is mainly caused by inappropriate pickup of lambda 2 or Q-iso-surface.

2. **The role of linearly unstable modes**: Although the linear modes are well-understood, Liu et al. [23] believe that the modes may be still linear only when its magnitude is smaller than 2%. When the perturbation is greater than 2%, there are no linear modes at all since the base flow has been changed. How can we still find T-S modes when we do not have Blasius base flow? The authors' DNS shows we do not have Blasius velocity profile at very early transition stage and the inflection points are developed as the vorticity rolls up. The significant perturbation growth and vortex structure formation are all nonlinear. The flight environment cannot keep the inflow perturbation to be smaller than 2%. Further

study found that although the linear unstable modes are important, they are small, cannot form vortex, and cannot cause the flow transition either by absolute instability, convectional instability, or mode interaction. The complex turbulence structure cannot be formed by those mode interactions or resonance, but sophistic vortex development, step-by-step [23]. Actually, the flow transition is vorticity redistribution and increment and the vorticity is given by the original Blasius solution for the flat plate case. Actually, the linearized N-S solution departs from DNS solution at very early stages (Figure 2)

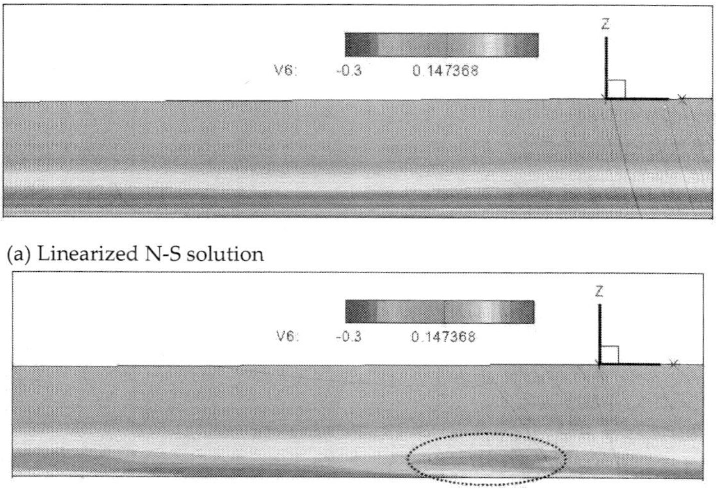

(a) Linearized N-S solution

(b) DNS solution

Figure 2. Comparison of linearized solution and DNS solution at early stage of flow transition

3. **Nature of turbulence**: Liu et al. [23] believe that the flow transition is caused by the flow inherent property that fluid cannot tolerate high shear, and shear must transfer to rotation. The role of perturbation (linear modes or others) is to trigger vorticity rollup, which leads to flow transition. Similar to linear modes, any perturbation, like gust, dust, sands, mosquito, fly, roughness, blowing, can trigger the vorticity rollup and then cause the transition. Most transition researchers believe turbulence is caused by unstable modes growth and their interaction or resonance to vortex breakdown. However, that is not the case and how can vortex break down like resonance to produce turbulence with coherent structure? There are some arguments that we have no exact definition on "vortex" and "breakdown." However, no matter how one defines "vortex," there is no possibility that turbulence is generated by "vortex breakdown."

4. **Λ-vortex to hairpin vortex**: Hama et al. [8] and Moin et al. [30] believe the Λ -vortex becomes hairpin vortex through a self-deformation mechanism. First, there must be no self-deformation. Deformation is a motion and any motion must be driven by some force and cannot be "self-deformed." Liu et al. [23] believe Λ -vortex root and vortex ring are generated separately and independently through different mechanisms and vortex ring is not part of Λ -vortex. Correctly understanding the hairpin vortex formation is key to understanding turbulence. According to Liu et al., the hairpin vortex has ring and legs. The legs are generated by vorticity rollup, which can generate low speed zones above the legs through vortex rotation (ejections), and the ring is generated by shear layer instability (K-H type.) In many turbulence textbooks (e.g., [7]) Λ -vortex is defined as a vortex tube. The authors believes it is a serious misunderstanding that vortex is defined as a vorticity tube without vorticity line leakage, like stream tube; Λ -vortex is a rotation core which is open for vorticity lines to come in and come out (Figure 3).

5. **Vortex reconnection**: Based on current theory, since the hairpin vortex tube (see [7]) has to be stretched, it will have to break down as the leg is placed on the wall surface where velocity is zero and the ring head is located almost near the inviscid area where the streamwise velocity is near one unit (the dimensional speed could be 170 m/s if M=0.5.) Some literatures suggest the hairpin vortex will break down and reconnect. The vortex tube breakdown concept directly violates the Helmholtz vorticity flux conservation law, which states that vortex tube foot must lie on the boundary and cannot lie down inside the flow field. There is no mechanism to support either vortex breakdown or reconnection. Liu et al. believe it is a serious mistake to consider "vortex" as a "vortex tube" (e.g., Figure 4 in the book, *Turbulence*, Davidson, 2004) and consider "vortex" as a congregation of vorticity lines with a rotation core, but, in general, is not a vortex tube. Actually, we never have vortex tube which is laid on the wall and vortex is never attached to the wall.

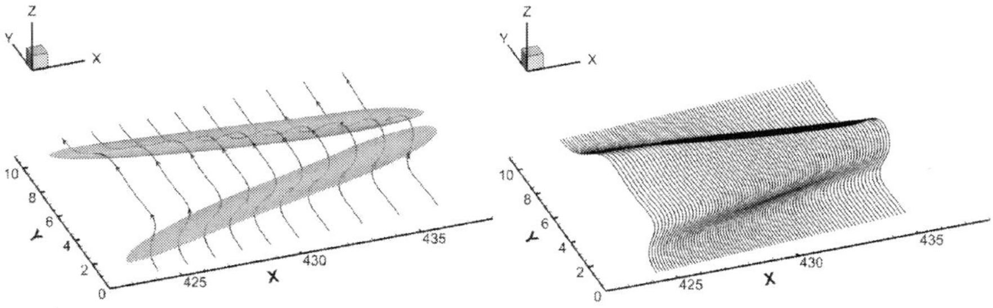

Figure 3. Λ-vortex (green part) is not a vortex tube as many vortex filaments (solid lines) penetrate the vortex

6. **Vortex attachment and detachment**: Some literatures suggest that the vortex is originally attached on the wall and then detached from the wall. It is really hard to believe how the vortex leg is originally linked with the wall surface and then detached from the wall. Liu et al. [23] believe the vortex (rotational core) is never attached on the wall. There is no

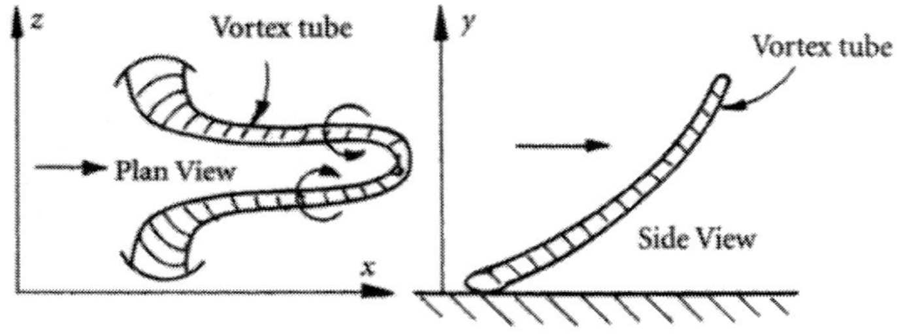

Figure 4. Vortex tube must break down to violate vorticity flux conservation law (copied from Davidson's book, *Turbulence*, published by Oxford University Press, 2004, see [7])

mechanism to support the switch from attachment to detachment, which would directly violate the vorticity flux conservation law and make the vorticity tube end inside the flow field. Actually, Λ - vortex, not vorticity tube, is never attached on the wall (Figure 5).

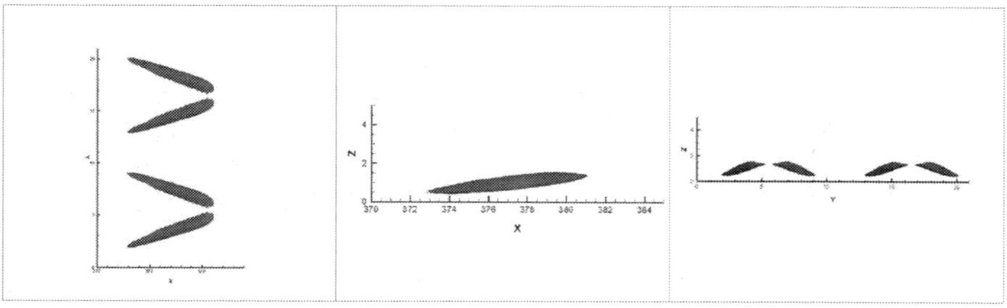

Figure 5. Projection of Λ -vortex on x-y plane; x-z plane; y-z plane at t = 3.07T

7. **Vortex ring auto-generation:** Some literatures suggest that the multiple vortex rings in a vortex package are auto-generated. We must be very careful when using the term "auto-generation" since everything must be generated under certain mechanism and, in general, cannot be auto-generated in this world. Liu et al. [23] believe all vortex rings are generated by the shear layer instability. There is no exception. There is a vortex rollup (Figure 6), which forms a low speed zone in the middle of the Λ -vortex (Figure 7). The low-speed zone will form a strong shear and the shear layer will further develop vortex rings through Kelvin-Helmholtz-type instability (Figure 8.)

8. **Turbulence bursting and intermittence:** Bursting means sudden appearance of physical quantity fluctuation and sudden increases of friction and Reynolds stress. These were explained as some unstable modes' sudden growth and breakdown.

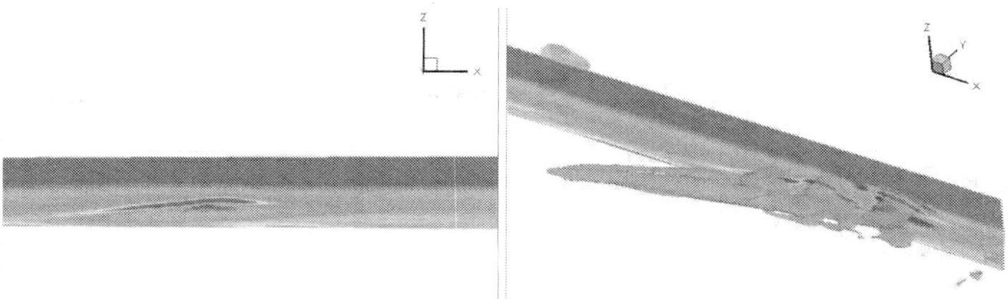

Figure 6. Vorticity rollup and vortex ring formation

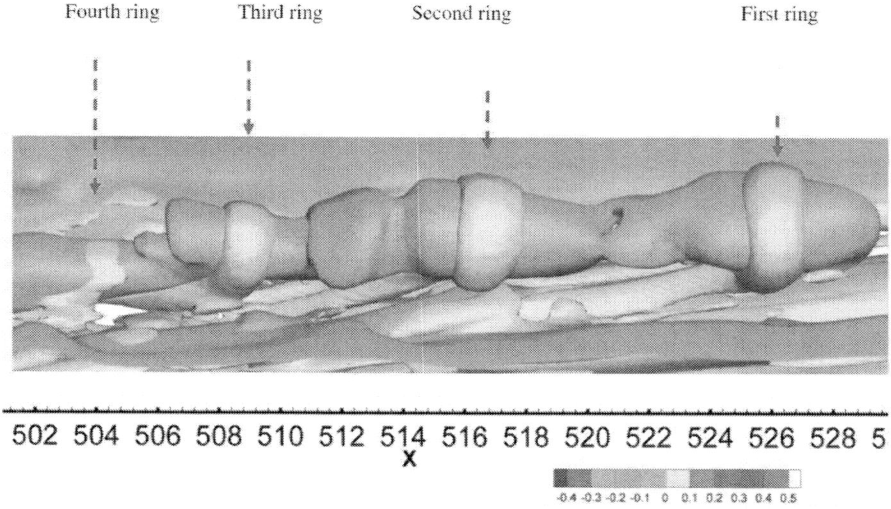

Figure 7. Low speed zone and vortex ring formation

However, that is not the case. Actually, this is caused by small length scale generation and fast motion of vortex rings, which has low speed in the ring center and low pressure in the rotation core center. The motion of vortex rings and vortex packages will cause the fluctuation. Turbulence cannot suddenly appear (bursting) and suddenly disappear (intermittence). Note that the vortex rings are moving with a self-rotation speed of around 10,000 circles per second. Even for a steady flow of such a vortex ring motion, it will cause a misunderstanding that the flow is strongly fluctuated in time if we installed a fixed probe inside the flow field. Turbulence bursting and intermittence are two important concepts in classical turbulence theory [33] but, unfortunately, they are basically misunderstandings. Turbulence cannot suddenly

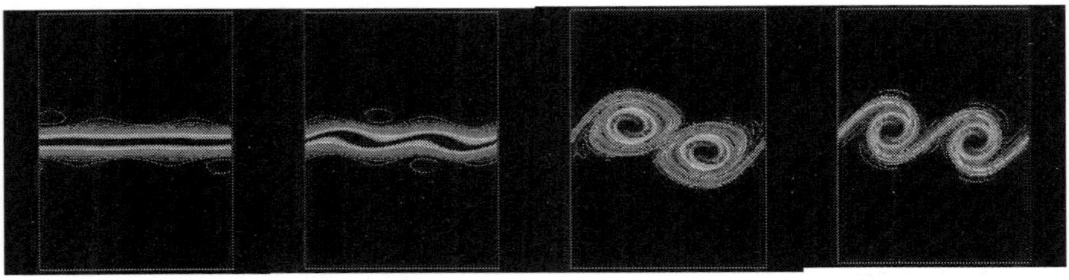

Figure 8. Kevin-Helmholtz instability

appear (bursting) and then suddenly disappear (intermittency). If turbulence can, there must be some superpower to manipulate the flow, which is impossible. These concepts are mistakenly formed by detection of flow fluctuations through fixed probes. Since the fast-rotating vortex rings have strong velocity gradient with low speed zones in the ring center and low pressure in the rotational core centers, the vortex package motion will show strong velocity and pressure fluctuations when the vortex package passes through these probes. Therefore, the velocity and pressure fluctuation, Reynolds stress, and surface friction will be quickly increased. After the vortex package left, the flow will recover and be quiet again. People usually call these "turbulence bursting and intermittency," which is really an incorrect understanding of the vortex package motion. The fluctuation including the frequencies and energy spectrum is determined by the vortex package structure and the vortex moving speed. Both "bursting" and "intermittency" can be reproduced by DNS for vortex package moving. Liu et al. [23] believe the turbulence generation cannot burst but is a very well-organized flow activity with four steps: Λ -vortex root formation, vortex ring formation, sweeps, and small vortices generation. The intermittence is a misunderstanding of vortex package self-motion (rotation) and relative motion. Turbulence cannot suddenly appear (burst) and then suddenly disappear (intermittence). The "turbulence bursting" and "intermittence" must be studied deeply and current misunderstandings must be clarified. The turbulence bursting and intermittency can be reproduced by DNS through a fixed-position probe (Figure 9), which agrees very well with the experiment by Borodulin and Kachanov [1]. This clearly shows that the fluctuation is caused by uneven vortex package movement. It means that fluctuation (turbulence) is vortex package motion.

9. **Richardson's [31] eddy cascade** (Figure 10a) which was described by a poem: "Big whirls have little whirls that feed on their velocity, and little whirls have lesser whirls and so on to viscosity." This has been accepted by the turbulence community for a long time. However, no one was able to observe such a cascade. It is really suspicious as the experimental tool is so advanced today (the resolution of Cai's experiment is around 1 μm) but we still cannot observe such a cascade. We have to be suspicious that there is no such a cascade. Actually, we see the large vortex induce smaller vortices through sweep in our DNS (Figure 10b).

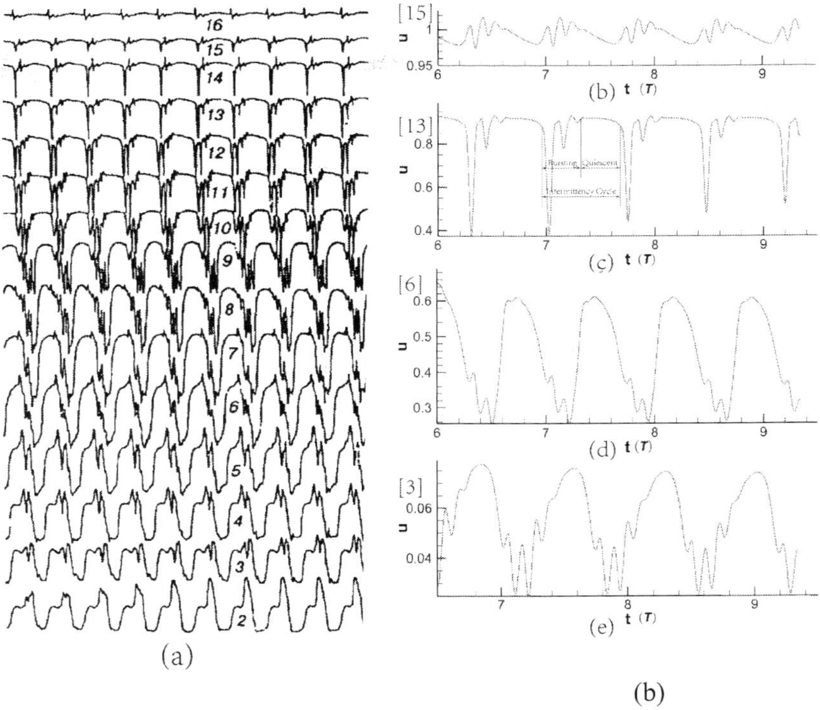

Figure 9. Comparison of experiment with DNS for turbulence intermittency

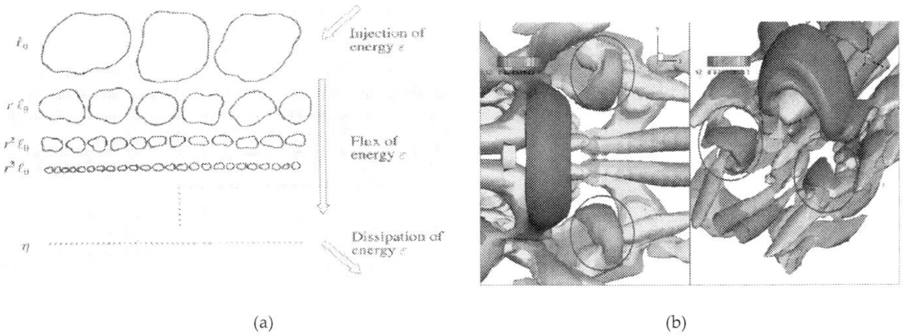

Figure 10. Richardson's eddy cascade (1928) and DNS observation on smaller vortices

10. **Energy transfer route:** Kolmogorov [11] believed that the large eddy passes energy to smaller vortices through "vortex breakdown" with very short eddy turnover time, which is *l*/*u* (see [7]), but no one was able to detect such a breakdown. Liu et al. [23] believe the large vortex passes the high energy through sweeps to generate positive spikes and further small vortices. Cai's experiment shows that when many small vortices appear, the large vortices are still alive and do not die as mothers. There is no "vortex breakdown" in any sense.

11. **Bypass transition and free stream turbulence**: If turbulence is generated by linear mode growth, nonlinear interaction, resonance, and vortex breakdown, current transition and turbulence theories have no ways to explain why we have "bypass transition" and "free stream turbulence." Liu et al. [23] believe flow transition is caused by the inherent property of fluids, that flow cannot tolerate shear and shear must transfer to rotation. In other words, laminar flow (shear-dominant) must transfer to turbulent flow (rotation-dominant) since the turbulent (rotation-dominant) state is a stable state. There is no necessary condition to have receptivity and linear growth for flow transition, and even no necessity to have Λ -vortex either.

These are just some sample questions to show that the classical or current turbulence theories are not able to give any convincing answer. Liu et al. believe one counterexample is good enough to overthrow a theory and we really do not need a second example. The current transition and turbulence theories have too many self-contradictions, definitely more than one or two.

Wallace [35] pointed out in his review paper: "... there has been remarkable progress in turbulent boundary layer research in the past 50 years, particularly in understanding the structural organization of the flow. Consensus exists that vortices drive momentum transport but not about the exact form of the vortices or how they are created and sustained." The authors have conducted a new high-order DNS with large number of grids to study the "turbulence generation and sustenance" and give exact form of the vortices or how they are created and sustained.

Many people misunderstand vorticity as rotation, vortex as vortex tube, and vorticity line as vortex line. Actually, as addressed above, vortex is a rotating core which consists of vortex lines with leaking, but vortex tube is a tube with vorticity lines without leakage, according to the definition in Davidson's book. Therefore, vorticity does not mean rotation, vortex is not vortex tube, and vorticity line is not vortex line. On the other hand, vortex line is part of vorticity lines and rotating vorticity is part of vorticity.

3. Liu's new findings

After 25 years of effort made by Liu and his students, the following new findings by high-order DNS were made:

- Mechanism of spanwise vorticity rollup

- Mechanism of transfer from flow shear to rotation
- Mechanism of spanwise vortex formation and role of the linear unstable modes
- Mechanism of vortex root formation
- Mechanism of first ring-like vortex formation
- Mechanism of multiple vortex ring formation
- Mechanism of second sweep formation
- Mechanism of low-speed zone and high shear layer formation
- Mechanism of positive spike formation
- Mechanism of secondary and tertiary vortex formation
- Mechanism of U-shaped vortex formation
- Mechanism of small length vortices generation
- Mechanism of multiple-level high shear layer formation
- Mechanism of energy transfer paths from the large length scales to the small ones
- Mechanism of symmetry loss and flow chaos
- Mechanism of thickening of turbulence boundary layer
- Mechanism of high surface friction of turbulent flow

4. Highlights of our new observations

4.1 People in general accept that turbulence is generated by some unstable modes that experience linear growth, nonlinear interaction or resonance, and vortex breakdown to turbulence. However, it is not the case. Linear modes can grow and trigger the flow transition, but flow transition is caused by the inherent property of fluid that "shear must transform to rotation."

4.2 Rotational and non-rotational vorticity and Helmholtz velocity decomposition revisited

Vorticity does not mean rotation as many people are confused by the two concepts. Actually, fluid particle motion can only be decomposed as symmetric tenser and non-symmetric tenser. The latter represents vorticity but not rotation. The vorticity can be further decomposed to rotational vorticity and non-rotational vorticity, which is different from Helmholtz decomposition of fluid particle motion.

$$\nabla\times\vec{V}=\Omega\vec{R}+\left(\nabla\times\vec{V}-\Omega\vec{R}\right)\approx\Omega\nabla\times\vec{V}+\left(1-\Omega\right)\nabla\times\vec{V}$$
$$\textit{In general } \vec{R}\neq\nabla\times\vec{V}$$

4.3 Although vorticity must keep conservation, once vorticity rolls up from the wall, the shear must transfer to rotation (non-rotational vorticity becomes rotational vorticity and deformation disappears) inside the flow field. On the other hand, laminar flow, which is dominated by shear and instability, must transfer to turbulent flow, which is dominated by rotation and stability. When the fluid particle becomes rotation like a rigid body, there is no deformation and thus no dissipation, which is the least energy-consuming state.

4.4 The nature of flow transition is that fluids inside the flow field cannot tolerate the shear, and shear must transfer to rotation when the Reynolds number is large enough. Flow transition is the inherent property of fluid flow.

4.5 Although the linearly unstable modes are important for flow transition, they are small, cannot form vortex, and cannot cause flow transition directly through either absolute instability or convective instability. They can only stimulate vorticity "rollup," which could cause flow transition. All linear unstable modes play a same role to push the vorticity up from the wall (rollup.) The flow trend to change shear to rotation will occur inside the flow field.

4.6 Linear mode suppression has been developed for several decades with little success. Now we understand that these efforts are unsuccessful since any factor that can cause vorticity rollup, like gust, dust, noise, mosquito, fly, can lead to failure of unstable mode suppression. The key issue is to avoid vorticity "rollup" and shear layer formation.

4.7 Because the linear unstable modes are small and cannot form vortex (negligible), mainly the vorticity which made turbulence structure is originated from the original wall boundary Blasius solution with absolute vorticity increment by stretching and tilting. Vorticity does not mean rotation and should be further decomposed to rotational vorticity and non-rotational vorticity.

4.8 The analytic linear solution becomes discrepant from DNS at the very beginning. They do not agree with each other even in the very early stages.

4.9 Vortex and hairpin vortex

The theory that Λ -vortex self-deforms to hairpin vortex, as given by Hamma [8] and Moin et al. [30] has been accepted by the turbulence community for many years; however, it is unfounded since deformation is a motion and any motion must be driven by force and cannot self-deform. Considering Λ -vortex as a vortex tube without vorticity line leakage is a serious mistake in turbulence study. The Λ -vortex root and ring head are formed by totally different mechanisms and vortex ring is not part of Λ -vortex (Figure 11.) A low-speed zone is formed above the lambda-vortex to form a high shear due to the vortex root rotation. The first vortex ring is generated by the high shear layer (K-H type) instability near the tip of the Λ -structure. Multiple vortex rings are all formed by shear layer instability, which is generated by momentum deficit (Figure 12).

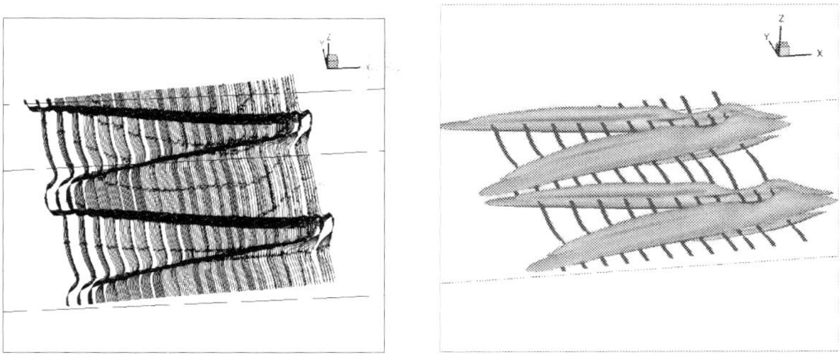

(a) Vorticity (b) Vortex

Figure 11. Vorticity and vortex of Λ -structure

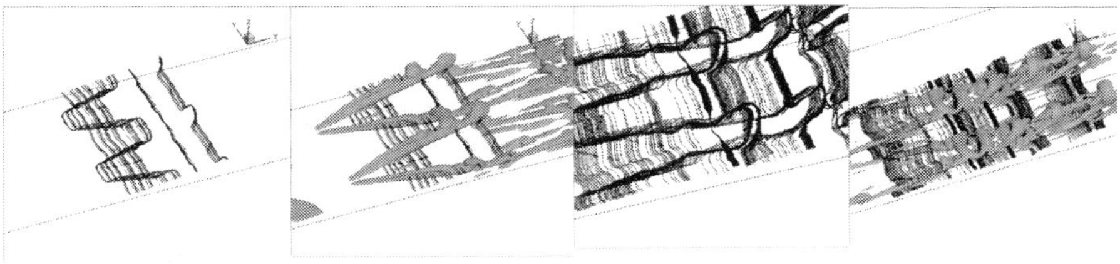

(a) Vorticity (b) Vortices (c) Vorticity (d) Vortices

Figure 12. Vorticity and vortices for multiple vortex rings

4.10 Vortex breakdown

"Vortex breakdown" is caused by faked visualization by using improper lambda2 values or Q-iso-surface. We can make various "vortex breakdowns" by using the same DNS data set with different lambda2 values (Figure 13). However, the vortex structures are very stable when they travel.

4.11 Energy transfer path from larger vortex to smaller vortices

The small vortices still need energy to survive, although rotation is the most stable state due to the minimized deformation and dissipation. In fact, the high energy is brought down to the lower boundary layer through fast large vortex rotation by multiple-level sweeps. Without these sweeps, all small vortices (turbulence) would dissipate quickly. Large vortex passed high energy through strong rotation (sweeps in particular) but definitely not through "vortex breakdown."

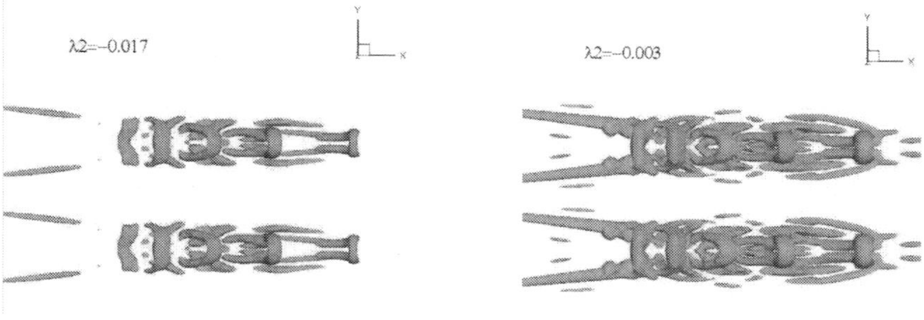

(a) Fake vortex breakdown (lambda2 = -0.0017) . (b) No breakdown (lambda = -0.003)

Figure 13. Fake "vortex breakdown" made by choosing inappropriate lambda2 values

4.12 Highlights of our new observations – chaos

The loss of flow symmetry begins from the middle of the multiple-level vortex structure, which shows it is an inherent instability of the multiple-level vortex structure. There is no proof that the chaos is caused by strong environmental disturbances and/or nonperiodic spanwise boundary conditions.

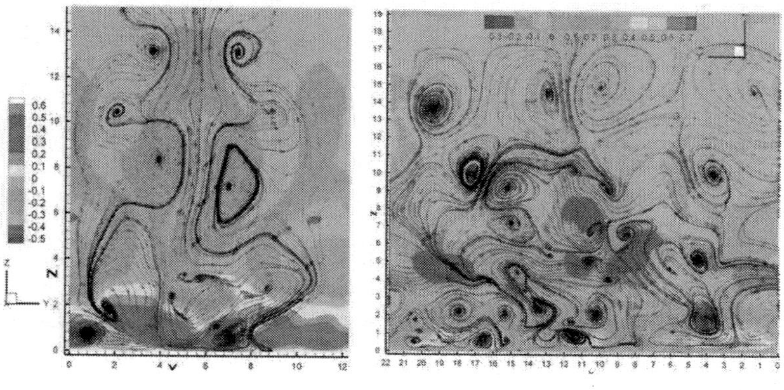

Figure 14. Asymmetry starts in the middle of the multi-level vortex package

5. Some conceptual mistakes in fundamental fluid dynamics

The authors believe some people have made at least several conceptual misunderstandings:

1. Considering vortex as "vortex tube": This is a major source of confusion.

2. Considering vorticity as rotation: They are two different concepts.

3. Considering hairpin vortex is self-deformed from Lambda vortex: There is no self-deformation.

4. Considering multiple vortex rings are auto-generated: Anything must be generated by certain mechanisms.

5. Considering vortex was first attached on the wall and then detached from the wall: Our textbooks say vortex legs can only lie down on the wall, but, in fact, vortex never attaches on the wall.

6. Considering small vortices are generated by large vortex breakdown: This is hypothesized by Richardson and Kolmokorov with no proof. The experiment conducted by Cai shows the large vortex is still there and even does not deform when a large number of small vortices are generated.

7. Considering vortex breaks down and then reconnects: Vortex ring and legs are never linked and they are separated as two different parts.

8. Considering turbulence is generated by unstable modes' linear growth, interaction, resonance, and breakdown either by absolute instability or convective instability: Linear modes can only trigger the vorticity rollup, but not part of the transition or part of the vortex structure.

9. Misunderstanding the uneven vortex package structure and package motion as "bursting" and "intermittency": This is caused by observation of fluid particle motion in a fixed-frame Euler system. Turbulence looks like velocity and pressure fluctuation, but fluctuation is really caused uneven vortex package motion.

10. Not realizing the vortex ring has a very fast-rotating core (e.g., around 10,000 circles/second) with large gradient in velocity and pressure.

11. Considering turbulent flow is a random motion: Turbulent flow cannot be random since fluid motion must follow conservation of mass, momentum, and energy.

Of course, these misunderstandings are hard to avoid as our pioneering scientists living in the 19th and early 20th centuries had neither computers nor high-resolution experimental instruments. They mainly presented hypotheses and assumptions, which must be reexamined. However, it would be an unpardonable mistake if we accept these hypotheses and teach our kids with the wrong concepts generation by generation without careful analyses by DNS and experiments.

6. Liu's new theory on flow transition and turbulence generation

By using high-order DNS in LBLT, Liu has revealed many new mechanisms, some of which are directly against the classical theory:

6.1. Nature of turbulence generation:

1. Fluids cannot tolerate high shear and shear must transfer to rotation and form a very fast rotation core (Dr. Cai has given his experimental observation)
2. Turbulence is not generated by "vortex breakdown" but "vortex buildup"
3. "Shear layer instability" is the "mother of turbulence"
4. Turbulence small scales are generated by multiple level shear layers which are generated by multiple level sweeps, ejections, negative and positive spikes.
5. Large vortices provide energy to smaller vortices through fast rotation which causes strong sweeps.

6.2. Nature of the flow transition

1. Flow transition is mainly caused by vorticity rollup from the wall.
2. Flow transition is a vorticity redistribution and increment from near wall to whole boundary layer.
3. Flow transition is a process of non-rotational vorticity transferring to rotational vorticity. Actually, laminar flow dominated by shear (non-rotational vorticity) is an unstable state but turbulent flow dominated by rotational vorticity is a stable state (rigid rotating vortex ring will have no deformation and then no energy dissipation). Of course, there are still shear layers between vortex rings and energy is still needed to keep these small vortex rings alive. The energy is provided by large vortex through rotation, sweeps in particular.

7. Conclusion

The classical and current turbulence theories are filled with many self-contradictions and, therefore, must be revisited. DNS and high-resolution experiment will pave ways to provide correct concepts and theories for turbulence study. Turbulence consists of rotations with different sizes of vortices. Laminar flow is dominated by shear which is unstable and turbulence is dominated by rotation which is stable. The rigid rotation of fluids has least energy consumption since it has no deformation (energy dissipation).

Flow transition is not a process of "vortex breakdown" but "vortex buildup." Vorticiy has rotational and non-rotational parts. Flow transition is a process of transformation of non-rotational vorticity to rotational vorticity, while the shear is gradually reduced but rotation is strengthened.

Flow transition is a fluid inherent property that shear must transfer to rotation when Reynolds number is large, i.e., transfer to a stable state.

The role of unstable modes is to stimulate the vorticity rollup from the wall. All high shear layers are produced by the vortex structure with multiple-level vortex rings, multiple-level sweeps and ejections, and multiple-level positive and negative spikes near the laminar sublayers. The vortex rotation generates low-speed zones which cause the high shear.

"Vortex breakdown" never happens. "Turbulence" is not generated by "vortex breakdown" but by positive spikes and consequent high shear layers.

There is a universal mechanism for turbulence generation and sustenance – the energy is brought by large vortex structure through fast rotation with multiple-level sweeps.

Flow disordering is caused by the instability of multiple-level vortex packages.

Acknowledgements

Liu's new turbulence theory was presented by Chaoqun Liu, with help received from many of his students. The authors are grateful to participants including Zhining Liu, Hua Shan, Li Jiang, Lin Chen, Xiaobin Liu, Ping Lu, Yonghua Yan, and Yiqian Wang.

This work was originally supported by AFOSR Grant FA9550-08-1-0201 (United States), supervised by Dr. John Schmisseur and the Department of Mathematics at University of Texas at Arlington. The authors are grateful to Texas Advanced Computing Center (TACC) for providing computation hours. This work is accomplished by using Code DNSUTA, and was released by Dr. Chaoqun Liu at University of Texas at Arlington in 2009.

Author details

Chaoqun Liu* and Shuhyi Chen

*Address all correspondence to: cliu@uta.edu

University of Texas at Arlington, Arlington, TX, USA

References

[1] Borodulin, V.I., Kachanov, Y.S. Formation and development of coherent structures in a transitional boundary layer. *J Appl Mech Tech Phys*, 1995;36(4): 532-564.

[2] Chen, L., Liu, X., Oliveira, M., Tang, D., Liu, C. Vortical structure, sweep and ejection events in transitional boundary layer, *Sci China, Series G, Physics, Mechanics, Astronomy*, 2009;39(10): 1520-1526.

[3] Chen, L., Liu, X., Oliveira, M., Liu, C. DNS for ring-like vortices formation and roles in positive spikes formation, AIAA Paper 2010-1471, Orlando, FL, USA, January 2010a.

[4] Chen, L., Tang, D., Lu, P., Liu, C. Evolution of the vortex structures and turbulent spots at the late-stage of transitional boundary layers, *Sci China, Physics, Mechanics, Astronomy*, January 2010b;53(1): 1–14.

[5] Chen, L., Liu, C. Numerical study on mechanisms of second sweep and positive spikes in transitional flow on a flat plate, *J Comp Fluids*, 2011a;40:28-41.

[6] Chen, L., Liu, X., Tang, D., Liu, C. Evolution of the vortex structures and turbulent spots at the late-stage of transitional boundary layers. *Sci China, Physics, Mechanics Astronomy*, 2011b;54(5): 986-990.

[7] Davidson, P.A., *Turbulence: An Introduction for Scientists and Engineers*. Page 123, Oxford University Press. ISBN 9780198529491, 2004.

[8] Hama, F.R., Nutant, J. Detailed flow-field observations in the transition process in a thick boundary layer. *Proc. 1963 Heat Transfer & Fluid Mech. Inst.*, Palo Alto, CA: Stanford Univ. Press, 1963; pp. 77-93.

[9] Jeong J., Hussain F. On the identification of a vortex, *J Fluid Mech* 1995, 285:69-94.

[10] Jiang, L., Chang, C. L. (NASA), Choudhari, M. (NASA), Liu, C. Cross-Validation of DNS and PSE Results for Instability-Wave Propagation, AIAA Paper 2003-3555, The 16th AIAA Computational Fluid Dynamics Conference, Orlando, Florida, June 23-26, 2003.

[11] Kolmogorov, A.N. (1941). The local structure of turbulence in incompressible viscous fluid for very large Reynolds numbers. *Proc Royal Soc Lond, Series A: Math Phys Sci* 1991;434: 9-13. Bibcode 1991RSPSA.434....9K. doi:10.1098/rspa.1991.0075.

[12] Liu, C., Liu, Z. Multigrid mapping and box relaxation for simulation of the whole process of flow transition in 3-D boundary layers. *J Comput Phys*, 1995;119: 325-341.

[13] Liu, Z., Xiong, G., Liu, C., Direct numerical simulation for the whole process of transition on 3-D airfoils. AIAA paper, 1996; AIAA 96-2081.

[14] Liu, C., Liu, Z. Direct Numerical Simulation for Flow Transition Around Airfoils, Proceedings of First AFOSR International Conference on DNS/LES, Louisiana Tech University, Ruston, Louisiana, August 4-8, 1997.

[15] Liu, C., Chen, L. Study of mechanism of ring-like vortex formation in late flow transition, AIAA Paper 2010-1456, Orlando, FL, 2010a.

[16] Liu, X., Chen, L., Oliveira, M., Tang, D., Liu, C. DNS for late stage structure of flow transition on a flat-plate boundary layer, AIAA Paper 2010-1470, Orlando, FL, January 2010b.

[17] Liu, C., Chen, L. Study of mechanism of ring-like vortex formation in late flow transition, AIAA Paper 2010-1456, Orlando, FL, January 2010c.

[18] Liu, X., Chen, Z., Liu, C. Late-stage vortical structures and eddy motions in transitional boundary layer status, *Chinese Phys Lett*, 2010d;27(2): 024706-1-4.

[19] Liu, C., Chen, L., Lu, P. New findings by high order DNS for late flow transition in a boundary layer. *J Model Simul Engin*, 2011a; 2011(721487): 1-16.

[20] Liu, C., Chen, L. Parallel DNS for vortex structure of late stages of flow transition. *J Comput Fluids*, 2011b;45: 129-137.

[21] Liu, C. Numerical and theoretical study on "Vortex Breakdown," *Int J Comput Math*, 2011c;88(17): 3702-3708.

[22] Liu, C., Chen, L., Lu, P., Liu, X. Study on multiple ring-like vortex formation and small vortex generation in late flow transition on a flat plate, theoretical and numerical fluid dynamics. 2013; 27(1): 41-70.

[23] Liu, C., Yan, Y., Lu, P. Physics of turbulence generation and sustenance in a boundary layer, *Comput Fluids*, 2014;102 (2014): 353-384.

[24] Lu, P., Liu, C. Numerical study of mechanism of u-shaped vortex formation, AIAA Paper 2011-0286, Orlando, Florida, USA, January 4-7, 2011.

[25] Lu, P., Wang, Z., Chen, L., Liu, C. Numerical study on U-shaped vortex formation in late boundary layer transition. *Comput Fluids*. 2011a;55: 36-47.

[26] Lu, P., Liu, C. Numerical study on mechanism of small vortex generation in boundary layer transition. AIAA Paper 2011-0287, 2011b, Orlando, Florida, USA, January 4-7, 2011.

[27] Lu, P., Liu, C. DNS study on mechanism of small length scale generation in late boundary layer transition. *Physica D: Nonlinear Phen*, 2011c;241 (2012): 11-24.

[28] Lu, P., Thampa, M, Liu, C. Numerical Study on Randomization in Late Boundary Layer Transition, AIAA 2012-0748, 2012, Nashville, TN, January 2012.

[29] Marshak, A. *3D Radiative Transfer in Cloudy Atmospheres*. Springer. ISBN 9783540239581, 2005; p. 76.

[30] Moin, P., Leonard, A. Kim, J. Evolution of curved vortex filament into a vortex ring. *Phys Fluids*, 29(4): 955-963.

[31] Richardson, L. F., *Weather Prediction by Numerical Process* (Cambridge Univ. Press, 1922).

[32] Rist, U., et al. Turbulence mechanism in Klebanoff transition: a quantitative comparison of experiment and direct numerical simulation. *J Fluid Mech*, 2002; 459:217-243.

[33] Schlichting, H., Gersten, K. *Boundary Layer Theory*, Springer, 8th revised edition, 2000, ISBN 3-540-66270-7.

[34] *USA Today*, Turbulence theory gets a bit choppy, September 10, 2006. http://www.usatoday.com/tech/science/columnist/vergano/2006-09-10-turbulence_x.htm.

[35] Wallace, J.M. Highlights from 50 years of turbulent boundary layer research. *J Turbulence*, 2013;13(53):1–70.

[36] Wu, X., Moin, P. Direct numerical simulation of turbulence in a nominally zero-pressure gradient flat-plate boundary layer, *JFM*, 630: 5-41, 2009.

[37] Yan, Y., Liu, C. Shear Layer Stability Analysis in Later Boundary Layer Transition and MVG controlled flow, AIAA2013- 0531, January 6-10, 2013.

[38] Yan, Y., Chen, C., Fu, H., Liu, C. DNS study on λ-vortex and vortex ring formation in flow transition at Mach number 0.5. *J Turbulence*, 2014;5(1): 1-21.

Chapter 2

Modeling of Wave Propagation from Arbitrary Depths to Shallow Waters – A Review

José Simão Antunes do Carmo

Additional information is available at the end of the chapter

http://dx.doi.org/10.5772/61866

Abstract

Numerical models are a useful instrument for studying complex superposition of wave–wave and wave–current interactions in coastal and estuarine regions and to investigate the interaction of waves with complex bathymetries or structures built in nearshore areas. Moreover, since their applications are significantly less expensive and more flexible than the construction of physical models, they are a convenient tool to support design. The ability of the standard Boussinesq and Serre or Green and Naghdi equations to reproduce these nonlinear processes is well known. However, these models are restricted to shallow water conditions, and addition of other terms of dispersive origin has been considered since the 1990s, particularly for approximations of the Boussinesq-type. To allow applications in a greater range of h_0 / λ, other than shallow waters, where h_0 is the water depth at rest and λ is the wavelength, a new set of Serre-type equations, with additional terms of dispersive origin, is developed and tested with the available data and with a numerical solution of a Boussinesq-type model, also improved with dispersive characteristics. Explicit and implicit methods of finite–difference are implemented to solve both approximations of Boussinesq and Serre types with improved dispersive characteristics. A finite element method is also implemented to solve an extension of a Boussinesq-type model that takes into account wave–current interactions. Application examples to solve real-world problems are shown and discussed. The performances of both 1HD models are compared with experimental data of very demanding applications, namely: (*i*) a highly nonlinear solitary wave propagating up a slope and reflecting from a vertical wall and (*ii*) a periodic wave propagating in intermediate-depth waters upstream a trapezoidal bar, followed by very shallow waters over the bar, and again in intermediate-depth water conditions downstream.

Keywords: Extended Boussinesq equations, Extended Serre equations, Dispersive characteristics, Intermediate waters, Numerical methods

1. Introduction

In recent decades, significant advances have been made in developing mathematical and numerical models to describe the entire phenomena observed in shallow water conditions. Indeed, not only our understanding of the phenomena has significantly improved but also the computational capabilities that are available have also increased considerably. In this context, we can now use more powerful and more reliable tools in the design of structures commonly used in coastal environments.

By the end of the 1970s, due to the lack of sufficiently deep knowledge, but above all for lack of computing power, the use of the linear wave theory for the simulation of phenomena, such as refraction and diffraction of waves, was common practice. In the 1980s, other models that take into account not only the refraction but also the diffraction process have been proposed and commonly used by [12] Berkhoff *et al.*, 1982, [21] Kirby and Dalrymple, 1983, [13] Booij, 1983, [20] Kirby, 1984, and [15] Dalrymple, 1988, among many others. However, as they are based on the linear theory, those models should not be utilized in shallow water conditions.

As noted in [6], even at that time models based on the Saint-Venant equations (see also [32] Saint-Venant, 1871) *"were frequently used in practical applications. However, as has been widely demonstrated, in shallow water conditions and for some types of waves, models based on a non-dispersive theory, of which the Saint-Venant model is an example, are limited and are not usually able to compute satisfactory results over long periods of analysis"* (see also [33]). Nowadays, it is generally accepted that for practical applications the combined gravity wave effects in shallow water conditions must be taken into account. In addition, the refraction and diffraction processes, the swelling, reflection and breaking waves, all have to be considered.

Also according to [6] *"a number of factors have made it possible to employ increasingly complex mathematical models"*. Indeed, not only has there been great improvement in our theoretical knowledge of the phenomena involved but also the numerical methods have been used more efficiently. The great advances made in computer technology, especially since the 1980s, improving information processing and enabling large amounts of data to be stored, have made possible the use of more mathematical models, of greater complexity and with fewer restrictions. Indeed, only models of order σ^2 ($\sigma = h_0 / \lambda$, where h_0 and λ represent, respectively, depth and wavelength characteristics) or greater, of the Boussinesq or Serre types [14, 35], are able to describe the entire dispersive and nonlinear interaction process of generation, propagation and run-up of waves resulting from wave–wave and wave–current interactions. It is also worth to point out that in more complex problems, such as wave generation by seafloor movement, propagation over uneven bottoms, and added breaking effects, high-frequency waves can arise as a consequence of nonlinear interaction.

In the past few years, the possibility of using more powerful computational facilities, and the technological evolution and sophistication of control systems have required thorough theoretical and experimental research designed to improve the knowledge of coastal hydrodynamics. Numerical methods aimed for the applications in engineering fields that are more sophisticated and with a higher degree of complexity have also been developed.

In Section 2, the general shallow water wave theory is used to develop different mathematical approaches, which are nowadays the basis of the most sophisticated models in hydrodynamics and sedimentary dynamics. Extensions of the equations for intermediate water of the more general approaches (order σ^2) are presented in Section 3. Numerical formulations of the models and appropriate application examples are presented in Section 4. Firstly, a submerged dissipation platform to protect the Bugio lighthouse, situated at the Tagus estuary mouth, Lisbon, is designed and tested numerically. The second application is a real-world problem concerning coastal protection, using a submerged structure to force the breaking waves offshore. The third example shows the agitations established in the port of Figueira da Foz, Portugal, for different sea states, with the objective of designing new protective jetties, or extend the existing ones. Finally, the fourth example shows the performance of the extended Serre equations for the propagation of a wave in very demanding conditions. Conclusions and future developments are given in Section 5.

2. Mathematical formulations

We start from the fundamental equations of the Fluid Mechanics, written in Euler's variables, relating to a three-dimensional and *quasi*-irrotational flow of a perfect fluid [Euler equations, or Navier-Stokes equations with the assumptions of non-compressibility ($d\rho / dt = div\vec{v} = 0$), irrotationality ($rot\vec{v} = 0$) and perfect fluid (dynamic viscosity, $\mu = 0$)]:

$$\begin{aligned}
&u_x + v_y + w_z = 0 \\
&u_t + uu_x + vu_y + wu_z = -p_x/\rho \\
&v_t + uv_x + vv_y + wv_z = -p_y/\rho \\
&w_t + uw_x + vw_y + ww_z = -p_z/\rho - g \\
&u_z = w_x;\ v_z = w_y;\ v_x = u_y
\end{aligned} \tag{1}$$

with $p = 0$ at $z = \eta(x, y, t)$, $w = \eta_t + u\eta_x + v\eta_y$ at $z = \eta(x, y, t)$, and $w = \xi_t + u\xi_x + v\xi_y$ at $z = -h_0 + \xi(x, y, t)$. In these equations ρ is density, t is time, g is gravitational acceleration, p is pressure, η is free surface elevation, ξ is bottom, and u, v, w are velocity components. Defining the dimensionless quantities $\varepsilon = a / h_0$ and $\sigma = h_0 / \lambda$, in which a is a characteristic wave amplitude, h_0 represents water depth, and λ is a characteristic wavelength, we proceed with suitable nondimensional variables:

$$x' = x/\lambda,\ y' = y/\lambda,\ z' = z/h_0,\ \eta' = \eta/a,\ \xi' = \xi/h_0,\ t' = t\sqrt{gh_0}/\lambda = tc_0/\lambda,\ p' = p/(\rho g h_0),$$
$$u' = u/\left(a\sqrt{g/h_0}\right) = uh_0/(ac_0),\ v' = v/\left(a\sqrt{g/h_0}\right) = vh_0/(ac_0), w' = w\lambda/\left(ah_0\sqrt{g/h_0}\right) = w\lambda/(ac_0),$$

where c_0 represents critical celerity, given by $c_0 = (gh_0)^{1/2}$, and, as above, η is free surface elevation, ξ represents bathymetry, u, v and w are velocity components, and p is pressure. In dimensionless variables, without the line over the variables, the fundamental equations and the boundary conditions are written [5]:

a. Fundamental equations

$$
\begin{aligned}
&u_x + v_y + w_z = 0 && \text{a} \\
&\varepsilon u_t + \varepsilon^2 u u_x + \varepsilon^2 v u_y + \varepsilon^2 w u_z = -p_x && \text{b} \\
&\varepsilon v_t + \varepsilon^2 u v_x + \varepsilon^2 v v_y + \varepsilon^2 w v_z = -p_y && \text{c} \\
&\varepsilon \sigma^2 w_t + \varepsilon^2 \sigma^2 u w_x + \varepsilon^2 \sigma^2 v w_y + \varepsilon^2 \sigma^2 w w_z = -p_z - 1 && \text{d} \\
&u_z = \sigma^2 w_x;\ v_z = \sigma^2 w_y;\ v_x = u_y && \text{e}
\end{aligned}
\qquad (2)
$$

b. Boundary conditions

$$
\begin{aligned}
&w = (1/\varepsilon)\xi_t + u\xi_x + v\xi_y,\ z = -1 + \xi && \text{a} \\
&w = \eta_t + \varepsilon u \eta_x + \varepsilon v \eta_y,\ z = \varepsilon\eta && \text{b} \\
&p = 0,\ z = \varepsilon\eta && \text{c}
\end{aligned}
\qquad (3)
$$

Integrating the first equation 2(a) between the bed $-1+\xi$ and the free surface $\varepsilon\eta$, taking into account 3(a) and 3(b), yields the continuity equation (4):

$$\left[\eta - (1/\varepsilon)\xi\right]_t + \left[(1 - \xi + \varepsilon\eta)\bar{u}\right]_x + \left[(1 - \xi + \varepsilon\eta)\bar{v}\right]_y = 0 \qquad (4)$$

where the bar over the variables represents the average value along the vertical. Then, accepting the fundamental hypothesis of the shallow water theory, $\sigma = h_0/\lambda << 1$, and developing the dependent variables in power series of the small parameter σ^2, that is

$$f = \sum_{i=0}^{\infty} (\sigma^2)^i f_i,\ \text{for } f = (u, v, w, p, \eta, \xi, A) \qquad (5)$$

where $A = u_x + v_y$, from continuity 2(a), and with 3(a) and 3(b), the following equations are obtained:

$$w_0 = -(z + 1 - \xi_0)A_0 + w_0^* \qquad (6)$$

$$w_0^{**} = -(1 + \varepsilon\eta_0 - \xi_0)A_0 + w_0^* \qquad (7)$$

where the simple and double asterisk represent the variables values at the bottom and at the surface, respectively. Of 2(e) we obtain, successively [34]:

$$\begin{aligned} u_0 &= u_0(x,y,t) \\ v_0 &= v_0(x,y,t) \end{aligned} \tag{8}$$

$$\begin{aligned} u_1 &= -(1/2)(z+1-\xi_0)^2 A_{0x} + (z+1-\xi_0)(\xi_{0x}A_0 + w^*_{0x}) + u^*_1 \\ v_1 &= -(1/2)(z+1-\xi_0)^2 A_{0y} + (z+1-\xi_0)(\xi_{0y}A_0 + w^*_{0y}) + v^*_1 \end{aligned} \tag{9}$$

so that the average values of the horizontal components of the velocity, on the vertical, are given by:

$$\begin{aligned} \bar{u} &= u_0 + \sigma^2 u^*_1 - (\sigma^2/6)(1+\varepsilon\eta_0 - \xi_0)^2 A_{0x} \\ &\quad + (\sigma^2/2)(1+\varepsilon\eta_0 - \xi_0)(\xi_{0x}A_0 + w^*_{0x}) + O(\sigma^4) \\ \bar{v} &= v_0 + \sigma^2 v^*_1 - (\sigma^2/6)(1+\varepsilon\eta_0 - \xi_0)^2 A_{0y} \\ &\quad + (\sigma^2/2)(1+\varepsilon\eta_0 - \xi_0)(\xi_{0y}A_0 + w^*_{0y}) + O(\sigma^4) \end{aligned} \tag{10}$$

On the other hand, taking into account that,

$$f = f_0 + O(\sigma^2) \text{ for } f = (u,v,\eta,\xi,w^*) \tag{11}$$

from (5) and (9) we obtain:

$$\begin{aligned} u^{**} &= \bar{u} - (\sigma^2/3)(1+\varepsilon\eta - \xi)^2 \bar{A}_x \\ &\quad + (\sigma^2/2)(1+\varepsilon\eta - \xi)(\xi_x \bar{A} + w^*_x) + O(\sigma^4) \\ v^{**} &= \bar{v} - (\sigma^2/3)(1+\varepsilon\eta - \xi)^2 \bar{A}_y \\ &\quad + (\sigma^2/2)(1+\varepsilon\eta - \xi)(\xi_y \bar{A} + w^*_y) + O(\sigma^4) \end{aligned} \tag{12}$$

Representing by $\Gamma = w_t + \varepsilon u w_x + \varepsilon v w_y + \varepsilon w w_z$ the vertical acceleration of the particles, we get $\Gamma = w_{0t} + \varepsilon u_0 w_{0x} + \varepsilon v_0 w_{0y} + \varepsilon w_0 w_{0z} + O(\sigma^2)$, and from (6), (7) and (11) the following approach is obtained:

$$\begin{aligned} \Gamma &= -(z+1-\xi)\left[\bar{A}_t + \varepsilon\bar{u}\bar{A}_x + \varepsilon\bar{v}\bar{A}_y - \varepsilon\bar{A}^2\right] \\ &\quad + \left[w^*_t + \varepsilon\bar{u}w^*_x + \varepsilon\bar{v}w^*_y\right] + O(\sigma^2) \end{aligned} \tag{13}$$

in which the terms within the first parenthesis (straight parenthesis) represent the vertical acceleration when the bottom is horizontal, and the terms within the second parenthesis (straight parenthesis) represent the vertical acceleration along the real bottom. It should be noted that equation 2(d) can be written as:

$$\varepsilon\sigma^2\Gamma = -p_z - 1 \tag{14}$$

where, by vertical integration between the bottom and the surface, the pressure p on the surface is obtained as:

$$\begin{aligned} p_x^{**} &= \varepsilon\eta_x\left(\varepsilon\sigma^2\Gamma^{**} + 1\right) \\ p_y^{**} &= \varepsilon\eta_y\left(\varepsilon\sigma^2\Gamma^{**} + 1\right) \end{aligned} \tag{15}$$

which, along with 2(b) and 2(c), allow us to obtain [34]:

$$\begin{aligned} \left(u_t + \varepsilon u u_x + \varepsilon v u_y + \varepsilon w u_z\right)^{**} + \eta_x\left(1 + \varepsilon\sigma^2\Gamma^{**}\right) = 0 \\ \left(v_t + \varepsilon u v_x + \varepsilon v v_y + \varepsilon w v_z\right)^{**} + \eta_y\left(1 + \varepsilon\sigma^2\Gamma^{**}\right) = 0 \end{aligned} \tag{16}$$

or even, given that $(f_s)^{**} = f_s^{**} - \varepsilon (f_z)^{**}\eta_s$, where $f = (u, v)$ and $s = (x, y, t)$:

$$\begin{aligned} u_t^{**} + \varepsilon u^{**}u_x^{**} + \varepsilon v^{**}u_y^{**} + \eta_x\left(1 + \varepsilon\sigma^2\Gamma^{**}\right) = 0 \\ v_t^{**} + \varepsilon u^{**}v_x^{**} + \varepsilon v^{**}v_y^{**} + \eta_y\left(1 + \varepsilon\sigma^2\Gamma^{**}\right) = 0 \end{aligned} \tag{17}$$

By developing expressions (17) in second approach (order 2 in σ^2), the following equations of motion (18) are obtained (for details see [34]):

$$\begin{aligned}
&\bar{u}_t + \varepsilon\bar{u}\bar{u}_x + \varepsilon\bar{v}\bar{u}_y + \eta_x \\
&\quad + \sigma^2\left\{\left[(2/3)(\varepsilon\eta - \xi)_x + (1/2)\xi_x\right]P + (1/3)(1 + \varepsilon\eta - \xi)P_x\right\} \\
&\quad + \sigma^2\left[\varepsilon\eta_x Q + (1/2)(1 + \varepsilon\eta - \xi)Q_x\right] + \sigma^4 = 0 \\
&\bar{v}_t + \varepsilon\bar{u}\bar{v}_x + \varepsilon\bar{v}\bar{v}_y + \eta_y \\
&\quad + \sigma^2\left\{\left[(2/3)(\varepsilon\eta - \xi)_y + (1/2)\xi_y\right]P + (1/3)(1 + \varepsilon\eta - \xi)P_y\right\} \\
&\quad + \sigma^2\left[\varepsilon\eta_y Q + (1/2)(1 + \varepsilon\eta - \xi)Q_y\right] + \sigma^4 = 0 \\
&P = (1 + \varepsilon\eta - \xi)\left(\varepsilon\bar{A}^2 - \varepsilon\bar{u}\bar{A}_x - \varepsilon\bar{v}\bar{A}_y - \bar{A}_t\right) \\
&Q = w_t + \varepsilon\bar{u}w_x + \varepsilon\bar{v}w_y \\
&w = (1/\varepsilon)\xi_t + \bar{u}\xi_x + \bar{v}\xi_y \\
&\bar{A} = \bar{u}_x + \bar{v}_y
\end{aligned} \tag{18}$$

where, likewise, the bar over the variables represents the average value along the vertical. In dimensional variables and with a solid/fixed bottom ($\xi_t = 0$), the complete set of equations is written, in second approach:

$$\begin{aligned}
&h_t + (h\bar{u})_x + (h\bar{v})_y = 0 \\
&\bar{u}_t + \bar{u}\bar{u}_x + \bar{v}\bar{u}_y + g\eta_x \\
&\quad + \left[(2/3)h_x + (1/2)\xi_x\right]P + (1/3)hP_x + h_x Q + (1/2)hQ_x = 0 \\
&\bar{v}_t + \bar{u}\bar{v}_x + \bar{v}\bar{v}_y + g\eta_y \\
&\quad + \left[(2/3)h_y + (1/2)\xi_y\right]P + (1/3)hP_y + h_y Q + (1/2)hQ_y = 0 \\
&P = h\left(\bar{A}^2 - \bar{u}\bar{A}_x - \bar{v}\bar{A}_y - \bar{A}_t\right) \\
&Q = w_t + \bar{u}w_x + \bar{v}w_y \\
&w = \bar{u}\xi_x + \bar{v}\xi_y \\
&\bar{A} = \bar{u}_x + \bar{v}_y
\end{aligned} \tag{19}$$

where $h = h_0 - \xi + \eta$ is total water depth. The one-dimensional form (1HD) of the equation system (19) is written, also with a fixed bottom:

$$\begin{aligned}
&h_t + (\bar{u}h)_x = 0 \\
&h\bar{u}_t + h\bar{u}\bar{u}_x + gh\eta_x + \left[h^2(P/3 + Q/2)\right]_x + \xi_x h(P/2 + Q) = 0 \\
&P = -h\left(\bar{u}_{xt} + \bar{u}\bar{u}_{xx} - \bar{u}_x^2\right) \\
&Q = \xi_x\left(\bar{u}_t + \bar{u}\bar{u}_x\right) + \xi_{xx}\bar{u}^2
\end{aligned} \tag{20}$$

Assuming additionally a relative elevation of the surface due to the waves ($\varepsilon = a/h_0$) having a value close to the square of the relative depth ($\sigma = h_0/\lambda$), i.e. $O(\varepsilon) = O(\sigma^2)$, from the system of equations (18), and at the same order of approximation, the following approach is obtained, in dimensional variables:

$$\begin{aligned}
&h_t + (h\bar{u})_x + (h\bar{v})_y = 0 \\
&\bar{u}_t + \bar{u}\bar{u}_x + \bar{v}\bar{u}_y + g\eta_x - \left[(1/6)\xi_x\right]P + (1/3)h_r P_x + (1/2)h_r Q_x = 0 \\
&\bar{v}_t + \bar{u}\bar{v}_x + \bar{v}\bar{v}_y + g\eta_y - \left[(1/6)\xi_y\right]P + (1/3)h_r P_y + (1/2)h_r Q_y = 0
\end{aligned} \tag{21}$$

where $h_r = h_0 - \xi$ is the water column height at rest, P and Q are given by $P = -(h_0 - \xi)(\bar{u}_x + \bar{v}_y)_t$ and $Q = (\bar{u}\xi_x + \bar{v}\xi_y)_t$. The momentum equations are written as:

$$\begin{aligned}&\overline{u}_t+\overline{u}\overline{u}_x+\overline{v}\overline{u}_y+g\eta_x+(1/6)h_r\xi_x\left(\overline{u}_x+\overline{v}_y\right)_t-(1/3)h_r^2\left(\overline{u}_x+\overline{v}_y\right)_{xt}\\&\quad+(1/3)h_r\xi_x\left(\overline{u}_x+\overline{v}_y\right)_t+(1/2)h_r\left(\overline{u}\xi_x+\overline{v}\xi_y\right)_{xt}=0\end{aligned}\tag{22}$$

$$\begin{aligned}&\overline{v}_t+\overline{u}\overline{v}_x+\overline{v}\overline{v}_y+g\eta_y+(1/6)h_r\xi_y\left(\overline{u}_x+\overline{v}_y\right)_t-(1/3)h_r^2\left(\overline{u}_x+\overline{v}_y\right)_{yt}\\&\quad+(1/3)h_r\xi_y\left(\overline{u}_x+\overline{v}_y\right)_t+(1/2)h_r\left(\overline{u}\xi_x+\overline{v}\xi_y\right)_{yt}=0\end{aligned}\tag{23}$$

with $\xi_t=0$, the complete system of equations (24) is obtained:

$$\begin{aligned}&h_t+\left(h\overline{u}\right)_x+\left(h\overline{v}\right)_y=0\\&\overline{u}_t+\overline{u}\overline{u}_x+\overline{v}\overline{u}_y+g\eta_x-(1/3)h_r^2\left(\overline{u}_{xxt}+\overline{v}_{xyt}\right)+h_r\xi_x\overline{u}_{xt}\\&\qquad+(1/2)h_r\left(\xi_{xx}\overline{u}_t+\xi_y\overline{v}_{xt}+\xi_x\overline{v}_{yt}+\xi_{xy}\overline{v}_t\right)=0\\&\overline{v}_t+\overline{u}\overline{v}_x+\overline{v}\overline{v}_y+g\eta_y-(1/3)h_r^2\left(\overline{u}_{xyt}+\overline{v}_{yyt}\right)+h_r\xi_y\overline{v}_{yt}\\&\qquad+(1/2)h_r\left(\xi_{yy}\overline{v}_t+\xi_y\overline{u}_{xt}+\xi_x\overline{u}_{yt}+\xi_{xy}\overline{u}_t\right)=0\end{aligned}\tag{24}$$

Further simplifying the equations of motion (18), retaining only terms up to order 1 in σ, i.e., neglecting all terms of dispersive origin, this system of equations is written in dimensional variables:

$$\begin{aligned}&h_t+\left(h\overline{u}\right)_x+\left(h\overline{v}\right)_y=0\\&\overline{u}_t+\overline{u}\overline{u}_x+\overline{v}\overline{u}_y+g\eta_x=0\\&\overline{v}_t+\overline{u}\overline{v}_x+\overline{v}\overline{v}_y+g\eta_y=0\end{aligned}\tag{25}$$

Approaches (19), (24) and (25) are known as Serre equations, or Green & Naghdi, Boussinesq and Saint-Venant, respectively, in two horizontal dimensions (2HD models). The classical Serre equations (19) [17] are fully nonlinear and weakly dispersive. Boussinesq equations (24) only incorporate weak dispersion and weak nonlinearity and are valid only for long waves in shallow waters. As for the Boussinesq-type models, also Serre's equations are valid only for shallow water conditions.

3. Derivation of higher-order equations

3.1. Weakly nonlinear approaches with improved dispersive performance

3.1.1. Nwogu's approach

To allow applications in a greater range of h_0/λ, other than shallow waters, [27] introduced higher-order dispersive terms into the governing equations to improve linear dispersion properties. By redefining the dependent variable, [30] achieved the same improvement without the need to add such terms to the equations.

Following [36], the extended Bussinesq equations obtained by [30] are derived in this section, using the nondimensionalised scaled equation system (2) – (3) as the starting point, rather than the procedure presented by Nwogu. For simplicity reasons, and without loss of generality, the one-dimensional extended Boussinesq equation system is derived. For consistency with the previous work (Section 2), the continuity equation 2(a) in two vertical dimensions (x, z) is integrated through the depth:

$$\int_{-1+\xi}^{z} w_z dz = -\int_{-1+\xi}^{z} u_x dz \tag{26}$$

$$w\big|_z - w\big|_{-1+\xi} = -\left(\int_{-1+\xi}^{z} u dz\right)_x + u\big|_{-1+\xi}\,\xi_x - u\big|_z\, z_x \tag{27}$$

Denoting $w\,|_z$ by w, with $\xi_t = 0$ and using Leibnitz' rule, the boundary condition 3(a) at $z = -1 + \xi$ gives:

$$w = -\left(\int_{-1+\xi}^{z} u dz\right)_x \tag{28}$$

Substituting (28) in the irrotationality equation 2(e), $u_z = \sigma^2 w_x$, yields:

$$u_z = \sigma^2 w_x = -\sigma^2\left(\int_{-1+\xi}^{z} u dz\right)_{xx} \tag{29}$$

Considering a Taylor series expansion (30) of $u(x, z, t)$, about $z = z_\alpha$, where the horizontal velocity $u_\alpha\,(x, t)$ denotes the velocity at depth z_α, $u\,(x, z_\alpha, t)$, this Taylor expansion is integrated through the depth from $-1 + \xi$ to z, yielding (31) (for details see [36]):

$$u = u_\alpha + (z - z_\alpha) u_z\big|_{z=z_\alpha} + \frac{(z - z_\alpha)^2}{2} u_{zz}\bigg|_{z=z_\alpha} + \cdots \tag{30}$$

$$\begin{aligned}\int_{-1+\xi}^{z} u dz = (z + 1 - \xi) u_\alpha + \left[\frac{(z - z_\alpha)^2}{2} - \frac{(1 - \xi + z_\alpha)^2}{2}\right] u_z\bigg|_{z=z_\alpha} \\ + \left[\frac{(z - z_\alpha)^3}{6} + \frac{(1 - \xi + z_\alpha)^3}{6}\right] u_{zz}\bigg|_{z=z_\alpha} + \cdots\end{aligned} \tag{31}$$

Substituting (31) in equation (29) gives:

$$u_z = -\sigma^2 \left\{ (z+1-\xi)u_\alpha + \left[\frac{(z-z_\alpha)^2}{2} - \frac{(1-\xi+z_\alpha)^2}{2} \right] u_z \bigg|_{z=z_\alpha} + \cdots \right\}_{xx} \tag{32}$$

Differentiating equation (32) with respect to z, noting from (29) that $u_z = O(\sigma^2)$:

$$\begin{aligned} u_{zz} &= -\sigma^2 \left\{ u_\alpha + (z-z_\alpha) u_z \big|_{z=z_\alpha} + \frac{(z-z_\alpha)^2}{2} u_{zz} \big|_{z=z_\alpha} + \cdots \right\}_{xx} \\ &= -\sigma^2 u_{\alpha xx} + O(\sigma^4) \end{aligned} \tag{33}$$

Differentiating equation (33) with respect to z and noting that both u_z and u_{zz} are $O(\sigma^2)$:

$$u_{zzz} = -\sigma^2 \left\{ u_z \big|_{z=z_\alpha} + (z-z_\alpha) u_{zz} \big|_{z=z_\alpha} + \cdots \right\}_{xx} = O(\sigma^4) \tag{34}$$

Repeated differentiation of this expression will produce expressions for the higher derivatives of u with respect to z and show them to be of $O(\sigma^4)$ order or greater. Substituting equations (33) and (34) back in equation (32):

$$\begin{aligned} u_z &= -\sigma^2 \left\{ (1-\xi+z_\alpha)u_\alpha - \frac{(1-\xi+z_\alpha)^2}{2} u_z \bigg|_{z=z_\alpha} + O(\sigma^2) \right\}_{xx} \\ &= -\sigma^2 \left[(1-\xi+z_\alpha)u_\alpha \right]_{xx} + O(\sigma^4) \end{aligned} \tag{35}$$

Substituting equations (33), (34) and (35) in the Taylor series expansion (30) produces an expression for the horizontal velocity component u:

$$u = u_\alpha - \sigma^2 \left\{ (z-z_\alpha)\left[(1-\xi+z_\alpha)u_\alpha\right]_{xx} + \frac{(z-z_\alpha)^2}{2}(u_\alpha)_{xx} \right\} + O(\sigma^4) \tag{36}$$

Substituting the horizontal velocity (36) in equation (28) leads to an expression for the vertical velocity component w:

$$w = -\Big\{(1-\xi+z_\alpha)u_\alpha + \sigma^2\Big[\frac{(1-\xi+z_\alpha)^2}{2}\big[(1-\xi+z_\alpha)u_\alpha\big]_{xx} - \frac{(1-\xi+z_\alpha)^3}{6}(u_\alpha)_{xx}\Big]\Big\}_x + O(\sigma^4) \quad (37)$$

Using the velocities u (36) and w (37) in the vertical momentum, equation 2(d) yields:

$$\varepsilon\sigma^2\Big\{-\big[(z+1-\xi)u_\alpha\big]_{xt} + O(\sigma^2)\Big\} + p_z + 1 + O(\varepsilon^2\sigma^2) = 0 \quad (38)$$

This can be rearranged to give an expression for the pressure p:

$$-p_z = 1 - \varepsilon\sigma^2\big[(z+1-\xi)u_\alpha\big]_{xt} + O(\varepsilon\sigma^4, \varepsilon^2\sigma^2) \quad (39)$$

Integrating through the depth from z to the free surface $\varepsilon\eta$:

$$-p\big|_{\varepsilon\eta} + p\big|_z = \varepsilon\eta - z + \varepsilon\sigma^2\Big\{\Big[\frac{z^2}{2} + (1-\xi)z\Big]u\Big\}_{xt} + O(\varepsilon\sigma^4, \varepsilon^2\sigma^2) \quad (40)$$

Using the free surface boundary condition 3(c) for the pressure, and denoting $p\,|_z$ simply as p gives:

$$p = \varepsilon\eta - z + \varepsilon\sigma^2\Big\{\Big[\frac{z^2}{2} + (1-\xi)z\Big]u\Big\}_{xt} + O(\varepsilon\sigma^4, \varepsilon^2\sigma^2) \quad (41)$$

Substituting the equations for the horizontal velocity component (36), the vertical velocity component (37) and the pressure (41) in the horizontal momentum equation 2(b) gives the Boussinesq momentum equation:

$$u_{\alpha t} - \sigma^2\Big\{(z-z_\alpha)\big[(1-\xi+z_\alpha)\,u_\alpha\big]_{xx} + \frac{(z-z_\alpha)^2}{2}(u_\alpha)_{xx}\Big\}_t + \varepsilon u_\alpha u_{\alpha x} + \eta_x + \sigma^2\Big\{\Big[\frac{z^2}{2} + (1-\xi)z\Big]u_\alpha\Big\}_{xxt} = O(\varepsilon\sigma^2, \sigma^4) \quad (42)$$

or

$$u_{\alpha t}+\varepsilon u_{\alpha}u_{\alpha x}+\eta_x+\sigma^2\left\{z_\alpha\left[(1-\xi)u_\alpha\right]_{xx}+\frac{z_\alpha^2}{2}(u_\alpha)_{xx}\right\}_t=O\left(\varepsilon\sigma^2,\sigma^4\right) \tag{43}$$

The second equation of the Boussinesq system is developed by first integrating the continuity equation 2(a) through the depth $-1+\xi$:

$$\left(\int_{-1+\xi}^{\varepsilon\eta} u\,dz\right)_x-\varepsilon u\big|_{\varepsilon\eta}\eta_x+u\big|_{-1+\xi}\xi_x+w\big|_{\varepsilon\eta}-w\big|_{-1+\xi}=0 \tag{44}$$

Using Liebnitz' Rule and the kinematic boundary conditions at the bed $z=-1+\xi$ 3(a) and at the free surface $z=\varepsilon\eta$ 3(b) gives:

$$\eta_t+\left(\int_{-1+\xi}^{\varepsilon\eta} u\,dz\right)_x=0 \tag{45}$$

From the expression (36) for the horizontal velocity u:

$$\int_{-1+\xi}^{\varepsilon\eta} u\,dz=(1-\xi+\varepsilon\eta)u_\alpha-\sigma^2\left\{\left[\frac{(z-z_\alpha)^2}{2}\right]_{-1+\xi}^{\varepsilon\eta}\left[(1-\xi+z_\alpha)u_\alpha\right]_{xx}+\left[\frac{(z-z_\alpha)^3}{6}\right]_{-1+\xi}^{\varepsilon\eta}u_{\alpha xx}\right\}+O\left(\sigma^4\right) \tag{46}$$

or

$$\int_{-1+\xi}^{\varepsilon\eta} u\,dz=(1-\xi+\varepsilon\eta)u_\alpha-\sigma^2\left\{\left[\frac{z_\alpha^2}{2}-\frac{(1-\xi+z_\alpha)^2}{2}\right]\left[(1-\xi+z_\alpha)u_\alpha\right]_{xx}+\left[-\frac{z_\alpha^3}{6}+\frac{(1-\xi+z_\alpha)^3}{6}\right]u_{\alpha xx}\right\}+O\left(\varepsilon\sigma^2,\sigma^4\right) \tag{47}$$

yielding:

$$\int_{-1+\xi}^{\varepsilon\eta} u\,dz = (1-\xi+\varepsilon\eta)u_\alpha - \sigma^2\left\{\left[-\frac{(1-\xi)^2}{2} + (1-\xi)z_\alpha\right]\left[(1-\xi+z_\alpha)u_\alpha\right]_{xx} + \left[\frac{(1-\xi)^3}{6} + \frac{(1-\xi)^2 z_\alpha}{2} + \frac{(1-\xi)z_\alpha^2}{2}\right](u_\alpha)_{xx}\right\} + O(\varepsilon\sigma^2,\sigma^4) \tag{48}$$

and thus

$$\int_{-1+\xi}^{\varepsilon\eta} u\,dz = (1-\xi+\varepsilon\eta)u_\alpha + \sigma^2\left\{\left[z_\alpha + \frac{1-\xi}{2}\right](1-\xi)\left[(1-\xi)u_\alpha\right]_{xx} + \left[\frac{z_\alpha^2}{2} - \frac{(1-\xi)^2}{6}\right](1-\xi)(u_\alpha)_{xx}\right\} + O(\varepsilon\sigma^2,\sigma^4) \tag{49}$$

Substituting equation (49) in the free surface equation (45) gives the Boussinesq continuity equation:

$$\eta_t + \left[(1-\xi+\varepsilon\eta)u_\alpha\right]_x + \sigma^2\left\{\left(z_\alpha + \frac{1-\xi}{2}\right)(1-\xi)\left[(1-\xi)u_\alpha\right]_{xx} + \left(\frac{z_\alpha^2}{2} - \frac{(1-\xi)^2}{6}\right)(1-\xi)(u_\alpha)_{xx}\right\}_x = O(\varepsilon\sigma^2,\sigma^4) \tag{50}$$

Returning to dimensional variables, unscaled form, equations (43) and (50) are written:

$$u_{\alpha t} + u_\alpha u_{\alpha x} + g\eta_x + z_\alpha (h_r u_{\alpha t})_{xx} + \frac{z_\alpha^2}{2}(u_\alpha)_{xxt} = 0 \tag{51}$$

$$\eta_t + \left[(h_0+\eta)u_\alpha\right]_x + \left[\left(z_\alpha + \frac{h_r}{2}\right)h_r (h_r u_\alpha)_{xx} + \left(\frac{z_\alpha^2}{2} - \frac{h_r^2}{6}\right)h_r (u_\alpha)_{xx}\right]_x = 0 \tag{52}$$

Setting the arbitrary depth $z_\alpha = \alpha h_0$, where $-1 \le \alpha \le 0$, the system (51)–(52) is rewritten:

$$u_{\alpha t} + u_\alpha u_{\alpha x} + g\eta_x + \alpha h_r\left[h_r (u_\alpha)_t\right]_{xx} + \frac{\alpha^2}{2}h_r^2 (u_\alpha)_{xxt} = 0 \tag{53}$$

$$\eta_t + \left[(h_r + \eta)\, u_\alpha \right]_x + \left[\left(\alpha + \frac{1}{2} \right) h_r^2 (h_r u_\alpha)_{xx} + \left(\frac{\alpha^2}{2} - \frac{1}{6} \right) h_r^3 (u_\alpha)_{xx} \right]_x = 0 \tag{54}$$

In two dimensions, the equation system (53)–(54) is written:

$$\begin{aligned} &\eta_t + \nabla \cdot \left[(h_r + \eta)\, \boldsymbol{u}_\alpha \right] \\ &\quad + \nabla \left\{ \left(\alpha + \frac{1}{2} \right) h_r^2 \nabla \left[\nabla \cdot (h_r \boldsymbol{u}_\alpha) \right] + \left(\frac{\alpha^2}{2} - \frac{1}{6} \right) h_r^3 \nabla (\nabla \cdot \boldsymbol{u}_\alpha) \right\} = 0 \end{aligned} \tag{55}$$

$$\boldsymbol{u}_{\alpha t} + (\boldsymbol{u}_\alpha \cdot \nabla) \boldsymbol{u}_\alpha + g \nabla \eta + \alpha h_r \nabla \left[\nabla \cdot (h_r \boldsymbol{u}_{\alpha t}) \right] + \frac{\alpha^2}{2} h_r^2 \nabla (\nabla \cdot \boldsymbol{u}_{\alpha t}) = 0 \tag{56}$$

where ∇ represents the two-dimensional gradient operator with respect to the horizontal coordinates (x, y) $(\nabla = \partial / \partial x, \partial / \partial y)$, and the velocity vector $\boldsymbol{u}(x, y, t) = (u, v)$ represents the two-dimensional velocity field at depth $z = \alpha h$.

3.1.2. Beji and Nadaoka's, and Liu and Sun's approaches

Starting from the standard Boussinesq equations and adopting the methodology introduced by [27] [11] presented a new approach that allows for applications until values of h_0 / λ to the order of 0.25, and still with acceptable errors in amplitude and phase velocity up to values of h_0 / λ near 0.50. By an addition and subtraction process, using the approximation $u_t = -g\ \eta_x$ and considering a dispersion parameter β in the momentum equation of the 1HD system (24), Beji and Nadaoka obtained an improved set of Boussinesq equations for variable depth, with β value obtained by comparison of the dispersion relation of the linearized form of the resulting equation with a second-order Padé expansion of the linear dispersion relation $\omega^2 / gk = \tanh(kh_r)$. In order to improve dispersion and linear shoaling characteristics in the Beji and Nadaoka's equations, [23] introduced two tuning parameters, α and γ, so that $\beta = 1.5\alpha - 0.5\gamma$. The nonlinearity in the previous Boussinesq-type models was improved by Liu and Sun adding higher-order terms accurate to the order of $O(\varepsilon\sigma^2)$. The 1HD standard Boussinesq equations and the approaches of Beji and Nadaoka and Liu and Sun are identified within the following system of equations (57) for water of variable depth:

$$\begin{aligned} &h_t + (hu)_x = 0 \\ &u_t + uu_x + g\eta_x - \{(1+\alpha)/2 - (1+\gamma)/6\} h_r^2 u_{xxt} \\ &\quad - (\alpha/2 - \gamma/6) g h_r^2 \eta_{xxx} + (1+\alpha) h_r \xi_x u_{xt} + \alpha g h_r \xi_x \eta_{xx} \\ &\quad + 0.5(1+\alpha) h_r \xi_{xx} u_t + 0.5 \alpha g h_r \xi_{xx} \eta_x = O(\varepsilon\sigma^2) \end{aligned} \tag{57}$$

After linearization of the equations system (57), [23] obtained the following dispersion relation (58):

$$\frac{\omega^2}{gk} = \frac{kh_r\left[1+\left(\alpha/2-\gamma/6\right)k^2h_r^2\right]}{1+\left(1+\alpha\right)k^2h_r^2/2-\left(1+\gamma\right)k^2h_r^2/6} \tag{58}$$

Comparing equation (58), written in terms of the phase speed (59)

$$C^2 = \frac{\omega^2}{k^2} = gh_r\left[\frac{1+\left(\alpha/2-\gamma/6\right)\left(kh_r\right)^2}{1+\left(1+\alpha\right)\left(kh_r\right)^2/2-\left(1+\gamma\right)\left(kh_r\right)^2/6}\right] \tag{59}$$

with the linear dispersion relation $\omega^2/gk = \tanh(kh_r)$, using the approach (60)

$$C_{Airy}^2 = \frac{\omega^2}{k^2} = \left(gh_r\right)\tanh\left(kh_r\right) = gh_r\left[\frac{1+\left(kh_r\right)^2/15}{1+2\left(kh_r\right)^2/5}\right] + O\left[\left(kh\right)^6\right] \tag{60}$$

allows to obtain the best values for the parameters α and γ: $\alpha = 0.1308$ and $\gamma = -0.0076$.

Considering appropriate values for the tuning parameters α and γ, we can identify within the equation system (57):

- The standard Boussinesq equations by setting $\alpha = \gamma = 0$
- The Beji and Nadaoka equations considering $\alpha = \gamma = 0.20$
- The Liu and Sun equations with $\alpha = 0.1308$ and $\gamma = -0.0076$

A visual comparison of numerical results of the extended Boussinesq approximation (57), with $\alpha = 0.1308$ and $\gamma = -0.0076$), shown in [1, 2] and [3], with a similar study performed by [37], using the extended Boussinesq (53)–(54) model (Nwogu's approach, with $\alpha = -0.531$) shows no relevant differences in the graphs. In this regard, it is worth remembering [4] *"although the methods of derivation are different, the resulting dispersion relations of these extended Boussinesq equations are similar, and may be thought of as a slight modification of the second-order Padé approximant of the full dispersion relation"*.

3.2. Fully nonlinear approaches with improved dispersive performance

Instead of using the horizontal velocity at a certain depth, other extensions of the Boussinesq equations have been made by using the velocity potential on an arbitrary depth, also with a tuning parameter. Wei *et al.* (1995) [39] used the Nwogu's approach to derive a Boussinesq-type model which retains full nonlinearity by including $O(\varepsilon\sigma^2)$ terms not considered in the

Nwogu's (53)–(54) system, and thus improving the nonlinearity to $O(\varepsilon)=1$. Wei *et al.*, and later [16] Gobbi *et al.*, derived a fourth-order Boussinesq model in which the velocity potential is approximated by a fourth-order polynomial in z. In terms of non-dimensional variables, both consider the boundary value problem for potential flow, given by:

$$\begin{aligned}
&\varphi_{zz} + \mu^2\nabla^2\varphi = 0\,, && -1+\xi \le z \le \varepsilon\eta \\
&\varphi_z + \mu^2\nabla h_{1-\xi}\bullet\nabla\varphi = 0\,, && z=-1+\xi \\
&\eta + \varphi_t + \frac{1}{2}\varepsilon\left[(\nabla\varphi)^2 + \frac{1}{\mu^2}(\varphi_z)^2\right] = 0\,, && z=\varepsilon\eta \\
&\eta_t + \varepsilon\nabla\varphi\bullet\nabla\eta - \frac{1}{\mu^2}\varphi_z = 0\,, && z=\varepsilon\eta
\end{aligned} \tag{61}$$

where, as above, z is the vertical coordinate starting at the still water level $h_0(x, y)$ and pointing upwards, scaled by a typical depth h_0, and η is the water surface displacement scaled by a representative amplitude a. The two dimensionless parameters ε and μ^2 are defined as $\varepsilon = a / h_0$ and $\mu^2=(k_0 h_0)^2$, with a representative wave number $k_0=2\pi / \lambda$, so $\mu^2=(2\pi\sigma)^2$. Time t is scaled by $[k_0(gh_0)^{1/2}]^{-1}$, and φ, the velocity potential, is scaled by $\varepsilon h_0(gh_0)^{1/2}$. We use the nondimensional water level $h_{1-\xi}$ instead of $1-\xi$. Integrating the first equation of (61) over the water column, and using the appropriate boundary conditions, the continuity equation is obtained:

$$\eta_t + \nabla\bullet M = 0 \tag{62}$$

where $M = \int_{-1+\xi}^{\varepsilon\eta} \nabla\varphi dz$. Retaining terms to $O(\mu^2)$, and denoting φ_α as the value of φ at $z=z_\alpha(x, y)$, an approximate velocity potential is given by:

$$\varphi = \varphi_\alpha(x,t) + \mu^2(z_\alpha - z)\nabla\bullet(h_{1-\xi}\nabla\varphi_\alpha) + \frac{\mu^2}{2}(z_\alpha^2 - z^2)\nabla^2\varphi_\alpha + O(\mu^4) \tag{63}$$

Substituting equation (63) into (62), a mass flux conservation equation is obtained [39]:

$$\begin{aligned}
\eta_t + \nabla\bullet\Bigg\{(h_{1-\xi}+\varepsilon\eta)\Bigg\{\nabla\varphi_\alpha + \mu^2\Bigg\{\nabla\left[z_\alpha\nabla\bullet(h_{1-\xi}\nabla\varphi_\alpha) + \frac{z_\alpha^2}{2}\nabla^2\varphi_\alpha\right] \\
+\frac{(h_{1-\xi}-\varepsilon\eta)}{2}\nabla\left[\nabla\bullet(h_{1-\xi}\nabla\varphi_\alpha)\right] - \frac{\left[h_{1-\xi}^2 - \varepsilon\eta h_{1-\xi} + (\varepsilon\eta)^2\right]}{6}\nabla^2\nabla\varphi_\alpha\Bigg\}\Bigg\}\Bigg\} = 0
\end{aligned} \tag{64}$$

Similarly, substituting (63) into the third equation of (61), a momentum equation is obtained in terms of the velocity potential. Then, given that at $z = z_\alpha$, $u_\alpha = \nabla \varphi_\alpha$, a fully nonlinear version of a Boussinesq type model in terms of η and u_α is written:

$$\begin{aligned}\eta_t + \nabla \cdot \left\{ \left(h_{1-\xi} + \varepsilon\eta \right) \left\{ \boldsymbol{u}_\alpha + \mu^2 \left\{ \left[\frac{z_\alpha^2}{2} - \frac{1}{6}\left(h_{1-\xi}^2 - \varepsilon\eta h_{1-\xi} + (\varepsilon\eta)^2 \right) \right] \nabla(\nabla \cdot \boldsymbol{u}_\alpha) \right.\right.\right. \\ \left.\left.\left. + \left[z_\alpha + \frac{1}{2}\left(h_{1-\xi} - \varepsilon\eta \right) \right] \nabla\left[\nabla \cdot \left(h_{1-\xi} \boldsymbol{u}_\alpha \right) \right] \right\} \right\} \right\} = O(\mu^4)\end{aligned} \tag{65}$$

$$\boldsymbol{u}_{\alpha t} + \varepsilon(\boldsymbol{u}_\alpha \cdot \nabla)\boldsymbol{u}_\alpha + \nabla\eta + \mu^2 \boldsymbol{R} + \varepsilon\mu^2 \boldsymbol{S} = O(\mu^4) \tag{66}$$

where

$$\boldsymbol{R} = \frac{1}{2} z_\alpha^2 \nabla(\nabla \cdot \boldsymbol{u}_{\alpha t}) + z_\alpha \nabla\left[\nabla \cdot \left(h_{1-\xi} \boldsymbol{u}_{\alpha t} \right) \right] - \nabla\left[\frac{1}{2}(\varepsilon\eta)^2 \nabla \cdot \boldsymbol{u}_{\alpha t} + \varepsilon\eta \nabla \cdot \left(h_{1-\xi} \boldsymbol{u}_{\alpha t} \right) \right] \tag{67}$$

$$\begin{aligned}\boldsymbol{S} = \nabla\left\{ (z_\alpha - \varepsilon\eta)(\boldsymbol{u}_\alpha \cdot \nabla)\left[\nabla \cdot \left(h_{1-\xi} \boldsymbol{u}_\alpha \right) \right] + \frac{1}{2}\left[z_\alpha^2 - (\varepsilon\eta)^2 \right](\boldsymbol{u}_\alpha \cdot \nabla)(\nabla \cdot \boldsymbol{u}_\alpha) \right\} \\ + \frac{1}{2} \nabla\left\{ \left[\nabla \cdot \left(h_{1-\xi} \boldsymbol{u}_\alpha \right) + \varepsilon\eta \nabla \cdot \boldsymbol{u}_\alpha \right]^2 \right\}\end{aligned} \tag{68}$$

It should be noted that Nwogu's approximation is recovered by neglecting higher-order terms. Numerical computations show that this model agrees well with solutions of the full potential problem over the range of relevant water depths.

A high-order, predictor–corrector, finite–difference numerical algorithm to solve this model was developed and is presented in [25] for the one- and two-layer models (see Sections 3.2.1 and 3.2.2). Considering one layer only, the corresponding numerical model that is the basis of the COULWAVE model is presented later, in Sections 4.1.3 and 4.1.4.

To improve the accuracy of numerical models, it has been common practice the use of high-order polynomials to approximate the vertical velocity dependence. However, this requires very elaborate and expensive numerical calculation procedures. A different approach is suggested by [26], which consists of using quadratic polynomials, matched at interfaces that divide the water column into layers. In this regard, it is worth mentioning [26] "*this approach leads to a set of model equations without the high-order spatial derivatives associated with high-order polynomial approximations*".

3.2.1. Mathematical model for one-layer

Defining the parameters $S_1 = \nabla . u_1$ and $T_1 = \nabla .[(1-\xi)\, u_1] + (1/\varepsilon)(\partial h_{1-\xi}/\partial t)$, the model uses the following approach for the continuity equation (to compute η values), in nondimensional variables:

$$\frac{1}{\varepsilon}\frac{\partial h_{1-\xi}}{\partial t} + \frac{\partial \eta}{\partial t} + \nabla \bullet \left[\left(\varepsilon\eta + h_{1-\xi}\right)\boldsymbol{u}_1\right] - \mu^2 \nabla \bullet \left\{ \left[\frac{\varepsilon^3\eta^3 + h_{1-\xi}^3}{6} - \frac{\left(\varepsilon\eta + h_{1-\xi}\right)k_1^2}{2} \right] \nabla S_1 + \left[\frac{\varepsilon^2\eta^2 - h_{1-\xi}^2}{2} - \left(\varepsilon\eta + h_{1-\xi}\right)k_1 \right] \nabla T_1 \right\} = O\left(\mu^4\right) \tag{69}$$

where u_1 = horizontal velocity vector, $k_1 = \alpha_1 h + \beta_1 \eta$, α_1 and β_1 are coefficients to be defined by the user. The index 1 means one-layer model. To compute the velocity components (u, v), the following approach of the momentum equation is solved, in nondimensional variables:

$$\begin{aligned} &\frac{\partial \boldsymbol{u}_1}{\partial t} + \varepsilon\, \boldsymbol{u}_1 \bullet \nabla \boldsymbol{u}_1 + \nabla \eta + \mu^2 \frac{\partial}{\partial t}\left\{ \frac{k_1^2}{2} \nabla S_1 + k_1 \nabla T_1 \right\} \\ &+ \varepsilon\mu^2 \left[\left(\boldsymbol{u}_1 \bullet \nabla k_1\right)\nabla T_1 + k_1 \nabla\left(\boldsymbol{u}_1 \bullet \nabla T_1\right) + k_1\left(\boldsymbol{u}_1 \bullet \nabla k_1\right)\nabla S_1 + \frac{k_1^2}{2}\nabla\left(\boldsymbol{u}_1 \bullet \nabla S_1\right) \right] + \varepsilon\mu^2 \left[T_1 \nabla T_1 - \nabla\left(\eta \frac{\partial T_1}{\partial t}\right) \right] \\ &+ \varepsilon^2\mu^2 \nabla\left(\eta S_1 T_1 - \frac{\eta^2}{2}\frac{\partial S_1}{\partial t} - \eta\, \boldsymbol{u}_1 \bullet \nabla T_1 \right) \\ &+ \varepsilon^2\mu^2 \nabla\left[\frac{\eta^2}{2}\left(S_1^2 - \boldsymbol{u}_1 \bullet \nabla S_1\right) \right] = O\left(\mu^4\right) \end{aligned} \tag{70}$$

The horizontal velocity vector is given as:

$$\boldsymbol{U}_1 = \boldsymbol{u}_1 - \mu^2 \left\{ \frac{z_1^2 - k_1^2}{2} \nabla S_1 + \left(z_1 - k_1\right)\nabla T_1 \right\} + O\left(\mu^4\right) \tag{71}$$

This one-layer model, often referred to as the "fully nonlinear, extended Boussinesq equations" in the literature (e.g. [38]), has been examined and applied to a significant extent. The weakly nonlinear version of (69)–(71) (i.e., assuming $O(\varepsilon) = O(\mu^2)$, thereby neglecting all nonlinear dispersive terms) was first derived by [30]. Through a linear and first-order nonlinear analysis of the model equations, Nwogu recommended that $z_1 = -0.531h$, and that value has been recommended and adopted by most researchers who use these equations. Refs. [22] and later [38] used the Nwogu's approach to derive a high-order model which retains "full nonlinearity" by including terms not considered in the Nwogu's approach. Comparing Liu's and Wei and Kirby's equations, there are some differences that can be attributed to the absence of some nonlinear dispersive terms in Wei and Kirby [19]. The above one-layer equations (69) and (70) are identical to those derived by [22].

3.2.2. *Mathematical model for two-layers*

Details about the two-layer mathematical model are presented in [25] and [24]. The model consists of a continuity equation, a momentum equation for the upper layer and a matching equation for the velocity in the lower layer. These equations are solved using an appropriate numerical method to compute values for the free surface elevation η and the velocity components (u, v). In dimensional variables, this set of equations is written ([24]):

$$\begin{aligned}&\frac{\partial\eta}{\partial t}+\nabla\cdot\left[(\eta-\zeta)\boldsymbol{u}_1+(\zeta+h_r)\boldsymbol{u}_2\right]\\&-\nabla\cdot\left\{\left[\frac{\zeta^3+h_r^3}{6}-\frac{(\zeta+h_r)k_2^2}{2}\right]\nabla S_2+\left[\frac{\zeta^2-h_r^2}{2}-(\zeta+h_r)k_2\right]\nabla T_2\right\}\\&-\nabla\cdot\left\{\left[\frac{\eta^3-\zeta^3}{6}-\frac{(\eta-\zeta)k_1^2}{2}\right]\nabla S_1+\left[\frac{\eta^2-\zeta^2}{2}-(\eta-\zeta)k_1\right]\nabla T_1\right\}=0\end{aligned}\tag{72}$$

$$\begin{aligned}&\frac{\partial\boldsymbol{u}_1}{\partial t}+\frac{1}{2}\nabla(\boldsymbol{u}_1\cdot\boldsymbol{u}_1)+g\nabla\eta\\&+\frac{\partial}{\partial t}\left[\frac{k_1^2}{2}\nabla S_1+k_1\nabla T_1-\nabla\left(\frac{\eta^2}{2}S_1\right)-\nabla(\eta T_1)\right]\\&+\nabla\left[\frac{\partial\eta}{\partial t}(T_1+\eta S_1)+(k_1-\eta)(\boldsymbol{u}_1\cdot\nabla)\,T_1+\frac{1}{2}(k_1^2-\eta^2)(\boldsymbol{u}_1\cdot\nabla)S_1\right.\\&\left.+\frac{1}{2}(T_1+\eta S_1)^2\right]-R_b+R_f+\nu_T\left[\nabla S_1-\nabla^2\boldsymbol{u}_1\right.\\&\left.-\nabla^2\left(\frac{k_1^2}{2}\nabla S_1+k_1\nabla T_1\right)+\nabla\left(\frac{\eta^2}{2}\nabla^2 S_1+\eta\nabla^2 T_1\right)\right]=0\end{aligned}\tag{73}$$

$$\boldsymbol{u}_2+\frac{k_2^2-\zeta^2}{2}\nabla S_2+(k_2-\zeta)\nabla T_2=u_1+\frac{k_1^2-\zeta^2}{2}\nabla S_1+(k_1-\zeta)\nabla T_1\tag{74}$$

where $S_1=\nabla\cdot\boldsymbol{u}_1$, $T_1=\zeta(S_2-S_1)+T_2$, $S_2=\nabla\cdot\boldsymbol{u}_2$, and $T_2=\nabla\cdot(h_r\boldsymbol{u}_2)$. R_b = breaking-related dissipation term, $R_f=[f/(h_r+\eta)]\boldsymbol{u}_b\,|\,\boldsymbol{u}_b\,|$ accounts for bottom friction, where $\boldsymbol{u}_b$ = velocity evaluated at the seafloor, and f = bottom friction coefficient, typically in the range of $10^{-3}-10^{-2}$, ν_T = constant eddy viscosity, $\nabla^2=(\partial^2/\partial x^2, \partial^2/\partial y^2)$, $k_1=-0.127h_r$ = evaluation level for the velocity u_1, $\zeta=-0.266h_r$ = layer interface elevation, s = evaluation level for the velocity u_2, and η = free surface elevation.

3.3. 1HD Serre's equations with improved dispersive performance

To allow applications in a greater range of h_0/λ, other than shallow waters, a new set of extended Serre equations, with additional terms of dispersive origin, is developed and tested in [1] and [2] by comparisons with the available test data. The equations are solved using an efficient finite–difference method, whose consistency and stability are tested in that work by comparison with a closed-form solitary wave solution of the Serre equations.

From the equation system (20), by adding and subtracting terms of dispersive origin, using the approximation $u_t=-g\,\eta_x$ and considering the parameters α, β and γ, with $\beta=1.5\alpha-0.5\gamma$, allows to obtain a new system of equations with improved linear dispersion characteristics:

$$\begin{aligned} & h_t+(uh)_x=0 \\ & u_t+uu_x+g(h+\xi)_x+(1+\alpha)(\Omega u_t-hh_xu_{xt})-(1+\beta)\frac{h^2}{3}u_{xxt} \\ & \quad +\alpha g\Omega(h+\xi)_x-\alpha ghh_x(h+\xi)_{xx}-\beta g\frac{h^2}{3}(h+\xi)_{xxx}-hh_xuu_{xx} \\ & \quad +\frac{h^2}{3}(u_xu_{xx}-uu_{xxx})+h(u_x)^2(h+\xi)_x+\xi_{xx}u^2(h+\xi)_x \\ & \quad +(\Omega+h\xi_{xx})uu_x+\frac{h}{2}\xi_{xxx}u^2=0 \end{aligned} \tag{75}$$

After linearization of the equation system (75), the dispersion relation (58) is obtained. As for the Boussinesq approach obtained by Liu and Sun, equating equation (58) with the linear dispersion relation $\omega^2/gk=\tanh(kh_r)$, using the approach (60), values of $\alpha=0.1308$ and $\gamma=-0.0076$ are obtained, so that $\beta=0.20$. It should be noted that the Serre's equation system (20) is recovered by setting $\alpha=\beta=0$.

4. Numerical formulations and applications

4.1. 2HD Boussinesq-type approaches

4.1.1. WACUP numerical model

An extension of the Boussinesq model (24) to take into account wave–current interactions has been derived and presented in [5]. This model is named WACUP (*a 2HD WAve Plus CUrrent*

Boussinesq-type model). With dimensional variables, taking mean quantities of the horizontal velocity components $U=(u+u_c)$ and $V=(v+v_c)$, where index c represents current and $u=U-u_c$ and $v=V-v_c$, the final set of these equations are written as follows:

$$h_t+(hU)_x+(hV)_y=0 \tag{76}$$

$$\begin{aligned}&U_t+UU_x+VU_y+g(h+\xi)_x-\frac{h_r^2}{3}\left(U_{xxt}+V_{xyt}\right)\\&-\frac{h_r^2}{3}\left[u_c\left(U_{xx}+V_{xy}\right)+v_c\left(U_{xy}+V_{yy}\right)\right]_x+h_r\left(\frac{1}{2}\xi_{tt}+u_c\xi_{xt}+v_c\xi_{yt}\right)_x\\&-v_l\left(U_{xx}+U_{yy}\right)-\frac{\tau_{s_x}}{\rho h}+\frac{\tau_{b_x}}{\rho h}=0\end{aligned} \tag{77}$$

$$\begin{aligned}&V_t+UV_x+VV_y+g(h+\xi)_y-\frac{h_r^2}{3}\left(U_{xyt}+V_{yyt}\right)\\&-\frac{h_r^2}{3}\left[u_c\left(U_{xx}+V_{xy}\right)+v_c\left(U_{xy}+V_{yy}\right)\right]_y+h_r\left(\frac{1}{2}\xi_{tt}+u_c\xi_{xt}+v_c\xi_{yt}\right)_y\\&-v_l\left(V_{xx}+V_{yy}\right)-\frac{\tau_{s_y}}{\rho h}+\frac{\tau_{b_y}}{\rho h}=0\end{aligned} \tag{78}$$

where τ_s and τ_b represent stresses on surface and at bottom, respectively.

The standard Boussinesq model (24) and the extended system of equations (76)–(78) are solved in [8] and [5], respectively, using an efficient finite element method for spatial discretization of the partial differential equations. Firstly, the (U, V) derivatives in time and third spatial derivatives are grouped in two equations. This means that an equivalent system of five equations is solved instead of the original equation system (76)–(78). The final equation system takes the following form:

$$h_t+hU_x+Uh_x+hV_y+Vh_y=0 \tag{79}$$

$$\begin{aligned}&r_t+u_cr_x+v_cr_y=-uU_x-vU_y-g(h+\xi)_x\\&+\frac{h_r^2}{3}\left[(u_c)_x\left(U_{xx}+U_{yy}\right)+(v_c)_x\left(V_{xx}+V_{yy}\right)\right]-h_r\left(\frac{1}{2}\xi_{tt}+u_c\xi_{xt}+v_c\xi_{yt}\right)_x\\&+\left[v_l+\frac{2}{3}h_r\left(\xi_t+u_c\xi_x+v_c\xi_y\right)\right]\left(U_{xx}+U_{yy}\right)+\frac{\tau_{s_x}}{\rho h}-\frac{\tau_{b_x}}{\rho h}=0\end{aligned} \tag{80}$$

$$s_t + u_c s_x + v_c s_y = -uV_x - vV_y - g(h+\xi)_y + \frac{h_r^2}{3}\left[(u_c)_y (U_{xx}+U_{yy}) + (v_c)_y (V_{xx}+V_{yy})\right] - h_r\left(\frac{1}{2}\xi_{tt} + u_c\xi_{xt} + v_c\xi_{yt}\right)_y + \left[v_l + \frac{2}{3}h_r(\xi_t + u_c\xi_x + v_c\xi_y)\right](V_{xx}+V_{yy}) + \frac{\tau_{s_y}}{\rho h} - \frac{\tau_{b_y}}{\rho h} = 0 \tag{81}$$

$$U - \frac{h_r^2}{3}(U_{xx}+U_{yy}) = r \tag{82}$$

$$V - \frac{h_r^2}{3}(V_{xx}+V_{yy}) = s \tag{83}$$

It should be noted that weakly vertical rotational flows were assumed, which strictly correspond to a limitation of the numerical method.

As the values of variables h, U, V, r and s are known at time t, we can use a numerical procedure based on the following steps to compute the corresponding values at time $t + \Delta t$ (for details see [5]):

1. The equation (79) allows us to predict the values of variable $h(h_p^{t+\Delta t})$, considering the known values of h, U and V at time t in the whole domain.

2. Equations (80) and (81) make it possible to predict the values of variables r ($r_p^{t+\Delta t}$) and s ($s_p^{t+\Delta t}$), taking into account the values of U^t, V^t, r^t, s^t and $\tilde{h}^{t+\Delta t} = 0.5h^t + 0.5h_p^{t+\Delta t}$ known for the whole domain.

3. Solutions of equations (82) and (83) give us the values of the mean-averaged velocity components U and V ($U^{t+\Delta t}$ and $V^{t+\Delta t}$), taking into account the predicted values of r and s ($r_p^{t+\Delta t}$ and $s_p^{t+\Delta t}$, respectively).

4. Equation (79) allows us to compute the depth h at time $t + \Delta t$ (values of $h^{t+\Delta t}$) considering the values of variables h^t, $U^{t+05\Delta t} = 0.5U^t + 0.5U^{t+\Delta t}$ and $V^{t+05\Delta t} = 0.5V^t + 0.5V^{t+\Delta t}$ known for the whole domain.

5. Equations (80) and (81) allow us to compute the values of variables r and s at time $t + \Delta t$ (values of $r^{t+\Delta t}$ and $s^{t+\Delta t}$), taking into account the values of r^t, s^t $h^{t+05\Delta t} = 0.5h^t + 0.5h^{t+\Delta t}$, $U^{t+05\Delta t} = 0.5U^t + 0.5U^{t+\Delta t}$ and $V^{t+05\Delta t} = 0.5V^t + 0.5V^{t+\Delta t}$ known for the whole domain.

The Petrov-Galerkin procedure is utilized to achieve solutions for the unknowns h, r and s. According to the weighted residual technique, minimization requires the "orthogonality" of the residual R_J to a set of weighting functions $W_{i,J}$, i.e.,

$$\int_{\Delta^e} W_{i,J} R_J d\Delta^e = 0,\ i = 1,\cdots,n \tag{84}$$

where the general form of the weighting functions applied to these equations is defined as:

$$W_{i,J} = N_i + \delta_{ui}\left(N_x\right)_i + \delta_{vi}\left(N_y\right)_i,\ i = 1,\cdots,n \tag{85}$$

and where the coefficients δ_{u_i}, and δ_{v_i} are functions of: (*i*) the local velocities U and V; (*ii*) the ratio of the wave amplitude to the water depth, and (*iii*) the element length. To illustrate this procedure, a complete solution of equations (79) (80) and (82) is presented here in detail. Introducing in equation (79), the approximated values are given by:

$$p \approx \hat{p} = \sum_{1=1}^{n} N_i p_i \tag{86}$$

the following residual R_{79} is obtained:

$$R_{79} = \hat{h}_t + \hat{h}\hat{U}_x + \hat{U}\hat{h}_x + \hat{h}\hat{V}_y + \hat{V}\hat{h}_y \tag{87}$$

The R_{79} error minimization leads to the following equation:

$$\begin{aligned}\int_{\Delta^e} W_{i,79} R_{79} d\Delta^e &= \int_{\Delta^e} W_{i,79}\left(\hat{h}_t + \hat{h}\hat{U}_x + \hat{U}\hat{h}_x + \hat{h}\hat{V}_y + \hat{V}\hat{h}_y\right) d\Delta^e \\ &= \int_{\Delta^e} W_{i,79}\left[\sum_{j=1}^{n} N_j\left(h_t\right)_j + \sum_{k=1}^{n}\left(N_x\right)_k U_k \sum_{j=1}^{n} N_j h_j + \sum_{k=1}^{n} N_k U_k \sum_{j=1}^{n}\left(N_x\right)_j h_j \right.\\ &\quad \left. + \sum_{k=1}^{n}\left(N_y\right)_k V_k \sum_{j=1}^{n} N_j h_j + \sum_{k=1}^{n} N_k V_k \sum_{j=1}^{n}\left(N_y\right)_j h_j\right] d\Delta^e = 0\end{aligned} \tag{88}$$

In the matrix form, this equation may be written as:

$$Ah_t + Bh = 0 \tag{89}$$

where

$$a_{i,j} = \int_{\Delta^e} W_{i,79} N_j d\Delta^e; \; i,j = 1,\cdots,n$$
$$b_{i,j} = \int_{\Delta^e} W_{i,79} \sum_{k=1}^{n} N_k U_k (N_x)_j d\Delta^e + \int_{\Delta^e} W_{i,79} \sum_{k=1}^{n} (N_x)_k U_k N_j d\Delta^e$$
$$+ \int_{\Delta^e} W_{i,79} \sum_{k=1}^{n} (N_y)_k V_k N_j d\Delta^e + \int_{\Delta^e} W_{i,79} \sum_{k=1}^{n} N_k V_k (N_y)_j d\Delta^e; \; i,j = 1,\cdots,n \tag{90}$$

The residual R_{80} is written as:

$$\begin{aligned} R_{80} &= \hat{r}_t + \hat{u}_c \hat{r}_x + \hat{v}_c \hat{r}_y + \hat{u}\hat{U}_x + \hat{v}\hat{U}_y \\ &+ g\left(\hat{h} + \hat{\xi}\right)_x - \frac{\hat{h}_r^2}{3}\left[(\hat{u}_c)_x\left(\hat{U}_{xx} + \hat{U}_{yy}\right) + (\hat{v}_c)_x\left(\hat{V}_{xx} + \hat{V}_{yy}\right)\right] \\ &+ \hat{h}_r\left[\frac{1}{2}\left(\hat{\xi}_{tt}\right)_x + (\hat{u}_c)_x\left(\hat{\xi}_t\right)_x + \hat{u}_c\left(\hat{\xi}_t\right)_{xx} + (\hat{v}_c)_x\left(\hat{\xi}_t\right)_y + \hat{v}_c\left(\hat{\xi}_t\right)_{xy}\right] \\ &- \nu_l\left(\hat{U}_{xx} + \hat{U}_{yy}\right) - \frac{\tau_{s_x}}{\rho h} + \frac{\tau_{b_x}}{\rho h} \end{aligned} \tag{91}$$

It should be noted that the term $\frac{2}{3}h_r(\xi_t + u_c\xi_x + v_c\xi_y)(V_{xx} + V_{yy})$ of equation (80) is of order σ^4 or greater. For this reason, it is not considerd in the numerical developments. Similarly, and for the same reason, considering $f = h_r^2/3$, all terms involving ∇f in the numerical developments are omitted.

The Green's theorem is used to solve the second derivatives present in equation (80) (residual R_{80}), and in equations (81) to (83), i.e., considering $\hat{p} = (\hat{U}, \hat{V})$, from $\hat{p}_{xx} + \hat{p}_{yy}$ we obtain:

$$\int_{\Delta^e} W_{i,l}\left(\hat{p}_{xx} + \hat{p}_{yy}\right) d\Delta^e = \int_{\Delta^e} W_{i,l} \sum_{j=1}^{n}\left[(N_{xx})_j + (N_{yy})_j\right] p_j \, d\Delta^e \tag{92}$$

and so

$$\int_{\Delta^e} W_{i,l} \sum_{j=1}^{n} [(N_{xx})_j + (N_{yy})_j] \, p_j \, d\Delta^e = -\int_{\Delta^e} (W_x)_{i,l} \sum_{j=1}^{n} (N_x)_j p_j \, d\Delta^e$$
$$- \int_{\Delta^e} (W_y)_{i,l} \sum_{j=1}^{n} (N_y)_j p_j \, d\Delta^e + \oint_{\Gamma^e} W_{i,l} \sum_{j=1}^{n} (N_n)_j p_j \, d\Gamma^e; \; p = (U, V) \tag{93}$$

Retaining terms up to order σ^3, the R_{80} error minimization leads to:

$$\begin{aligned}\int_{\Delta^e} W_{i,80} R_{80} d\Delta^e &= \int_{\Delta^e} \Bigg\{ W_{i,80} \Bigg[\sum_{j=1}^{n} N_j (r_t)_j + \sum_{k=1}^{n} N_k (u_c)_k \sum_{j=1}^{n} (N_x)_j r_j \\ &+ \sum_{k=1}^{n} N_k (v_c)_{k_x} \sum_{j=1}^{n} (N_y)_j r_j + \sum_{k=1}^{n} N_k u_k \sum_{j=1}^{n} (N_x)_j U_j + \sum_{k=1}^{n} N_k v_k \sum_{j=1}^{n} (N_y)_j U_j \\ &+ g \sum_{j=1}^{n} (N_x)_j (h+\xi)_j \Bigg] - \sum_{l=1}^{n} N_l \frac{(h_r^2)_l}{3} \sum_{k=1}^{n} (N_x)_k \left[(u_c)_k U_{x^2y^2} + (v_c)_k V_{x^2y^2} \right] \\ &+ W_{i,80} \Bigg[\sum_{k=1}^{n} N_k \frac{(h_r)_k}{2} \sum_{j=1}^{n} (N_x)_j (\xi_{tt})_j + \sum_{l=1}^{n} N_l (h_r)_l \sum_{k=1}^{n} (N_x)_k (u_c)_k \sum_{j=1}^{n} (N_x)_j (\xi_t)_j \\ &\quad + \sum_{l=1}^{n} N_l (h_r)_l \sum_{k=1}^{n} (N_x)_k (v_c)_k \sum_{j=1}^{n} (N_y)_j (\xi_t)_j \Bigg] \\ &+ \sum_{l=1}^{n} N_l (h_r)_l \left[\sum_{k=1}^{n} N_k (u_c)_k (\tilde{\xi}_t)_{xx} + \sum_{k=1}^{n} N_k (v_c)_k (\tilde{\xi}_t)_{xy} \right] - \nu_l U_{x^2y^2} \\ &- W_{i,80} \sum_{k=1}^{n} N_k \left[(\tau_{s_x})_k - (\tau_{b_x})_k \right] \Big/ \sum_{j=1h}^{n} N_j h_j \Bigg\} d\Delta^e = 0\end{aligned} \tag{94}$$

where

$$(\tilde{\xi}_t)_{xx} = -(W_x)_{i,80} \sum_{j=1}^{n} (N_x)_j (\xi_t)_j, \; (\tilde{\xi}_t)_{xy} = -(W_x)_{i,80} \sum_{j=1}^{n} (N_y)_j (\xi_t)_j \tag{95}$$

and

$$p_{x^2y^2} = -\left[(W_x)_{i,J} \sum_{j=1}^{n} (N_x)_j + (W_y)_{i,J} \sum_{j=1}^{n} (N_y)_j \right] p_j, \; p = (U, V) \tag{96}$$

In the matrix form, this equation may be written as:

$$Ar_t + Br = C \tag{97}$$

where

$$a_{i,j} = \int_{\Delta^e} W_{i,80} N_j d\Delta^e, \; i,j = 1,\cdots,n \tag{98}$$

$$b_{i,j} = \int_{\Delta^e} W_{i,80} \left[\sum_{k=1}^{n} N_k (u_c)_k (N_x)_j + \sum_{k=1}^{n} N_k (v_c)_k (N_y)_j \right] d\Delta^e, \; i,j = 1,\cdots,n \tag{99}$$

$$
\begin{aligned}
c_i = -\int_{\Delta^e} \Bigg\{ & W_{i,80} \left[\sum_{k=1}^{n} N_k u_k \sum_{j=1}^{n} (N_x)_j U_j + \sum_{k=1}^{n} N_k v_k \sum_{j=1}^{n} (N_y)_j U_j \right. \\
& \left. + g \sum_{j=1}^{n} (N_x)_j (h+\xi)_j \right] - \sum_{l=1}^{n} N_l \frac{(h_r^2)_l}{3} \sum_{k=1}^{n} (N_x)_k \left[(u_c)_k U_{x^2y^2} + (v_c)_k V_{x^2y^2} \right] \\
& + W_{i,80} \left[\sum_{k=1}^{n} N_k \frac{(h_r)_k}{2} \sum_{j=1}^{n} (N_x)_j (\xi_{tt})_j + \sum_{l=1}^{n} N_l (h_r)_l \sum_{k=1}^{n} (N_x)_k (u_c)_k \sum_{j=1}^{n} (N_x)_j (\xi_t)_j \right. \\
& \qquad \left. + \sum_{l=1}^{n} N_l (h_r)_l \sum_{k=1}^{n} (N_x)_k (v_c)_k \sum_{j=1}^{n} (N_y)_j (\xi_t)_j \right] \\
& + \sum_{l=1}^{n} N_l (h_r)_l \left[\sum_{k=1}^{n} N_k (u_c)_k (\tilde{\xi}_t)_{xx} + \sum_{k=1}^{n} N_k (v_c)_k (\tilde{\xi}_t)_{xy} \right] - v_l U_{x^2y^2} \\
& - W_{i,80} \sum_{k=1}^{n} N_k \left[(\tau_{s_x})_k - (\tau_{b_x})_k \right] \Big/ \sum_{j=1h}^{n} N_j h_j \Bigg\} d\Delta^e = 0 \, , \; i = 1, \ldots . n
\end{aligned}
\tag{100}
$$

The residual R_{82} is written as:

$$
R_{82} = \hat{U} - \frac{\hat{h}_r^2}{3} \left(\frac{\partial^2 \hat{U}}{\partial x^2} + \frac{\partial^2 \hat{U}}{\partial y^2} \right) - \hat{r} \tag{101}
$$

According to Galerkin's procedure, after using integration by parts (or Green's theorem) to reduce the second derivatives, the R_{82} error minimization leads to the following equation:

$$
\begin{aligned}
\int_{\Delta^e} W_{i,82} R_{82} d\Delta^e = \int_{\Delta^e} \Bigg\{ & W_{i,82} \sum_{j=1}^{n} N_j U_j \\
& + \sum_{k=1}^{n} N_k \frac{(h_r^2)_k}{3} \left[(W_x)_{i,82} \sum_{j=1}^{n} (N_x)_j + (W_y)_{i,82} \sum_{j=1}^{n} (N_y)_j \right] U_j \\
& - W_{i,82} \sum_{j=1}^{n} N_j r_j \Bigg\} d\Delta^e - \oint_{\Gamma^e} N_k \frac{(h_r^2)_k}{3} W_{i,82} (N_n)_j U_j \, d\Gamma^e
\end{aligned}
\tag{102}
$$

The last term of (102) can be written as:

$$
\begin{aligned}
\oint_{\Gamma^e} N_k \frac{(h_r^2)_k}{3} W_{i,82} (N_n)_j U_j \, d\Gamma^e = & \oint_{\Gamma_i^e} N_p \frac{(h_r^2)_p}{3} W_{q,82} (N_n)_r U_r \, d\Gamma_i^e \\
& + \oint_{\Gamma_e^e} N_p \frac{(h_r^2)_p}{3} W_{q,82} U_n \, d\Gamma_e^e
\end{aligned}
\tag{103}
$$

where $p, q=1, \cdots, n^e$, n^e being the number of nodes in the corresponding element side coincident with the boundary domain, Γ_i^e represents the element sides within the domain, with the corresponding integral null because the resulting element contributions are equal, but with opposite signals, and Γ_e^e represents the element sides coincident with the boundary domain. Accordingly, an equivalent form of equation (102) may be written as follows:

$$\begin{aligned}\int_{\Delta^e} W_{i,82} R_{82} d\Delta^e = &\int_{\Delta^e} \Big\{ W_{i,82} \sum_{j=1}^{n} N_j \\ &+ \sum_{k=1}^{n} N_k \frac{h\left(\frac{2}{r}\right)_k}{3} \Big[(W_x)_{i,82} \sum_{j=1}^{n} (N_x)_j + (W_y)_{i,82} \sum_{j=1}^{n} (N_y)_j \Big] \Big\} U_j \, d\Delta^e \\ &= \int_{\Delta^e} W_{i,82} \sum_{j=1}^{n} N_j r_j d\Delta^e + \oint_{\Gamma^e} N_p \frac{\left(h_r^2\right)_p}{3} W_{q,82} U_n \, d\Gamma_e^e \\ p, q = 1, &\cdots, n^e\end{aligned} \tag{104}$$

In the matrix form, this equation may be written as:

$$AU = B \tag{105}$$

where

$$a_{i,j} = \int_{\Delta^e} \left\{ W_{i,82} \sum_{j=1}^{n} N_j + \sum_{k=1}^{n} N_k \frac{\left(h_r^2\right)_k}{3} \left[\left(W_x\right)_{i,82} \sum_{j=1}^{n} \left(N_x\right)_j + \left(W_y\right)_{i,82} \sum_{j=1}^{n} \left(N_y\right)_j \right] \right\} d\Delta^e \tag{106}$$

$$\begin{aligned}b_i = \int_{\Delta^e} W_{i,82} \sum_{j=1}^{n} N_j r_j d\Delta^e + \oint_{\Gamma^e} N_p \frac{\left(h_r^2\right)_p}{3} W_{q,82} U_n \, d\Gamma^e \\ p, q = 1, \cdots, n^e.\end{aligned} \tag{107}$$

As recommended in [5] "*a suitable grid is normally crucial to the success of a finite element model. In our case, the following rules must be fulfilled for its generation.*

a. *Element side lower than the local depth.*

b. *Minimum of 20 to 25 elements per wave length.*

c. *Courant number always lower than one in the whole domain".*

4.1.2. Real case study using the WACUP model

The fortification of S. Lourenço da Cabeça Seca (lighthouse of Bugio) – Tagus estuary (Portugal) – has endured over the course of four centuries the continuous action of waves and currents, as well as bathymetric modifications resulting from the movement of significant quantities of sand in the area where it is located.

With the intention of preventing the destruction of this fortification, several studies were conducted to evaluate the best protection structure. The studies have led to a protective structure which consists of a circular dissipation platform, with a level of 2 m (HZ) and about 80 m radius (Figure 1).

Figure 1. Submerged dissipation platform to protect the Bugio lighthouse, situated at the mouth of the Tagus estuary.

The wave–current Boussinesq-type model WACUP was used to obtain the time-dependent hydrodynamic characteristics of the joint action of a relatively common wave over the high flow tide current. The wave characteristics are: wave height, $H = 3.5$ m, period $T = 12$ s and direction = 180°. Figure 2 shows the free surface level when the wave approaches the platform and its action on the Bugio lighthouse. Figure 3 shows a free surface water level, in a quasi-stationary state.

As can be seen, the wave action on the fortification was drastically reduced as a consequence of the wave breaking, refraction and diffraction on the dissipation platform constructed around the fortification.

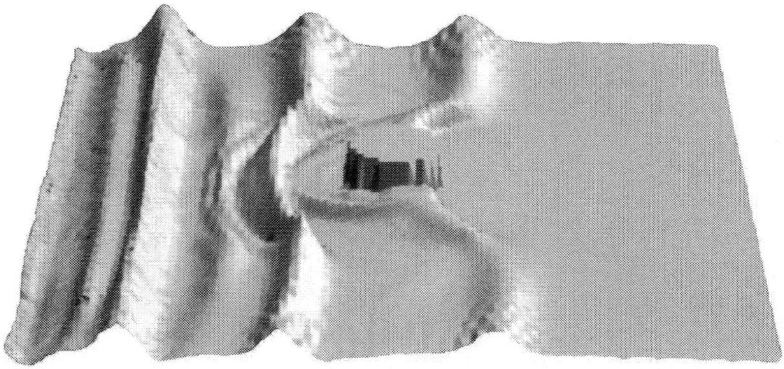

Figure 2. Perspective of the free surface obtained by simulation around the Bugio lighthouse, situated in the Tagus estuary, Portugal, in a transient condition.

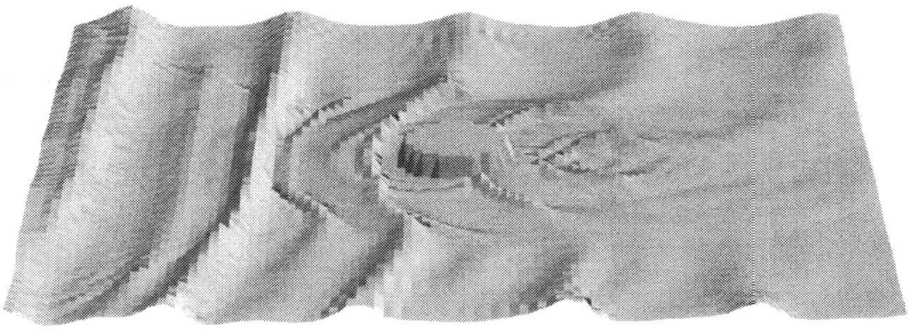

Figure 3. Protection of the Bugio lighthouse situated at the mouth of the Tagus estuary. Perspective of the free surface obtained with a numerical simulation, in a quasi-steady state.

4.1.3. COULWAVE one-layer model

A numerical scheme similar to that of [38] and [39] is utilized by [25], with the inclusion of extra nonlinear terms. Basically using the same high-order predictor–corrector scheme, [25] developed a numerical code (COULWAVE) based on Nwogu's equations for one and two layers. Parameterizations of bottom friction and wave breaking have been included in the code, as well as a moving boundary scheme to simulate wave runup and rundown. A finite–difference algorithm is used for the general one- and two-layer model equations.

According to [9], "*the equations are solved using a high-order predictor-corrector scheme, employing a third order in time explicit Adams-Bashforth predictor step, and a fourth order in time Adams-Moulton implicit corrector step*" [31]. Also in accordance with [9], "the implicit corrector step must be iterated until a convergence criterion is satisfied".

In order to solve numerically the nondimensional equations (69)–(71), these are previously rewritten in dimensional variables. Then, to simplify the predictor–corrector equations, the velocity time derivatives in the momentum equations are grouped into the dimensional form (for details see [25]):

$$U = u + \frac{k^2 - \eta^2}{2} u_{xx} + (k - \eta)(h_r u)_{xx} - \eta_x \left[\eta u_x + (h_r u)_x \right] \tag{108}$$

$$V = v + \frac{k^2 - \eta^2}{2} v_{yy} + (k - \eta)(h_r v)_{yy} - \eta_y \left[\eta v_y + (h_r v)_y \right] \tag{109}$$

where subscripts denote partial derivatives. For reasons of stability and less iterations required in the process of convergence, the nonlinear time derivatives, arisen from the nonlinear dispersion terms $\nabla\left[\eta(\nabla.(h_r u_\alpha)_t + h_{r_{tt}}/\varepsilon)\right]$ and $\nabla\left[(\eta^2/2)\nabla.u_{\alpha t}\right]$, can be reformulated using the relations:

$$\nabla\left[\eta\left(\nabla\bullet\left(h_r u_\alpha\right)_t+\frac{h_{r\,tt}}{\varepsilon}\right)\right]=\nabla\left[\eta\left(\nabla\bullet\left(h_r u_\alpha\right)+\frac{h_{r\,t}}{\varepsilon}\right)\right]_t-\nabla\left[\eta_t\left(\nabla\bullet\left(h_r u_\alpha\right)+\frac{h_{r\,t}}{\varepsilon}\right)\right] \tag{110}$$

$$\nabla\left(\frac{\eta^2}{2}\nabla\bullet u_{\alpha t}\right)=\nabla\left(\frac{\eta^2}{2}\nabla\bullet u_\alpha\right)_t-\nabla\left(\eta\eta_t\nabla\bullet u_\alpha\right) \tag{111}$$

The explicit predictor equations are:

$$\eta_{i,j}^{n+1}=\eta_{i,j}^{n}+\frac{\Delta t}{12}\left(23E_{i,j}^{n}-16E_{i,j}^{n-1}+5E_{i,j}^{n-2}\right) \tag{112}$$

$$U_{i,j}^{n+1}=U_{i,j}^{n}+\frac{\Delta t}{12}\left(23F_{i,j}^{n}-16F_{i,j}^{n-1}+5F_{i,j}^{n-2}\right)+2\left(F_1\right)_{i,j}^{n}-3\left(F_1\right)_{i,j}^{n-1}+\left(F_1\right)_{i,j}^{n-2} \tag{113}$$

$$V_{i,j}^{n+1}=V_{i,j}^{n}+\frac{\Delta t}{12}\left(23G_{i,j}^{n}-16G_{i,j}^{n-1}+5G_{i,j}^{n-2}\right)+2\left(G_1\right)_{i,j}^{n}-3\left(G_1\right)_{i,j}^{n-1}+\left(G_1\right)_{i,j}^{n-2} \tag{114}$$

where

$$\begin{aligned}E=&-h_{rt}-\left[\left(\eta+h_r\right)u\right]_x-\left[\left(\eta+h_r\right)v\right]_y\\&+\left\{\left(\eta+h_r\right)\left[\left(\frac{1}{6}\left(\eta^2-\eta h_r+h_r^2\right)-\frac{1}{2}k^2\right)S_x+\left(\frac{1}{2}\left(\eta-h_r\right)-k\right)T_x\right]\right\}_x\\&+\left\{\left(\eta+h_r\right)\left[\left(\frac{1}{6}\left(\eta^2-\eta h_r+h_r^2\right)-\frac{1}{2}k^2\right)S_y+\left(\frac{1}{2}\left(\eta-h_r\right)-k\right)T_y\right]\right\}_y\end{aligned} \tag{115}$$

$$\begin{aligned}F=&-\frac{1}{2}\left[\left(u^2\right)_x+\left(v^2\right)_x\right]-g\eta_x-kh_{rxtt}-k_t h_{rxt}+\left(Eh_{rt}+\eta h_{rtt}\right)_x\\&-\left[E\left(\eta S+T\right)\right]_x-\left[\frac{1}{2}\left(k^2-\eta^2\right)\left(uS_x+vS_y\right)\right]_x\\&-\left[\left(k-\eta\right)\left(uT_x+vT_y\right)\right]_x-\frac{1}{2}\left[\left(T+\eta S\right)^2\right]_x\end{aligned} \tag{116}$$

$$F_1=\frac{\eta^2-k^2}{2}v_{xy}-\left(k-\eta\right)\left(h_r v\right)_{xy}+\eta_x\left[\eta v_y+\left(h_r v\right)_y\right] \tag{117}$$

$$G = -\frac{1}{2}\left[\left(u^2\right)_y + \left(v^2\right)_y\right] - g\eta_y - kh_{rytt} - k_t h_{ryt} + \left(Eh_{rt} + \eta h_{rtt}\right)_y$$
$$-\left[E(\eta S + T)\right]_y - \left[\frac{1}{2}\left(k^2 - \eta^2\right)\left(uS_x + vS_y\right)\right]_y \tag{118}$$
$$-\left[(k-\eta)\left(uT_x + vT_y\right)\right]_y - \frac{1}{2}\left[(T + \eta S)^2\right]_y$$

$$G_1 = \frac{\eta^2 - k^2}{2} u_{xy} - (k-\eta)\left(h_r u\right)_{xy} + \eta_y\left[\eta u_x + \left(h_r u\right)_x\right] \tag{119}$$

and

$$S = u_x + v_y ; T = h_{rt} + \left(h_r u\right)_x + \left(h_r v\right)_y \tag{120}$$

All first-order spatial derivatives are differenced with fourth-order ($\Delta x^4 = \Delta y^4$) accurate equations, which are five-point differences. Second-order spatial derivatives are approximated with three-point centered finite-difference equations, which are second-order accurate. The fourth-order implicit corrector expressions for the free surface elevation, η, and the horizontal velocities, u and v, are:

$$\eta_{i,j}^{n+1} = \eta_{i,j}^{n} + \frac{\Delta t}{24}\left(9E_{i,j}^{n+1} + 19E_{i,j}^{n} - 5E_{i,j}^{n-1} + E_{i,j}^{n-2}\right) \tag{121}$$

$$U_{i,j}^{n+1} = U_{i,j}^{n} + \frac{\Delta t}{24}\left(9F_{i,j}^{n+1} + 19F_{i,j}^{n} - 5F_{i,j}^{n-1} + F_{i,j}^{n-2}\right) + \left(F_1\right)_{i,j}^{n+1} - \left(F_1\right)_{i,j}^{n} \tag{122}$$

$$V_{i,j}^{n+1} = V_{i,j}^{n} + \frac{\Delta t}{24}\left(9G_{i,j}^{n+1} + 19G_{i,j}^{n} - 5G_{i,j}^{n-1} + G_{i,j}^{n-2}\right) + \left(G_1\right)_{i,j}^{n+1} - \left(G_1\right)_{i,j}^{n} \tag{123}$$

As noted in [25], "*the system is solved by first evaluating the predictor equations, then u and v are solved via (108) and (109), respectively. Both (108) and (109) yield a diagonal matrix after finite differencing. The matrices are diagonal, with a bandwidth of three (due to three-point finite differencing), and the efficient Thomas algorithm can be utilized. At this point in the numerical system, we have predictors for η, u and v*". Next, the corrector expressions are evaluated, and again u and v are determined from (108) and (109).

Also in accordance with [25], "*the error is calculated, in order to determine if the implicit correctors need to be reiterated. The error criteria employed is a dual calculation, and requires that either*"

$$\max\left|\frac{w^{n+1}-w_*^{n+1}}{w^{n+1}}\right|<\frac{\varepsilon}{100} \text{ or } \frac{\sum\left|w^{n+1}-w_*^{n+1}\right|}{\sum\left|w^{n+1}\right|}<\varepsilon \tag{124}$$

In these expressions, w represents any of the variables η, u and v, and w_* is the previous iterations value. The value of the error is set to 10^{-6}. Linear stability analysis performed by [38, 18] and [40] show that $\Delta t<\Delta x/(2\,c)$ to ensure stability, where c is the celerity.

4.1.4. Research study using COULWAVE model

For the analysis concerning coastal protection, the mean currents around a submerged structure (artificial reef) are analysed. The output of the COULWAVE model corresponds to the velocity values at a depth 0.531 h_r below the water surface. The velocity at this depth is used to determine the velocity cells near the shoreline that could give an indication of the sediment transport. Divergent cells indicate erosion near by the shoreline and convergent cells indicate sedimentation.

The numerical simulations to study the 2HD behavior of the hydrodynamics around the reef have been done for four cases: two reef geometries, varying the reef angle (45° and 66°), and two wave conditions. The characteristics of the simulations for the different cases are described in Table 1.

	Reef angle (°)	H (m)	T (s)	Number of grid points per wavelength	Grid size (m)	Time step (s)
C1	45	4.0	15	60	2.77	0.11999
C2	66	4.0	15	70	2.37	0.10285
C3	45	1.5	9	43	2.14	0.09288
C4	66	1.5	9	43	2.14	0.09288

Table 1. Main characteristics of the simulations performed.

As recorded in [28], "*the computational domain is around 1870 m in the long-shore and 1670 m in the cross-shore direction with a constant node spacing of around $\Delta x=\Delta y$ =2.0 m and a Courant number of 0.5. The total simulation time was 800 s. A flat bottom is placed in front of the slope where waves are generated using the source function method ([39] Wei et al., 1995). The source function is located at x=80 m along the y direction*" (Figure 4).

Two sponge layers are used, one in front of the offshore boundary to absorb the outgoing wave energy, and the other on the beach, both with a width of 0.5 times the wavelength of the incident wave. The numerical results obtained by the model are the time series of the free surface elevation, the two velocity components, u and v, and the wave breaking areas (Figure 5).

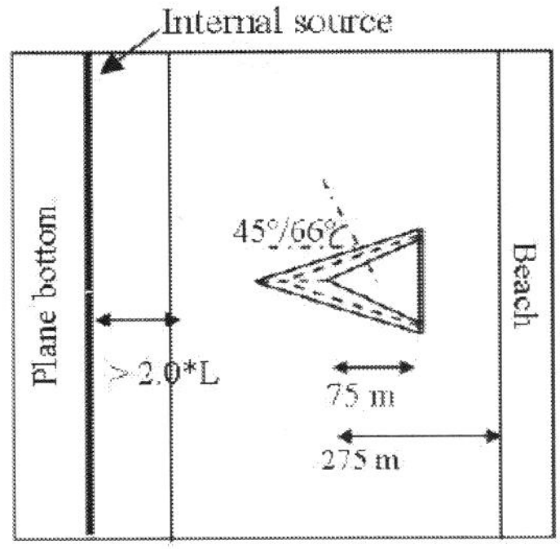

Figure 4. Schematic representation of the simulated geometry (not at scale) [28].

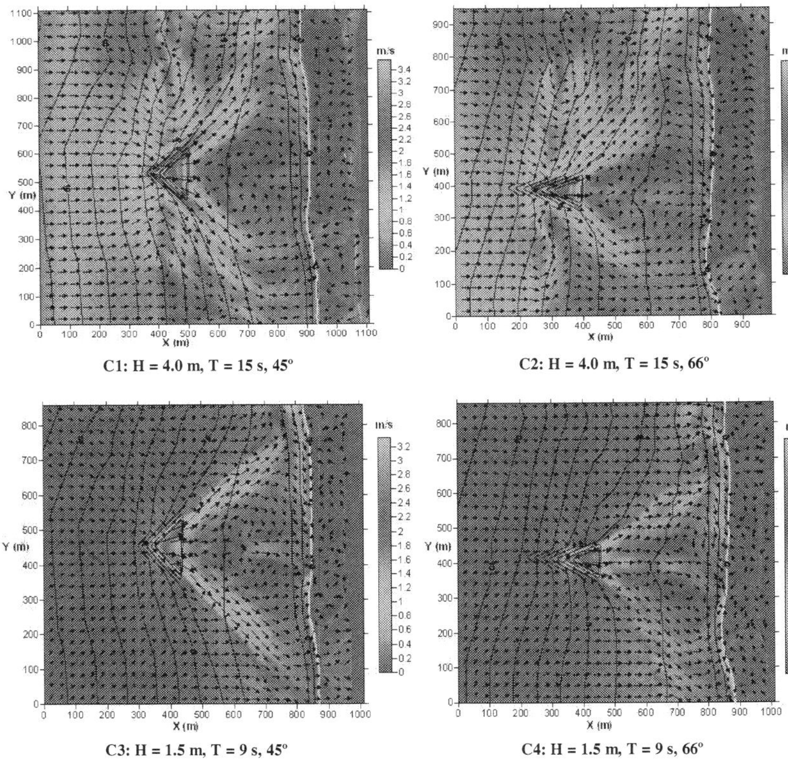

Figure 5. Velocity patterns of cases C1 to C4 [28].

Results of this simulation are described in [29], including the observed phenomena, such as *"along the reef, an increase of the wave height is observed, in opposition to the situation without the reef, owing to the decrease of the depth in the reef zone. Moreover, owing to the increase in wave heights, the wave breaking occurs earlier (and in general over the reef) in comparison with the situation without a reef. From Figure 5, it is clear that the presence of the reef significantly alters the wave heights, with the wave height increasing along the reef as a consequence of decreasing water depth"*.

Figure 5 also shows for all cases that convergent cells appear, indicating a possible sedimentation near the shoreline, suggesting that the chosen geometries (Figure 4) are advantageous for both coastal protection and improving surf conditions. Anyway, as referred in [28], a morphological study should be done in order to confirm these results.

4.2. 2HD Serre's standard model

4.2.1. Numerical formulation

The equation system (19) is solved in [7] using an explicit finite–difference method based on the MacCormack time-splitting scheme. For this purpose, the equations are written in the following form:

$$h_t + P_x + Q_y = 0 \tag{125}$$

$$\begin{aligned} P_t + (uP)_x + (vP)_y + \left\{\left[(g+\beta)/2 + \alpha/3\right]h^2\right\}_x \\ = -(g+\beta+\alpha/2)h\xi_x - \tau_{bx} + Rdiv(h\,grad\,u) \end{aligned} \tag{126}$$

$$\begin{aligned} Q_t + (uQ)_x + (vQ)_y + \left\{\left[(g+\beta)/2 + \alpha/3\right]h^2\right\}_y \\ = -(g+\beta+\alpha/2)h\xi_y - \tau_{by} + Rdiv(h\,grad\,v) \end{aligned} \tag{127}$$

where $P=hu$, $Q=hv$, $\alpha=d^2h/dt^2$; $\beta=d^2\xi/dt^2$ and the bottom friction terms, τ_{bx} and τ_{by}, are obtained through (128):

$$\tau_{bx} = g\frac{P\sqrt{P^2+Q^2}}{K^2h^{7/3}} \text{ and } \tau_{by} = g\frac{Q\sqrt{P^2+Q^2}}{K^2h^{7/3}} \tag{128}$$

In order to apply the MacCormack's method, equations (125) –(127) are split into two systems of three equations throughout the *Ox* and *Oy* directions. The corresponding operators *Lx* and *Ly* take the following form [7]:

Operator *Lx*

$$h_t + P_x = 0 \tag{129}$$

$$P_t + (uP)_x + \left\{\left[(g+\beta)/2 + \alpha/3\right]h^2\right\}_x = -(g+\beta+\alpha/2)h\xi_x - \tau_{bx} + R\,P_{xx} \tag{130}$$

$$Q_t + (uQ)_x = RQ_{xx} \tag{131}$$

Operator *Ly*

$$h_t + Q_y = 0 \tag{132}$$

$$P_t + (vP)_y = RP_{yy} \tag{133}$$

$$Q_t + (vQ)_y + \left\{\left[(g+\beta)/2 + \alpha/3\right]h^2\right\}_y = -(g+\beta+\alpha/2)h\xi_y - \tau_{by} + RQ_{yy} \tag{134}$$

Considering the generic variable F, the solution at time $(n+1)\Delta t$, for the computational point (i, j), is obtained from the known solution $F_{i,j}^{n}$ through the following symmetric application:

$$F_{i,j}^{n+1} = Lx\left(\frac{\Delta t}{4}\right)Ly\left(\frac{\Delta t}{4}\right)Lx\left(\frac{\Delta t}{4}\right)Ly\left(\frac{\Delta t}{4}\right)Ly\left(\frac{\Delta t}{4}\right)Lx\left(\frac{\Delta t}{4}\right)Ly\left(\frac{\Delta t}{4}\right)Lx\left(\frac{\Delta t}{4}\right)F_{i,j}^{n} \tag{135}$$

where each operator, *Lx* and *Ly*, is composed of a predictor–corrector sequence and n represents a generic time t. In the application (135) of eight predictor–corrector sequences, alternately backward and forward space differences are used. After each predictor and each corrector of the application F, the values of the velocities (u, v) are updated and the values of the vertical accelerations, α and β, are recalculated (for details see [7]).

4.2.2. Real case study

This model was applied to compute the agitation established in the port of Figueira da Foz, Portugal, for different sea states. This port is 2,250 m long and 400 m wide, approximately. Its average depth is of the order of 7 m, with approximately 12 m throughout the outer port basin.

Two simulations were performed, considering the state of rest as initial condition in both cases, although for different tidal heights. The first case corresponds to the entrance of a sinusoidal wave in the open sea (input boundary), with a mean wave height $H = 3.90$ m, period $T = 15$ s, wavelength $\lambda = 147$ m, and direction $\Phi = 260°$ W. The second case corresponds to a mean wave height $H = 4.80$ m, period $T = 17.5$ s, wavelength $\lambda = 173$ m, and direction $\Phi = 258.2°$ W. Figures 6 and 7 show perspective views of the free surface computed in the basin 269.1 s and 360 s after excitation. As can be seen, a zone with stronger agitation is observed in the outer harbor in the second case.

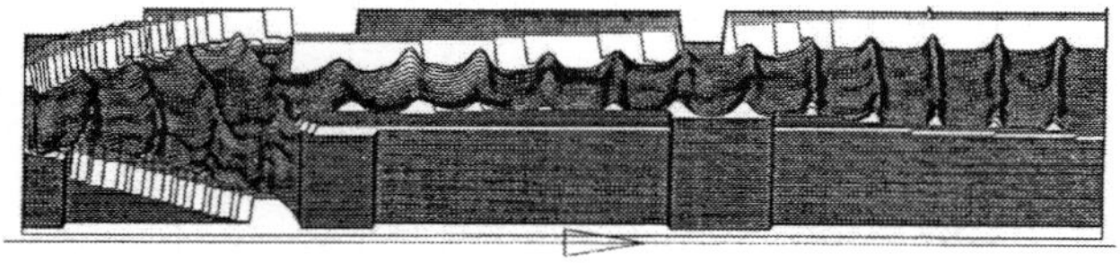

Figure 6. Port of Figueira da Foz. Perspective view of the free surface computed 269.1 s after excitation, for a mean wave height, $H = 3.90$ m, period $T = 15$ s, wavelength $\lambda = 147$ m and direction $\Phi = 283°$W. Tide height, 3.35 m (HZ).

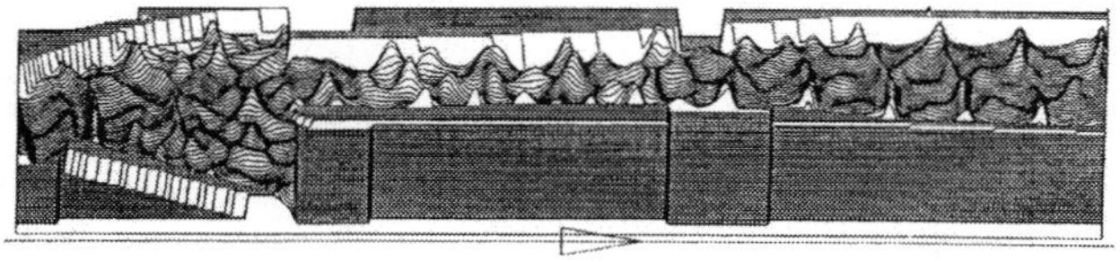

Figure 7. Port of Figueira da Foz. Perspective view of the free surface computed 360 s after excitation, for a mean wave height $H = 4.80$ m, period $T = 17.5$ s, wavelength $\lambda = 173$ m and direction $\Phi = 258.2°$ W. Tide height, 2.65 m (HZ).

4.3. 1HD Serre's extension model

4.3.1. Numerical solution

The equation system (75) is solved using an efficient finite–difference method, whose consistency and stability were tested in [1] and [2] by comparison with a closed-form solitary wave solution of the Serre equations. For this purpose, the terms containing derivatives in time of u are grouped. The final system of three equations is rewritten according to the following equivalent form (SERIMP model) [1, 2 3]:

$$
\begin{aligned}
& h_t + (uh)_x = 0 && \text{a} \\
& q_t + \left\{ uq - \frac{1}{2}\left[u^2 + (1+2\alpha) h^2 (u_x)^2 + (1+2\alpha)(\xi_x)^2 u^2 - h\,\xi_x (u^2)_x \right] \right\}_x && \\
& \quad + \left[g(1+\alpha\Omega) + \alpha h u u_{xx} \right] \eta_x - \alpha g h h_x \eta_{xx} - \beta g \frac{h^2}{3} \eta_{xxx} && \text{b} \\
& \quad - \frac{\alpha}{2}\left[(h\,\xi_{xx} u^2)_x + h_x \xi_{xx} u^2 - h\,\xi_{xx} u u_x \right] + \left(\alpha - \frac{\beta}{3} \right) h^2 u_x u_{xx} && \\
& \quad + \beta \frac{h^2}{3} u u_{xxx} + \tau_b / (\rho h) = 0 && \\
& \left[1 + (1+\alpha)\,\Omega \right] u - (1+\alpha)\, h h_x u_x - (1+\beta) \frac{h^2}{3} u_{xx} = q && \text{c} \\
& \Omega = \xi_x \eta_x + \frac{1}{2} h\, \xi_{xx} + (\xi_x)^2 && \text{d}
\end{aligned}
\tag{136}
$$

To compute the solution of equation system (75) (values of the variables h and u at time $t+\Delta t$), we use a numerical procedure based on the following scheme, itself based on the last equation system (136), for variables h, q and u. Knowing all values of h_i and u_i, $i=1, N$, in the whole domain at time $n\Delta t$, the equations (136c) and (136d) are used to obtain the first values of q_i and Ω_i in the whole domain. Then, we continue with the following steps, in which the index p means predicted values (see also [1, 2] and [3]):

1. The first equation (136a) is used to predict the variable values h_{pi} at time $t+\Delta t$ ($h_{pi}^{t+\Delta t}$), in the whole domain.

2. The second equation (136b) makes it possible to predict the variable values q_{pi} at time $t+\Delta t$ ($q_{pi}^{t+\Delta t}$), taking into account the values $\bar{h}_i^{t+\Delta t}=0.5\left(h_i^t+h_{pi}^{t+\Delta t}\right)$, namely for Ω_i in the whole domain.

3. The third equation (136c) makes it possible to compute the mean-averaged velocities $u_i^{t+\Delta t}$ at time $t+\Delta t$, taking into account the predicted values $h_{pi}^{t+\Delta t}$ and $q_{pi}^{t+\Delta t}$, namely for Ω_i in the whole domain.

4. The first operation (step 1) is repeated in order to improve the accuracy of the variable values h_i at time $t+\Delta t$ ($h_i^{t+\Delta t}$), using the values $\bar{u}_i^{t+\Delta t}=0.5\left(u_i^t+u_i^{t+\Delta t}\right)$ in the whole domain.

5. Finally, the second operation (step 2) is repeated in order to improve the accuracy of the variable values q_i at time $t+\Delta t$ ($q_i^{t+\Delta t}$), taking into account the values $\bar{h}_i^{t+\Delta t}=0.5\left(h_i^t+h_i^{t+\Delta t}\right)$ and $\bar{u}_i^{t+\Delta t}=0.5\left(u_i^t+u_i^{t+\Delta t}\right)$ in the whole domain.

At each interior point i, the first, second and third-order spatial derivatives are approximated through centered differences and the time derivatives are approximated using forward differences. The convective terms $(uh)_x$ and $(uq)_x$ in equations (136a) and 136b) are approximated through centered schemes in space and time for variables h and q. At each time t, these terms are written in the following form:

$$\left(uh\right)_x=u_i^t\left(\frac{h_{i+1}^t-h_{i-1}^t+h_{i+1}^{t+\Delta t}-h_{i-1}^{t+\Delta t}}{4\Delta x}\right)+\frac{1}{2}\left(h_i^t+h_i^{t+\Delta t}\right)\left(\frac{u_{i+1}^t-u_{i-1}^t}{2\Delta x}\right) \tag{137}$$

$$\left(uq\right)_x=u_i^t\left(\frac{q_{i+1}^t-q_{i-1}^t+q_{i+1}^{t+\Delta t}-q_{i-1}^{t+\Delta t}}{4\Delta x}\right)+\frac{1}{2}\left(q_i^t+q_i^{t+\Delta t}\right)\left(\frac{u_{i+1}^t-u_{i-1}^t}{2\Delta x}\right) \tag{138}$$

All finite–difference equations are implicit. Therefore, the solution of system (136) requires, in each time step, the computation of five three-diagonal systems of N-2 equations (steps 1 to 5), which are easily computed using the three-diagonal matrix algorithm (TDMA), also known as the Thomas algorithm. The stability condition to be observed can be expressed in terms of the Courant/CFL number, and is given by:

$$C_R = \sqrt{gh}\frac{\Delta t}{\Delta x} < 1.0 \tag{139}$$

4.3.2. Boundary conditions

We often prescribe an influx on the left boundary, usually a mono- or bi-chromatic wave flow. The initial condition for this (mono-chromatic) influx is:

$$\eta(0,t) = \frac{a}{2}\left\{\sin(\omega t)\tanh\left(t - \frac{\pi}{\omega}\right) + \sin(\omega t) - \frac{\cos(\omega t)}{\omega}\left[1 - \tanh^2\left(t - \frac{\pi}{\omega}\right)\right]\right\} \tag{140}$$

$$u(0,t) = \frac{a}{2}\frac{g}{\omega}\cos(\omega t)\left[\tanh\left(t - \frac{\pi}{\omega}\right) + 1\right] \tag{141}$$

where a is the wave amplitude and the *tanh* is used as a ramp function. It increases smoothly from 0 to 1, to create a smooth start. For t large, the boundary condition reduces to:

$$\eta(0,t) = \frac{a}{2}\sin(\omega t) \tag{142}$$

$$u(0,t) = \frac{g}{\omega}\cos(\omega t) \tag{143}$$

If we want to avoid reflections at the right boundary (output), the domain is extended with a damping region of length L_{damp}. In this case, terms like $-m(x)\,\eta$ and $-m(x)\,u$ may be added to the continuity (136a) and momentum (136b) equations, respectively. The length of the damped region is chosen such that we do not see any significant reflections.

4.3.3. Solitary wave travelling up a slope and reflection on a vertical wall

Experimental data and numerical results are available for a solitary wave propagating on the bathymetry shown in Figure 8 [1, 2]. It shows a constant depth before $x = 55$ m and a slope 1:50 between $x = 55$ m and $x = 75$ m. An impermeable vertical wall is placed at $x = 75$ m, corresponding to fully reflecting boundary conditions. A solitary wave of amplitude 0.12 m is initially centered at $x = 25$ m. The computational domain was uniformly discretized with a spatial step $\Delta x = 0.05$ m. A zero friction factor has been considered. Computations were carried out with a time step $\Delta t = 0.010$ s. Figure 9 compares numerical time series of surface elevation and test data at $x = 72.75$ m.

Figure 9 shows two peaks; the first one corresponding to the incident wave, and the second to the reflected wave. The extended Serre model predictions for both peaks agree well with the measurements. *RMSE* values equal to 0.0090 m and 0.0117 m were found in first and second peaks, respectively, for the wave height. Regarding the phase, there is a loss of approximately 0.05 s and of 0.10 s in those peaks (for details see [1] and [2]).

http://dx.doi.org/10.5772/61866

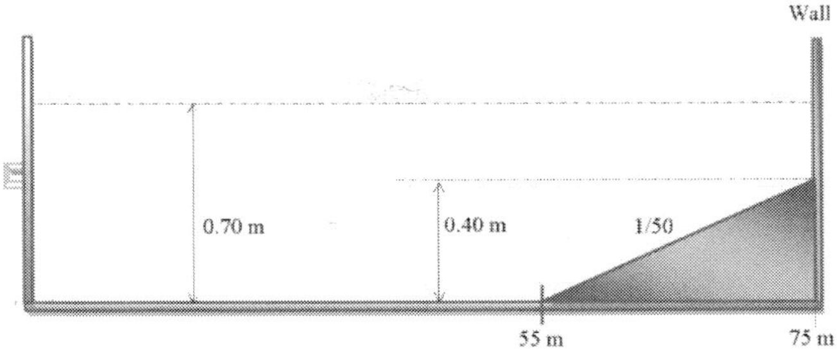

Figure 8. Bathymetry for a solitary wave travelling up a slope and its reflection on a vertical wall (not in scale).

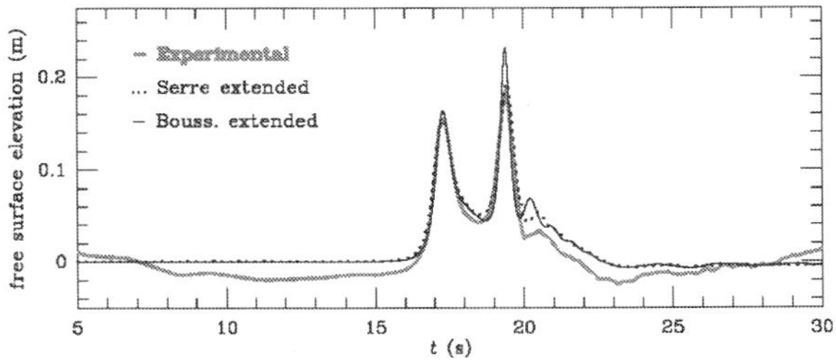

Figure 9. Solitary wave travelling up a slope and its reflection on a vertical wall. Free surface elevation in a depth gauge located at x = 72.75 m. Experimental (—); Serre extended (···); Boussinesq extended (—) [1, 2, 3].

Predictions of the extended Boussinesq equations for both peaks are less accurate. Particularly for the reflected peak, this is overestimated in about 20%. This result is not surprising, given the lower validity of the Boussinesq model for waves of higher relative amplitude. Indeed, this model assumes $O(\varepsilon)<<1$, contrary to the Serre model, which is $O(\varepsilon)=1$. A visual comparison of numerical results of the extended Boussinesq approximation with a similar study performed by [37], using the extended Boussinesq model developed by [30], shows no relevant differences in the graphs (see [1, 2] and [3]).

4.3.4. Periodic wave over an underwater bar

Beji and Battjes (1993) [10] conducted experiments in a flume of 0.80 m wide with a submerged trapezoidal bar. The up- and down-wave bottom slopes of the submerged bar are 1:20 and 1:10, respectively. Before and after the bar, the water depth is 0.40 m, with a reduction to 0.10 m above the bar, as shown in Figure 10. Experimental data obtained in this installation are available in the literature, and can be used for comparisons.

The measured data are compared with numerical results of the extended Boussinesq model (57), with $\alpha=0.1308$ and $\gamma=-0.0076$, and the extended Serre equations (75) (SERIMP model, in [2] and [3]), both improved with linear dispersive characteristics. Comparisons are made in

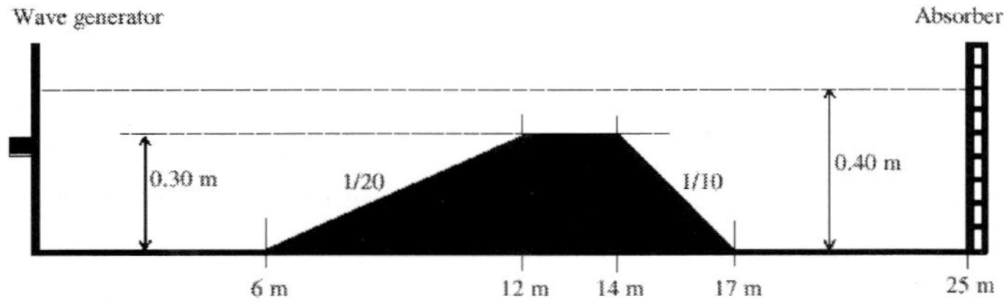

Figure 10. Bathymetry for a periodic wave propagating over a bar (not in scale) [2].

three wave gauges located at $x = 10.5$ m, $x = 13.5$ m and $x = 17.3$ m. For this purpose, a regular incident wave case with height 0.02 m, period $T = 2.02$ s and wavelength 3.73 m has been simulated. The computational domain was discretized with a uniform grid interval $\Delta x = 0.025$ m. A time step $\Delta t = 0.0010$ s was used. Globally, numerical results of the improved Serre and Boussinesq models agree quite well with the measured data (for details see [2] and [3]).

Following is presented a comparison of the standard Serre's model (20) with the extended Serre equations (75) (SERIMP model). The standard Serre's model (20) is only valid for shallow waters, thus under conditions up to $h_0 / \lambda = 0.05$. In this experiment, the dispersion parameter ($\sigma = h_0 / \lambda$) is greater than 0.05 (about 0.11) in front and behind the bar, and therefore affects the validity of the numerical outcomes. Due to the fact that over the bar there are very shallow water conditions ($\sigma \approx 0.03$) the standard Serre equations are used considering the input boundary located at section $x = 13.5$ m, where the input signal is known (measured data). In this way, results of the Serre's standard model are not influenced, as would happen, by changes arising from the wave propagation before the bar, under intermediate water depths.

Figure 11 shows a comparison of numerical results of the 1HD standard Serre's model (20) with the extended Serre equations (75), considering, in the first case, the input boundary at $x = 13.5$ m (gauge signal) (see [2]). The influence of additional terms of dispersive origin included in the extended Serre equations is clearly shown in Figure 11. The standard Serre model results (dashed line) are clearly of lesser quality. It should be noted that this application also demonstrates the good behavior of our numerical model to propagate a complex signal imposed at boundary.

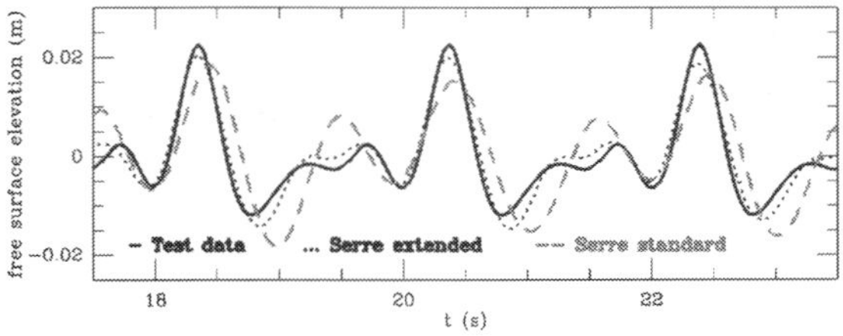

Figure 11. Periodic wave propagating over a bar. Comparison of test data (—) with numerical results of the extended Serre model (75) (···) and the standard Serre equations (20) (---) [2].

5. Conclusions and recommendations

This work presents some of the most recent methodologies to improve the linear dispersion characteristics of the classical Boussinesq equations for variable depth. A simple procedure to improve the linear dispersive performance of Serre's equations is presented as well.

Extensions of the classical Boussinesq and Serre equations for intermediate depths and deep-water applications are really important, and have been the subject of major developments in recent decades, since these are the conditions typically found in nearshore areas. Application examples to solve real-world problems clearly demonstrate the capabilities and potential of the Boussinesq- and Serre-type models with improved linear dispersion characteristics.

The influence of the dispersion characteristics is clearly evidenced by the generation and propagation of waves in intermediate water depths, as shown here in cases of very demanding applications over bottoms with considerable slopes. The overall agreement of the extended Serre model with improved dispersion characteristics is very good both in shallow water conditions as in intermediate water depths. The influence of additional terms of dispersive origin included in the extended Serre's equations is clearly shown.

This work also shows that current models of Boussinesq type with improved dispersive characteristics are accurate enough for applications in extensive real-world conditions.

Although the contribution of this work to improve the dispersive characteristics of the 1HD standard Serre equations (20) is clear, the author is aware of the need to further improve the dispersion properties of the equation system (75).

An extended version of the two-dimensional system of equations (19) with improved linear dispersion characteristics is being developed and will be published soon.

Author details

José Simão Antunes do Carmo*

Address all correspondence to: jsacarmo@dec.uc.pt

University of Coimbra, Department of Civil Engineering, Coimbra, Portugal

References

[1] Antunes do Carmo JS. Boussinesq and Serre type models with improved linear dispersion characteristics: applications. J Hydraulic Res 2013a;51(6):719–27:814090. DOI: 10.1080/00221686.814090.

[2] Antunes do Carmo JS. Extended Serre equations for applications in intermediate water depths. Open Ocean Eng J 2013b;6:16–25.

[3] Antunes do Carmo JS. Applications of Serre and Boussinesq type models with improved linear dispersion characteristics. In: Jesús MB, Irene A, Alberto P, José MG, Nuno S, Miguel S. (Eds.) Congreso de Métodos Numéricos en Ingeniería; 25–28 June; Bilbao, Spain. Barcelona, Spain: International Center for Numerical Methods in Engineering (CIMNE); 2013c. DOI: ISBN 978-84-941531-4-3.

[4] Antunes do Carmo JS. Closure to "Boussinesq- and Serre-type models with improved linear dispersion characteristics: applications". J Hydraulic Res 2015;53:284–5.

[5] Antunes do Carmo JS, Seabra-Santos FJ. On breaking waves and wave-current interaction on shallow water: a 2DH finite element model. Int J Num Method Fluids 1996;22:429–44.

[6] Antunes do Carmo JS, Seabra-Santos FJ. Wave-current interactions over bottom with appreciable variations in both space and time. Adv Eng Software 2010;41:295–305.

[7] Antunes do Carmo JS, Seabra-Santos FJ, Almeida AB. Numerical solution of the generalized Serre equations with the MacCormack finite-difference scheme. Int J Num Method Fluids 1993b;16:725–38. DOI: 10.1002/fld.1650160805.

[8] Antunes do Carmo JS, Seabra-Santos FJ, Barthélemy E. Surface waves propagation in shallow-water: a finite element model. Int J Num Method Fluids 1993a;16:447–59. DOI: 10.1002/fld.1650160602.

[9] Antunes do Carmo JS, Voorde MT, Neves MG. Enhancing submerged coastal constructions by incorporating multifunctional purposes. J Coastal Conserv 2011;15(4): 531–46.

[10] Beji S, Battjes JA. Experimental investigations of wave propagation over a bar. Coastal Eng 1993;19(1,2):151–62.

[11] Beji S, Nadaoka K. A formal derivation and numerical modelling of the improved Boussinesq equations for varying depth. Ocean Eng 1996;23(8):691–704.

[12] Berkhoff JCW, Booij N, Radder AC. Verification of numerical wave propagation models for simple harmonic linear water waves. Coastal Eng 1982;6:255–79.

[13] Booij N. A note on the accuracy of the mild-slope equation. Coastal Eng 1983;7:191–203.

[14] Boussinesq J. Théorie des ondes et des remous qui se propagent le long d'un canal rectangulaire horizontal. J Math Pures Appliq 1872;2(17):55–108.

[15] Dalrymple RA. Model for refraction of water waves. J Waterway Port CoastalOcean Eng 1988;114(4):423–35.

[16] Gobbi MFG, Kirby JT, Wei G. A fully nonlinear Boussinesq model for surface waves. Part 2. Extension to O(kh)4. J Fluid Mechanics 2000;405:181–210.

[17] Green AE, Naghdi PM. A derivation of equations for wave propagation in water of variable depth. J Fluid Mechanics 1976;78(2):237–46.

[18] Hsiao SC. Permeable effects on nonlinear water waves [thesis]. Cornell University: 2000.

[19] Hsiao SC, Liu PLF. Permeable effects on nonlinear water waves. In: Royal Society of London, editor. London. London:2002.

[20] Kirby JT. A note on linear surface wave-current interaction over slowly varying topography. J Geophys Res 1984;89(C):745–47.

[21] Kirby JT, Dalrymple RA. A parabolic equation for the combined refraction-diffraction of stokes waves by mildly varying topography. J Fluid Mechanics 1983;136:435–66.

[22] Liu PLF. Model equations for wave propagation from deep to shallow water. In: Liu PLF, (Ed.) Advances in Coastal Engineering. World Scientific. 1994, pp. 125–157.

[23] Liu ZB, Sun ZC. Two sets of higher-order Boussinesq-type equations for water waves. Ocean Eng 2005;32:1296–310.

[24] Lynett P. Nearshore wave modeling with high-order Boussinesq-type equations. J Waterway Port Coastal Ocean Eng 2006;132(5):348–57.

[25] Lynett P, Liu PLF. (Eds.) Modeling Wave Generation, Evolution, and Interaction with Depth Integrated, Dispersive Wave Equations COULWAVE Code Manual. 1st ed. Ithaca, NY: Cornell University Long and Intermediate Wave Modeling Package; 2002.

[26] Lynett P, Liu PLF. Linear analysis of the multi-layer model. Coastal Eng 2004;51:439–54. DOI: 10.1016/coastaleng.2004.05.004.

[27] Madsen PA, Murray R, Sørensen OR. A new form of the Boussinesq equations with improved linear dispersion characteristics. Coastal Eng 1991;15(4):371–88.

[28] Mendonça AM, Fortes CJ, Capitão R, Neves MG, Moura T, Antunes do Carmo JS. Wave hydrodynamics around a multi-functional artificial reef at Leirosa. J Coastal Conserv 2012a;16(4):543–53. DOI: 10.1007/s11852-012-0196-1.

[29] Mendonça AM, Fortes CJ, Capitão R, Neves MG, Antunes do Carmo JS, Moura T. Hydrodynamics around an Artificial Surfing Reef at Leirosa, Portugal. J Waterway Port Coastal Ocean Eng 2012b;138(3):226–35. DOI: 10.1061/(ASCE)WW.1943-5460.0000128.

[30] Nwogu O. Alternative form of Boussinesq equations for nearshore wave propagation. J Waterway Port Coastal Ocean Eng 1993;119:618–38.

[31] Press WH, Flannery BP, Teukolsky SA. Numerical Recipes. Cambridge University Press. 1989, pp. 569–72.

[32] Saint-Venant B. Theory of unsteady water flow, with application to river floods and to propagation of tides in river channels. Computes Rendus Acad. Sci., Paris. 1871;73:148–54, 237–40.

[33] Seabra Santos FJ. Contribution a l'ètude des ondes de gravité bidimensionnelles en eau peu profonde [thesis]. Université Scientifique et Médicale et Institut National Polutechnique de Grenoble, France:1985.

[34] Seabra Santos FJ. As aproximações de Wu e de Green & Naghdi no quadro geral da teoria da água pouco profunda (in Portuguese). In: APRH, editor. Simpósio Luso-Brasileiro de Hidráulica e Recursos Hídricos (4º SILUSBA); 14–16 June; Lisbon, Portugal, LNEC, Lisbon:1989. pp. 209–219.

[35] Serre F. Contribution à l'étude des écoulements permanents et variables dans les canaux. La Houille Blanche 1953;8:374–88.

[36] Walkley MA. A Numerical Method for Extended Boussinesq Shallow-Water Wave Equations [thesis]. Leeds, UK: University of Leeds, School of Computer Studies:1999.

[37] Walkley MA, Berzins M. A finite element method for the one-dimensional extended Boussinesq equations. Int J Num Method Fluids 1999;29:143–57.

[38] Wei G, Kirby JT. Time-dependent numerical code for extended Boussinesq equations. J Waterway Port Coastal Ocean Eng 1995;121(5):251–61.

[39] Wei G, Kirby JT, Grilli ST, Subramanya R. A fully nonlinear Boussinesq model for surface waves. I. Highly nonlinear, unsteady waves. J Fluid Mechanics 1995;294:71–92.

[40] Woo S-B. Finite element modeling of the fully-nonlinear extended Boussinesq equations [thesis]. Ithaca, NY: Cornell University:2002.

Chapter 3

CFD-Based Investigation of Wind-Strokes over Highway Bridge Section

Medzid Muhasilovic, Kenan Imsirpasic, Karel Ciahotny and Brano Sirok

Additional information is available at the end of the chapter

http://dx.doi.org/10.5772/61928

Abstract

There is an almost everlasting debate on the possibilities of the investigation tools for their applications in a prospective fashion (while solving the engineering tasks) and – against this fact – on the use of these engineering tools while correcting the existing technology problems. A unique chance to compare these needs of research in traffic intentions (while setting the modern road communications through southern Bosnia and Herzegovina) with the natural occurrences in the atmosphere (such is a strong north wind in this geographic region) offers us the highway section Pocitelj-Zvirovici. Exactly in such cases (and before the actual construction of this highway bridge) "for the sake" of prospective engineering, the CFD mechanism (the "toolkit" for performing the computational fluid dynamics) was applied to engage this atmospheric problem. Both as steady-state explorations (while applying the k-ε turbulence treatment) and as the time-dependent CFD-based mode, we explored the wind-strokes of 10 m/s, 20 m/s, 30 m/s and 40 m/s, expected but certainly unwanted strong gaseous flows over the bridge, detecting in this way the traffic safety edge points. The results coming from the explorations performed by the CFD tool are explained and discussed.

Keywords: Traffic safety, Highway bridge, Wind-stroke, CFD (computational fluid dynamics)

1. Introduction

While establishing the modern road and railway infrastructure that is not only impressive in the construction way but is also needed for accurate and important trafficking, one confronts the reality that is always surrounding such objects: the nature of our planet [1-4]. In spite of evaluations [5-8] and certain suggestions[9], every new object of traffic infrastructure that is exposed to rather strong atmospheric influences is presenting the safety risks and calls for exploring in a large-scale fashion. Such investigations, due to the ever-stronger software and hardware tools [10-12], are performed not only through the physical measuring [13] and scaled testing [14] but also more frequently by applying the CFD (computational fluid dynamics)-based approach[12]. The latter research mode [15] did find application in wind exploring [16, 17] and traffic safety [18] which is the research pathway of the work presented in this paper, offering very satisfying results accomplished in its attempts at "prospective engineering" for particular explored cases of fluid phenomena [19].

All of these research attempts that have been brought up into the CFD community do report on good capability of the numerical approaches used in handling the reactive flows in straight, enclosed traffic infrastructure. Besides the slight denivelation of a few percent, the geometry of the arbitrary objects of interest was relatively a simple one.

Therefore, the aim of the study is to explore the (accidental) wind-strokes over such a bridge that, as a segment of to-be-constructed highway, for sure turns up as an element of this modern traffic road communication and hence suggests some countermeasures serving the overall traffic safety.

2. Numerical approach

2.1. Treatment of turbulence: Mathematical model in this study

For turbulence-modelled conservation equations, for mass and momentum, employing a time-averaged k-ε turbulence model (a CFD mode that was applied in this study as well), the governing integro-differential equations must be discretised in space only [20, 21]. These equations, together with the equations of state for an ideal gas, form here a closed set of coupled equations. These are again discretised and solved on a three-dimensional, finite-volume Cartesian mesh. In choosing the numerical method, we [22] rely on the standard of the finite volumes [21, 23, 24]. The spatial discretisation of time-independent equations employs a segregated solution method. The linearised equations result in a system of linear equations for each cell in the computational domain, containing the unknown variable at the cell centre as well as the unknown values in surrounding neighbour cells. This mechanism for a scalar transport equation [22] is also used to discretise the momentum equations; in the same mode for the pressure field (if face mass fluxes were known) and the velocity field will be obtained in same way as well.

In cases where the pressure field and face mass fluxes are not known, FLUENT (the software package applied in this research attempt) uses a co-located scheme, whereby pressure and velocity are both stored at cell centres. A need for interfacial values includes an application of an interpolation scheme to compute pressure and velocity out of cell values. The integration over the arbitrary volume (a cell in a computational domain) can be performed and is discretised through an arbitrary surface of a face.

Executing these numerical steps, the equations can express the state for each other cell in the computational grid. This again will result in a set of algebraic equations with a sparse coefficient matrix. In this way, the segregated solver is handling "the updating" of a single variable

http://dx.doi.org/10.5772/61928

field, by considering all the cells of the domain at the same time, solving the governing equations sequentially (segregated one from another). Subsequently, the next field of another variable will be solved by again considering the entire cells at the same time. The computational loop for the converged solution had about 5,500 iterations.

3. Procedure of investigation

The estimation of the boundary conditions in this CFD-based investigation was supported by the experience of some previous studies and so were the bridge surrounding space characterised as open (pressure) atmospheric boundaries with minor pressure increase or pressure drop of 2 Pa, respectively. All zones formed around the road bridge an open (pressure) boundary, which was used for initialising the values for the velocity and pressure in the computational domain, while the global temperature was set to 293 K.

The bridge body and bridge road elements as well were presumed to be nonadiabatic in the area where the objects of interest (the investigated bridge crown) are situated. This decision was based on some reality-oriented investigation on modern bridge construction, denoting the thermal conductivity of a reinforced concrete to be λ =2,3 *W/mK*.

3.1. The explored object of interest

The cross-section shapes of this highway viaduct are distinguished as those between the major carrier pylons. Further, the bridge crown shapes are mounted onto the "bridge legs" of this traffic steady object. Standing with the angle of ca 3.1 °, the road treks of this bridge have the bow-length of 954 m and their arch radius is 983 m. Going partly over the river bed and partly over the terrain valley, the highway bridge "Pocitelj-Zvirovici" demonstrates its highest section to be 96 m. The wide range between the six major pylons is set to 147 m.

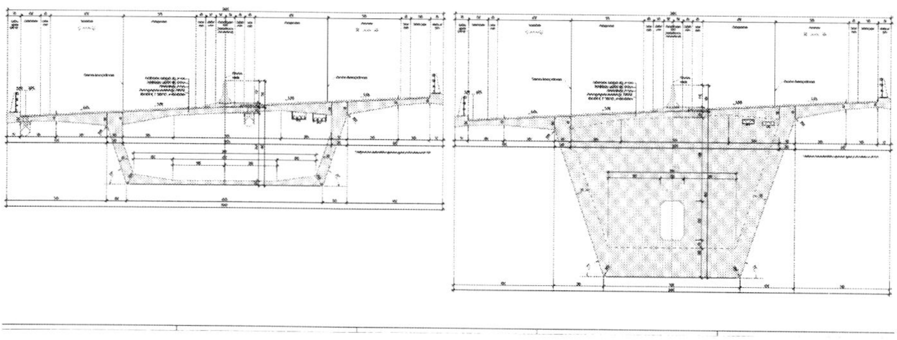

Figure 1. The cross section of the bridge crown in the free air as well as over the pylon

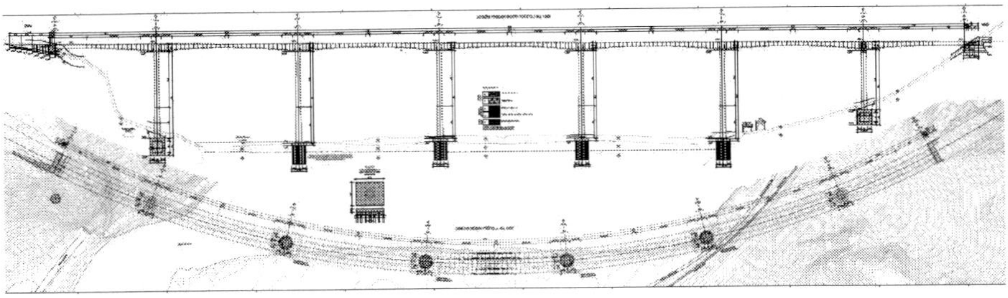

Figure 2. The constructive disposition of the highway bridge "Pocitelj-Zvirovici" in southern BH

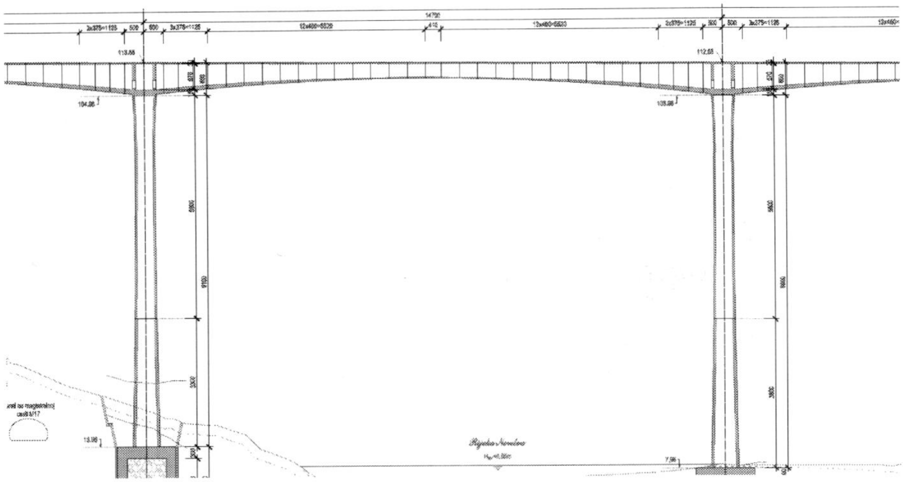

Figure 3. A side view of one of the bridge sections while crossing the river Neretva and following its shore terrain

3.2. Computational domain

The area in which the computation with applied mathematical model approach and additional numerical discretisation was performed is the very volume that a fluid can take without the "walls", where the solid body is the shape of the explored road bridge. Therefore, the computational domain of the section "Pocitelj-Zvirovici" was set to be 30m x 22m x 14,5 m.

3.2.1. The highway bridge "Pocitelj-Zvirovici"

The mesh of this computational domain (Figs. 4 and 5) is characterised through tetrahedral cells and hexahedral prisms of a random structure. In this case, a denser grid was also applied in the area around the zones where particular mechanical fluid phenomena are expected, having so more grid points to support the major occurrences.

Such an unstructured mesh (sized here to 350 mm) was installed in such zones of whole computational domain. However, the following parts of the 954-m-long bridge are also meshed with unstructured hexahedral cells in the explained way, having subsequent increased cell size up to 400 mm, 800 mm and 1,200 mm – as the distance from the bridge bottom towards the open space was growing.

The "Pocitelj-Zvirovici" bridge body and the road elements were in the computational domain defined as nonadiabatic walls. The fluid domain is air, with the ambient conditions and no fluid movement, onto which the wind-strokes were expected. The computational fluid sides were designed as opened pressure boundaries.

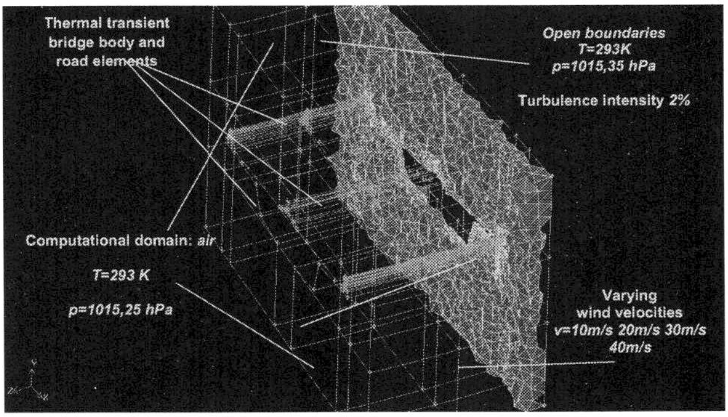

Figure 4. The unstructured hexahedral mesh is applied over the "Pocitelj-Zvirovici" highway bridge – here, at one fourth of its length, around the zone of the fuel pool

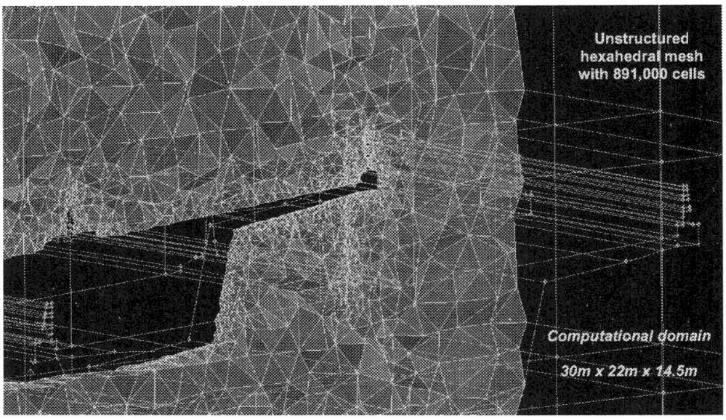

Figure 5. The meshing detail over the "Pocitelj-Zvirovici" highway bridge – here, demonstrating the solution around the road fence

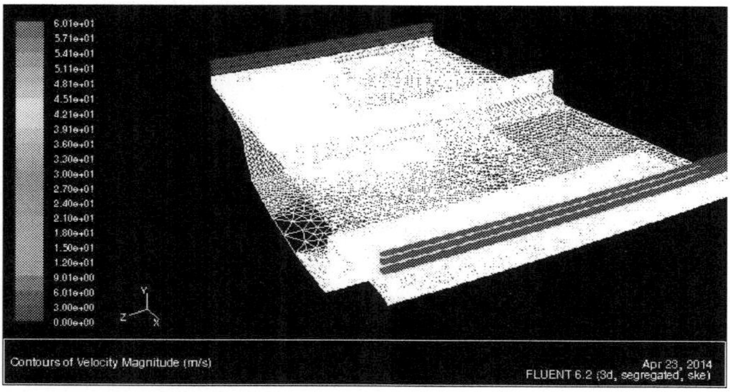

Figure 6. The meshing detail of the Bridge segment "Pocitelj-Zvirovici"

4. Discussion and outlook

Consulting the meteorological survey of the State Weather Service of Bosnia and Herzegovina[25], we performed several investigation scenarios by varying computationally aided, simulative conditions of the unwanted wind-strokes: from both south and north side of the highway bridge "Pocitelj-Zvirovici". Here (Fig. 4), we have chosen the sketch that presents CFD calculation of a (not seldom) 40 m/s stroke, from the more influencing north wind. Taken alone, this result (that was presented by the time-independent Reynolds-averaged Navier-Stokes numerical research) points up the need for a wind flow disturbing panel fences along the roadsides [26].

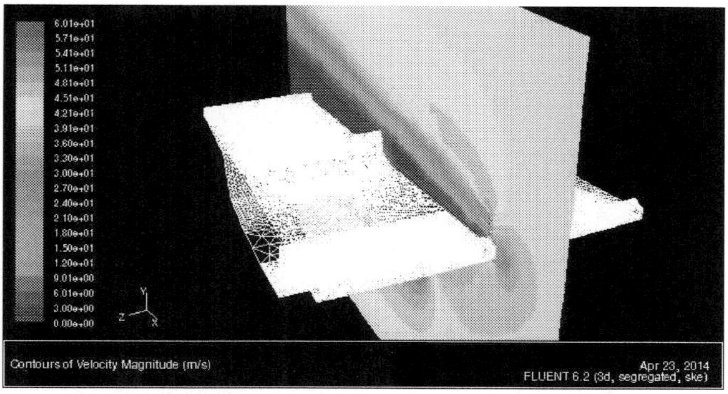

Figure 7. The computationally aided research of the wind-stroke, with the velocity of above 140 km/h: after the "Staupunkt" zones, the air, as every Newtonian fluid [6] that eventually runs around the obstacle and increases its propagation velocity

Even "weaker", sudden wind occurrences (out of 10 m/s – Fig. 8) that do turn up more frequently in the roses of wind for the geographic region, where this highway viaduct is to be constructed, do present the traffic hazard for the infrastructure element in this road communication. Our investigation performed in CFD mode does correlate with modern studies [27] and does report [28] on the importance [1] of taking into account security and safety.

During this research, we were able to perform the importance of a prospective engineering approach in treating the (gaseous) fluid phenomena in a large-scale mode. In this case, the traffic safety profits out of our findings – where again those results are "pushing" towards further prospective investigations. Exactly CFD-based research obligates to continuing investigation of the mentioned highway bridge, where the whole set of interacting answers and their questions could be satisfied: by applying another turbulence treatment (while using the *LES* model, for instance), by changing the regime of the numerical approach in another time fashion, by discretising the computational domain while using the other basic-cell shapes, and by applying different real-case scenarios based upon the atmospheric data from the statistics of the meteorology services.

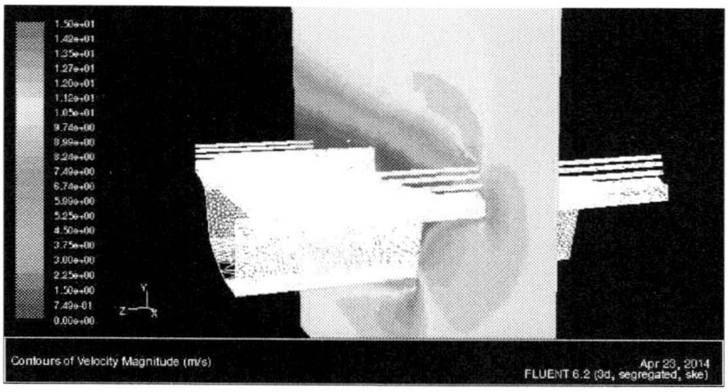

Figure 8. Even the "weaker" wind-stroke of some 36 km/h, after passing its "Staupunkt" zones, increases its velocity propagation and according to the CFD-based observation, reaches the wind fluctuation of above 50 km/h

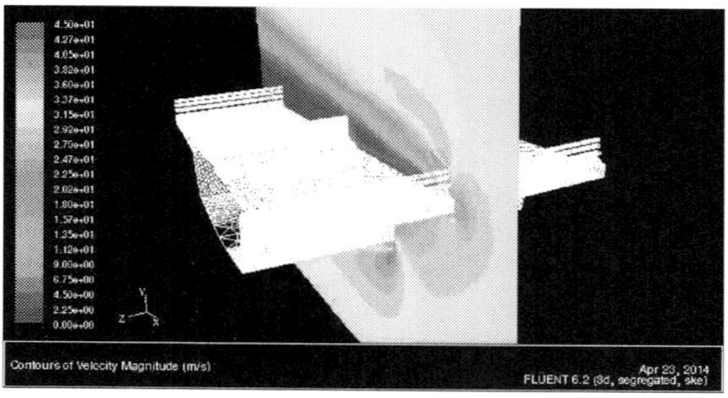

Figure 9. Averaged by its occurrence, the wind-stroke in a range of 70 km/h is due to the climatic changes, one of the frequent and unwanted air blows over the viaduct infrastructure throughout Europe

Figure 10. The mentioned dangerous increase of wind velocity after the obstacle in the form of traffic road security fence is visible

The safety issue caused by atmospheric occurrences varies in different parts of Europe (let alone in various parts of the world); what is inviting or even more is it is "provoking" us to further explore on these traffic safety issues. The landscape of the geographic regions where the future (both road and rail) infrastructure is to be constructed and to them belonging particular meteorological circumstances offers a unique perspective for each of these traffic objects. Modern hardware and software [29] developments [30] in the last 20 years [31] endorsed the computational modelling [32] into the modern and very reliable CFD toolkit [29] for scientific engineering and research [33].

Author details

Medzid Muhasilovic[1*], Kenan Imsirpasic[1], Karel Ciahotny[2] and Brano Sirok[3]

*Address all correspondence to: muhasilovic@gmail.com

1 IPSA-Institute, Sarajevo, Bosnia and Herzegovina

2 VSCHT, Vysoka Skola Chemicko Technologicka, Praha, Czech Republic

3 Univerza v Ljubljani, Fakulteta za Strojnistvo, Ljubljana, Slovenija

References

[1] H.K. Lorenzo Procino, G. Bartoli, A. Borsani, *Wind barriers on bridges: the effect of wall porosity*, in *BBAA VI International Colloquium on: Bluff Bodies Aerodynamics & Applications*. 2008. Milano, Italy.

[2] D. Hembre, C. Otto, J. Payette, G. Pingree, *The 50th anniversary of the Golden-Gate Bridge*, in *Laboratory for Construction Technology, Graduate School of Design* 1988, Harward University: Cambridge, Massatchussets, USA.

[3] P. Bowers, J. Boneck, *Charleston's bridges cross troubled waters*, in *Charleston City Paper*2012: Charleston, South Carolina, USA.

[4] E. Simiu, P. Vickery, A. Kareem, *Relation between Saffir–Simpson hurricane scale wind speeds and peak 3-s gust speeds over open terrain.* Journal of Structural Engineering, 2007. 133.

[5] A. Greeman, *Calls for wind shields after lorry falls off bridge*, in *New Civil Engineer, www.nce.co.uk*2005.

[6] Hideharu Yagi, Fumitake Yoshida, *Gas absorption by newtonian and non-newtonian fluids in sparged agitated vessels.* Industrial & Engineering Chemistry Process Design and Development, 1975. 14(4): p. 15.

[7] C. Airong, Y. Qingzhong., Z. Xigang, M. Rujin, Z. Zhiyong, *Aerodynamic problems of a super-long span cable-stayed bridge*, in *IABSE SYMPOSIUM*2005: Lisbon, Portugal.

[8] T. Martin, I.A. MacLeod, *The Tay Rail Bridge disaster revisited*, in *Proceedings of the Institution of Civil Engineers* 2004. p. 5.

[9] D.E. Newland, *Vibration of the London Millennium Footbridge: Part 1 - Cure*, in *Department of Engineering* 2002, University of Cambridge: CAMBRIDGE CB2 1PZ, UK.

[10] S. Muzaferija, D. Gosman, *Finite-volume CFD procedure and adaptive error control strategy for grids of arbitrary topology.* Journal of Computational Physics, 1997. 138(2).

[11] S. Svaic, I. Boras, M. Andrassy, *A numerical approach to hidden defects in thermal nondestructive testing.* Journal of Mechanical Engineering, 2007. 53(3): p. 165.

[12] L. H. Cheng, T.H. Ueng, C. W. Liu, *Simulation of ventilation and fire in the underground facilities.* Fire Safety Journal, 2001. 36(6): p. 597–619.

[13] J. Modic, *Porocilo o meritvah pri pozernem preizkusu v cestnem tunelu SENTVID*, 2008, University in Ljubljana, Mechanical ENgineering: Ljubljana, Slovenia.

[14] L. Zhoul, Y. Ge., *Wind tunnel test for vortex-induced vibration of vehicle-bridge system section model.* Journal of the Brazilian Society of Mechanical Sciences and Engineering, 2008. 30(2).

[15] A. Kareem, *Numerical simulation of wind effects: a probabilistic perspective.* Journal of Wind Engineering and Industrial Aerodynamics, 2008. 96: p. 25.

[16] T. Kijewski-Correa, e.a., *Validating wind-induced response of tall buildings: synopsis of the Chicago full-scale monitoring program.* Journal of Structural Engineering, 2006. 132(10).

[17] D.-K. Kwon, T. Kijewski-Correa, A. Kareem, *e-Analysis of high-rise buildings subjected to wind loads.* Journal of Structural Engineering, 2008. 134: p. 1139.

[18] X. Chen, A. Kareem, *Identification of critical structural modes and flutter derivatives for predicting coupled bridge flutter.* Journal of Wind Engineering and Industrial Aerodynamics, 2008. 96: p. 14.

[19] X. Chen, A. Kareem, *Revisiting multimode coupled bridge flutter: some new insights.* Journal of Structural Engineering, 2006. 132(10).

[20] http://www.fluent.com.

[21] H. K. Versteeg, W. Malalasekera, *An Introduction to Computational Fluid Dynamics,* 1995, London: Longman Group Ltd.

[22] M. Muhasilovic, M. Deville, *Tunnel-curvatire´s influence on the propagation of the consequences of large-scale accidental fire - a CFD-investigation.* Turkish Journal of Engineering and Environmental Sciences, 2007. 31: p. 391.

[23] Hirsch, C., *Numerical Computation of Internal and External Flows.* Vol. I. 1988, Chichester Brisbane Toronto New York: John Wiley & Sons. p. 515.

[24] M. Peric, J.H. Ferziger, *Computational Methods for Fluid Mechanics,* 2001, Berlin: Springer Verlag. p. 423.

[25] http://www.fhmzbih.gov.ba/.

[26] K. Ilic, *Projekt i Izvedba Viadukta "Hreljin" na Autocesti Rijeka-Zagreb,* in *Inzenjerski Projektni Zavod, Zagreb* 2008.

[27] H. Kozmar, L. Procino, G. Bartoli, A. Borsani, *Wind barriers on bridges: effects of wind incidence angle on flow field characteristics,* in *The Seventh Asia-Pacific Conference on Wind Engineering* 2009: Taipei, Taiwan.

[28] S. James, *Forth Replacement Crossing Study - Windshielding on Bridge Options,* 2007, Jacobs UK Ltd.

[29] S. R. Lee, H.S. Ryou, *A numerical study on smoke movement in longitudinal ventilation tunnel fires for different aspect ratio.* Building and Environment, 2006. 41(6): p. 719–725.

[30] M. Deville, A.D., *COSMASE,* 1996: Lausanne, Switzerland.

[31] J. S. M. Lee, W.K. Chow., *Numerical studies on performance evaluation of tunnel ventilation safety systems.* Tunneling and Underground Space Technology, 2003. 18(5): p. 435–452.

[32] S. Muzaferija, V. Seidl, A. Kneer *Parallel multidimensional calculation of steady-state and time-dependent flows with combustion,* in *Euro-Par '97: the Third International Euro-Par Conference on Parallel Processing* 1997.

[33] C. C. Hwang, J.C. Edwards, *The critical ventilation velocity in tunnel fires - a computer simulation.* Fire Safety Journal, 2005. 40(3): p. 213–244.

Chapter 4

Computational Fluid Dynamics Achievements Applied to Optimal Crop Production in a Greenhouse

Jorge Flores-Velázquez, Abraham Rojano, Adriana Rojas-Rishor and Waldo Ojeda Bustamante

Additional information is available at the end of the chapter

http://dx.doi.org/10.5772/61006

Abstract

Computational fluid dynamics has been successfully used in protected agriculture to simulate greenhouse weather as physical processes. The variables involved are velocity, wind direction related to either absolute or relative humidity, temperature as well as deficit vapor pressure, and carbon dioxide, among others. The research evolution is changing from the traditional validation of new designs and management to testing efficient production with less environmental pollution. This work points out this kind of assessment based on the physical principles of conservation of mass, momentum, and energy. Constitutive relationships like Darcy-Forchheimer porosity model in the momentum equation as well as the geometry and physical properties of the materials involved are needed to fulfill the particular solutions of temperature, wind, and humidity. This chapter is enhanced by the effect of solar radiation in more processes like crop transpiration with dynamical meshes and condensation.

Keywords: Computational fluid dynamics, greenhouse, climate variables, crop

1. Introduction

Intensive production systems are established in order to obtain the highest yields in the evolution of agriculture as the primary sector. One of the techniques of crop production is protected agriculture. Greenhouse has helped to increase production, improve quality, and provide sustained and profitable yields. By confining the environment in a greenhouse, the local atmosphere must fit well for cultivation. Making these adjustments sometimes cannot be easy, because the components of the atmosphere (temperature, humidity, CO_2) are intrinsically linked. The Computational Fluid Dynamics (CFD) is a tool that provides alternatives for analysis and management of greenhouse environment from a numerical point of view. CFD has advantages over other simulators, because it allows spatial and temporal analysis of the dynamics of the air inside the greenhouse, by introducing local climatic variables as starting conditions. The mathematical approach is based on physical laws expressed as partial differential equations dealing with the dependent variables as unknowns. Therefore, solving this system of equations allows either to see specific details or monitoring of greenhouse environment.

2. Models applied to protected agriculture

The inside greenhouse climate depends mainly on local environmental factors, geometry, as well as the type of materials, but they all effect the global distributional variables like: inner air temperature, exterior temperature, relative humidity, solar radiation, and wind speed and direction, further the case of the crop in the greenhouse changes transpiration rate, steam and carbon dioxide. The mentioned variables are complex factors which force solutions for a better approximation.

2.1. Lumped mechanistic models

Mechanistic models are based on physical equations. A general representation of this model is shown in Eq. (1). The main equation calculates the temperature inside the greenhouse depending on the factors described on the right side:

$$V_{grh}\rho_{air}cp_{air}\frac{dT_{in}}{dt} = Q_{sun} - Q_{vent} - Q_{conv} + Q_{sol} \tag{1}$$

Depending on the components of the greenhouse, a more complete equation can be formulated as follows:

$$\begin{aligned} V_{grh}\rho_{air}cp_{air}\frac{dT_{in}}{dt} &= \varepsilon * A_{cl} * Rad_{out} - \\ &- F_v * \rho_{air} * cp_{air}\left(T_{in} - T_{out}\right) - \alpha_{cl}\left(T_{in} - T_{out}\right) + \alpha_{soil}\left(T_{in} - T_{soil}\right) \end{aligned} \tag{2}$$

In Eq. (1) and (2) are avoided culture or transport of energy by radiation because it does not provide low temperature. Eq. (2) describes the energy representing the greenhouse environment taking into account the ground effect.

2.2. Black box model (ARX)

The black box model allows statistical description based on the outputs, given inputs on a limited range. The model only uses data obtained from direct measurements and is considered an empirical approach. So this system also provides a description of the climate of a greenhouse [1].

In general, a mathematical way to write the ARX model with one input and one output is represented in Eq. (3)

$$\begin{aligned} & y(t)+a_1 y(t-1)+.....+a_{na} y(t-na)= \\ & =b_1 u(t-nk)+b_2 u(t-nk-1)+......+b_{nb} u(t-nk-nb+1)+e(t) \end{aligned} \tag{3}$$

where

y (t) = output of the ARX model for t = t, t_{-1} t_{-na}...

u (t) = input of the ARX model for t = t_{-nk}, t_{-nk-1}.... $t_{-nk-nb+1}$

na = number of time steps out in the past

nb = number of time steps in the past entry

nk = entry delay, u(t), the output y(t)

2.3. Artificial intelligence techniques: artificial neural networks, fuzzy logic-based systems, evolutionary algorithms

Artificial neural networks are nonlinear empirical models, which have been successfully used for modeling greenhouse climate [2]. The nonlinear models are suitable candidates for dealing with complexities observed in the greenhouse behavior as moisture. However, in reference [3] mentions that require a large amount of data and time. Perhaps the best use of neural networks in the case of the greenhouse system is a combination of other alternatives: mechanistic models [4, 5]. Thus, either technical or artificial intelligence as systems based on neural networks and fuzzy logic (neural-fuzzy system) would be extended and promising future applications.

2.4. Distributed fluid dynamics models

CFD techniques treat the values of the dependent variables as primary unknowns in a finite number of points, then a set of algebraic equations derived from the fundamental equations applied to the domain are solved through predefined algorithms. Three fundamental physical principles supporting the known Navier–Stokes equations are mass, momentum, and energy conservation. These are obtained for balance on a defined volume control, as given by Eq. (4)

$$\frac{\partial(\rho\phi)}{\partial t}+\nabla\cdot\left(\rho\bar{u}\phi\right)=\nabla\cdot(\Gamma\nabla\phi)+S_{\varnothing} \tag{4}$$

where the four terms correspond to transient, convection, diffusion, and source, respectively. They represent the variable $\varnothing$ depending on mass, velocity, temperature, and chemical species distributed in all locations in a given time ($\varphi = \varphi$ (x, y, z, t)), in a Cartesian system with Euler viewpoint description. ρ denotes density (kg m^{-3}); t is time (s); ∇ is the gradient operator; ϕ denotes the state variable like air temperature (K), ammonia (NH_3), and mass of fraction (kg kg^{-1}); u is the wind speed (m s^{-1}); Γ is the diffusion coefficient (m^2 s^{-1}); and S_φ is the source term which represents the variation of the amount of material transported.

The computational studies of flow and heat transfer are mainly based on the solution of these equations together with the particular initial and boundary conditions that fulfill the problem. Commercial software regarding fluid dynamics have been designed with a user-friendly environment. To minimize the possibility of erroneously run away the desirable results by the numerical model is recommended to use prototype problems (benchmarks) [6, 7, 8].

2.4.1. Processes representation

Since the atmosphere inside the greenhouse is influenced by local environmental variables, roofing material, irrigation systems, crop type, among others, a model based on computational fluid dynamics is able to represent these variables [9]. A simulation of wind distribution using CFD in a greenhouse is shown in Figure 1.

Figure 1. Climate conditions in a greenhouse by using CFD

3. Computational fluid dynamics process simulation in a greenhouse

The CFD is based on the Navier–Stokes equations, expressed as a solution of partial differential equations for mass and momentum balances. Their deduction is typically extended through heat and mass balance contained in a control volume. Such equations can be expressed in a generalized form as Eq. 4 [10]. Based on the three basics principles of conservation, mass, momentum, and energy, they can be represented in a conservative way

Continuity

$$\frac{\partial \rho}{\partial t} + \nabla \cdot (\rho \mathbf{V}) = 0 \tag{5}$$

Equation 5 is valid for any infinitesimal element, where ρ is the density of fluid (kg cm^{-3}) and $\nabla \cdot \mathbf{V}$ represents the rate of change of volume, and $\frac{\partial \rho}{\partial t}$ is the density change with respect to time and **V** the flow velocity.

Momentum and energy equations are expressed for each component in Cartesian coordinates.

Momentum

X component:

$$\frac{\partial (u)}{\partial t} + \nabla \cdot (\rho u \mathbf{V}) = -\frac{\partial p}{\partial x} + \frac{\partial \tau_{xx}}{\partial x} + \frac{\partial \tau_{yx}}{\partial y} + \frac{\partial \tau_{zx}}{\partial z} + \rho f_x \tag{6}$$

Y component:

$$\frac{\partial (v)}{\partial t} + \nabla \cdot (\rho v \mathbf{V}) = -\frac{\partial p}{\partial y} + \frac{\partial \tau_{xy}}{\partial x} + \frac{\partial \tau_{yy}}{\partial y} + \frac{\partial \tau_{zy}}{\partial z} + \rho f_y \tag{7}$$

Z component:

$$\frac{\partial (w)}{\partial t} + \nabla \cdot (\rho w \mathbf{V}) = -\frac{\partial p}{\partial z} + \frac{\partial \tau_{xz}}{\partial x} + \frac{\partial \tau_{yz}}{\partial y} + \frac{\partial \tau_{zz}}{\partial z} + \rho f_z \tag{8}$$

Energy

$$\begin{aligned}
&\frac{\partial}{\partial t}\left[\rho\left(e + \frac{V^2}{2}\right)\right] + \nabla \cdot \left[\rho\left(e + \frac{V^2}{2}\right)\mathbf{V}\right] = \\
&= \rho q + \frac{\partial}{\partial x}\left(k\frac{\partial T}{\partial x}\right) + \frac{\partial}{\partial y}\left(k\frac{\partial T}{\partial y}\right) + \frac{\partial}{\partial z}\left(k\frac{\partial T}{\partial z}\right) - \\
&- \frac{\partial (up)}{\partial x} - \frac{\partial (vp)}{\partial y} - \frac{\partial (wp)}{\partial z} + \frac{\partial (u\tau_{xx})}{\partial x} + \\
&+ \frac{\partial (u\tau_{yx})}{\partial y} + \frac{\partial (u\tau_{zx})}{\partial z} + \frac{\partial (v\tau_{xy})}{\partial x} + \frac{\partial (v\tau_{yy})}{\partial y} + \\
&+ \frac{\partial (v\tau_{zy})}{\partial z} + \frac{\partial (w\tau_{xz})}{\partial x} + \frac{\partial (w\tau_{yz})}{\partial y} + \frac{\partial (w\tau_{zz})}{\partial z} + \rho \mathbf{f} \cdot \mathbf{V}
\end{aligned} \tag{9}$$

where q is the rate of heat volumetric flow (kJ/m^3 s), k is the thermal conductivity(W/(m °C)), ϱ is the fluid density(kg/m^3), e is the internal energy of the fluid element due to random molecular motion per unit mass (J), $V^2 / 2$ is the kinetic energy per unit mass (m^2/s^2); this is due to translational movement of the fluid element, T is the fluid temperature (°C), f is the force (N), and τ is the Reynolds stress, which represents the turbulent fluctuations of the fluid.

The equation for the internal energy e can be expressed as the static enthalpy h. Indeed, the total enthalpy is given by $h = h + V^2 / 2$.

The simulation is performed assuming steady-state conditions. Additionally, the greenhouse airflow patterns are assumed to be turbulent and the *k*-ε model is applied to solve for the kinetic energy (*k*) and the viscous dissipation rate of turbulent energy (*ε*). Several researchers have shown that this model makes good representation of the turbulent nature of fluid flow within controlled environments [11, 12, 13, 14] and it is one of the most commonly used [15].

Because low wind speeds prevailed in the experimental site through the year, the average wind speed and gust speed computed from the measured values were 1.89 and 4.34 m s^{-1}, respectively. Free and mixed convection may drive air exchange, instead of forced convection. Similar convection regimes were simulated in this study, as in previous studies performed on the same location [16].

3.1. Problem formulation

The simulation of physical and biological processes within the greenhouses is the challenge of the century from the numerical point of view. Although the theoretical basis had been established over the last centuries, advances in scalar or visualization representation were not given until this decade.

It is common to simulate single physical and biological processes, but is still pending models which combine, for example, the processes of photosynthesis and respiration, in order to improve the production of greenhouse crops.

3.1.1. Climate modeling

The analysis by CFD involving heat and mass transfer following the numerical solution of the fundamental equations of mass, energy, and momentum conservation, can deal with transient and distributed solutions, illustrating in space (2 or 3 dimensions) and time the variables of the climate and atmosphere, simultaneously.

Air movement will be based on a temperature gradient and consequently phenomenon of convection mass transport. Two models are essential for the analysis of this case; the flotation model for natural convection accounts for the variation of density as a function of temperature. Boussinesq model is used for suitable results and is given by Eq. (10) [17].

$$\rho = \rho_i 1 - \beta\left(T - T_i\right) \tag{10}$$

where ρ is the density (kg m^{-3}), β is the coefficient of thermal expansion (K^{-1}), T is the air temperature (K), and the subscript i means inside.

Finally, air movement assumes a mixture of liquid, steam, and nonconsumable gas, the standard equation that governs the mixture model and the mixing turbulence model flow for vapor mass fraction (f) can be written as Eq. (11) [18].

$$\frac{\partial}{\partial \varepsilon}(\rho f)+\nabla(\rho \vec{v}_v f)=\nabla(\gamma \nabla f)+R_e-R_c \tag{11}$$

Beginning from this statement, the air movement has been exposed. For instance, Figures 2, 3, and 4 show the dynamics of the air into the greenhouse, under different frames, cenital (Figure 2), tunnel (Figure 3), and screen house (Figure 4).

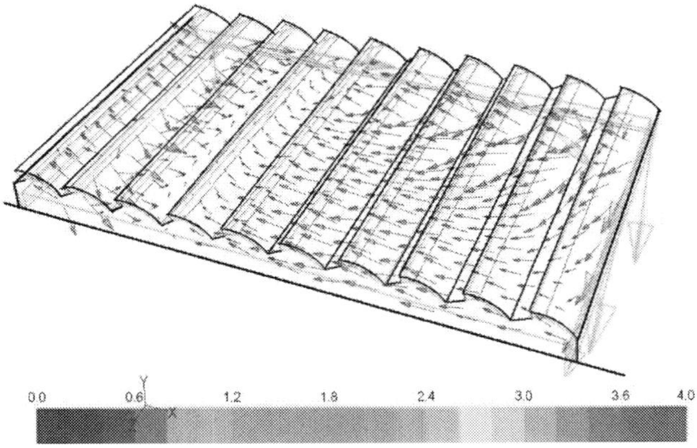

Figure 2. Wind behavior in a multi-spam (roof windows) frame

3.1.2. Mechanical ventilation

The mechanical ventilation is less used than natural ventilation, therefore investigation about this is scanty. For instance, the American Society of Agricultural Engineers (ASAE) provides the rules for the design of the mechanical ventilation system [19]; however, [20] there are limitations to these rules, such as the difficulty to incorporate approximate values for the transpiration coefficient in the calculus of the ventilation system.

Other studies like [21] and [22], show that the ventilator allows more control on the temperature in the greenhouse than the passive ventilation [23], and found a vertical gradient of temperature and humidity in the air much more homogeneous with mechanical ventilation. However, in other studies like [19], it was observed that even with mechanical ventilation there can be a notable lack of thermal and humidity homogeneity. Since they measured major

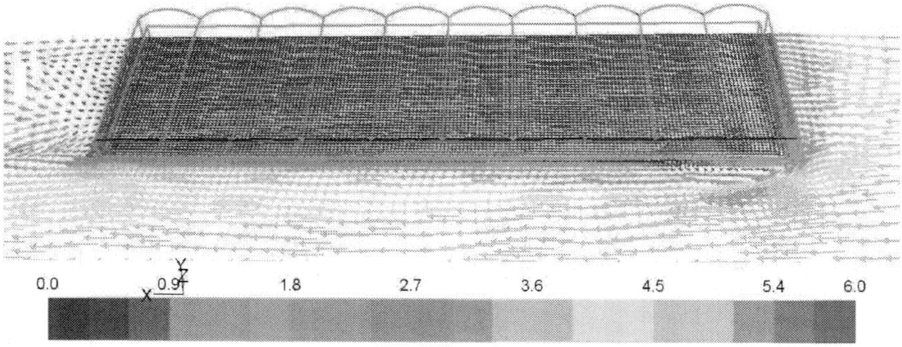

Figure 3. Wind velocity (m s^{-1}) in a tunnel frame

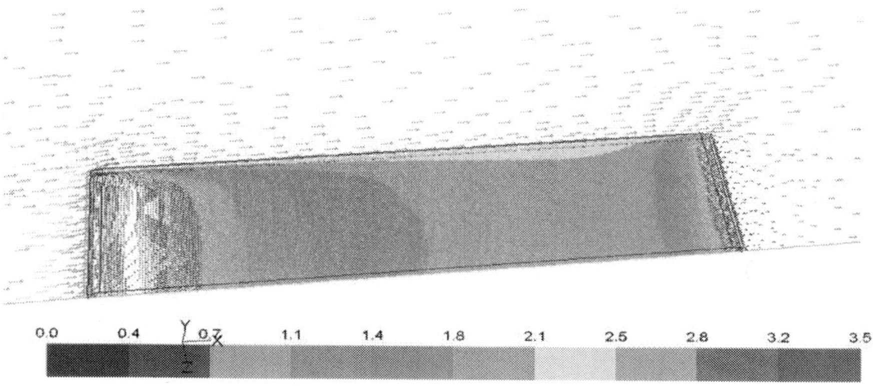

Figure 4. Wind velocity (m s^{-1}) and distribution in a screen house frame

temperatures in the high part of the crop for some time, the homogeneity can also be improved if roof windows are lightly open, especially when the greenhouse is big [24]. Both, mechanical and natural ventilation are processes with a high degree of difficulty in the analysis. So the investigation about the combination between natural and mechanical ventilation is very incipient and limited.

Nowadays, Computational Fluid Dynamics (CFD) is a strong tool for analysis, because it allows to study in detail the ventilation process and climatic variables inside of the greenhouse [24, 25]. Questions on the management of the system of ventilation to support sufficient traffic of the air and to achieve acceptable levels of heat transmission and mass between crop and the air can be solved by means of the application of this tool [13, 25–28]. An example of mechanical ventilation in a greenhouse is shown in Figure 5.

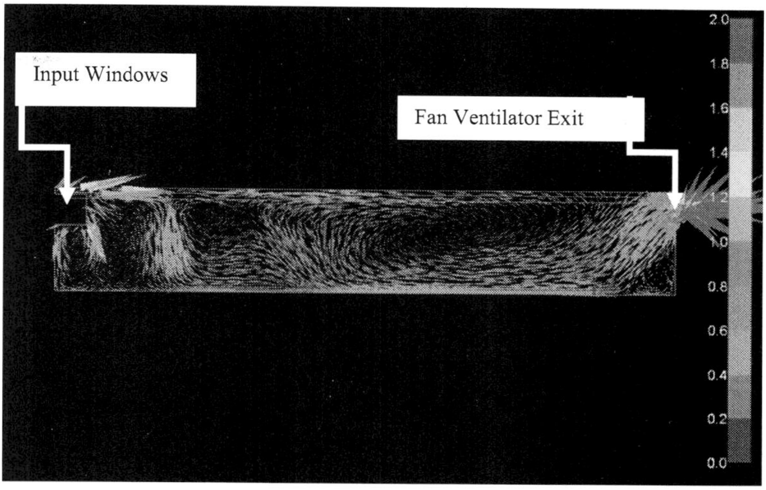

Figure 5. Wind vectors as a result of mechanical ventilation system (colored by magnitude)

3.1.3. Heating system

Greenhouse heating is one of the factors with a significant impact on crop production. From the environmental and economic points of view, the use of fossil fuels in agricultural production is costly; however, in regions with cold winters, the use of heating systems is essential to achieve sustainable production. A method to prevent the temperature going below the minimum threshold for crop production could be based on the forces of convection.

The heating source can be of various types such as collector walls or heating systems, based on water or gas driven through pipes. Heating pipes are effective devices used to keep the greenhouse warm and are implemented in many of the cold climate greenhouses to maintain a targeted temperature [29]. Their principles for heat transfer are convection and radiation; however there is a need to find ways to increase their efficiency. The position of the pipes in the greenhouse and the power of the heating system influence the spatial distribution of temperature and flow patterns due to convective forces.

Reference [29] reported that the best place to locate the pipes is at a half crop height, with the tubes near to the leaves. Other configurations have been studied by several researchers [30, 31], which highlighted the influence of the heating system on the cultivation in terms of convection and radiation. As a result of these studies, the advantages of placing a pipe heating system at the bottom of the crop without affecting the evapotranspiration from plants were discovered. Reference [32], reported that this type of heating system favors the removal of moisture. Humidity, as well as other factors, such as condensation [33] and cooling [34], especially in closed greenhouses [9], have been analyzed by computational fluid dynamics (CFD).

From the numerical point of view, reference [35] performed an experimental validation of a greenhouse model based on CFD to analyze its air thermal gradients. They found that the largest temperature differences occur in the zones close to the ground and under the roof. Conversely, air temperatures inside the greenhouse were more uniform and had a homogeneous distribution. Similarly, Tadj et al. [36] used ANSYS CFX to analyze the airflow and temperature patterns under three scenarios simulating several pipe locations and a tomato crop. They observed the same behavior of the air and temperature patterns, which consisted of strong thermal gradients near the floor and ceiling, and a homogeneous distribution in the rest of the greenhouse.

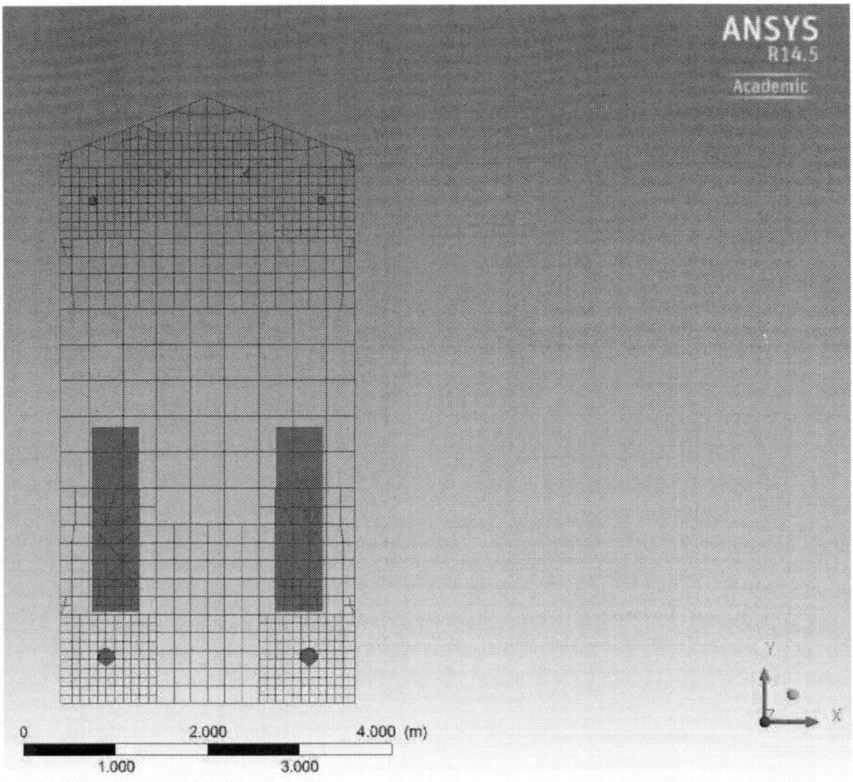

Figure 6. Meshing in a cabinet with heating system by pipes (low and upper)

In this chapter the feasibility of the natural convection phenomena as an energy transference process has been analyzed through simulations, produced in a Venlo type greenhouse with the presence of a fully developed tomato crop. Convection forces were assumed to be produced by the energy stored naturally in the soil mainly due to solar radiation and the energy produced from a pipe heating system (figure 6). The crop was simulated with the "single leaf" approach [37] by considering the plant canopy as a porous zone. CFD simulations were performed for the analysis considering the effects of an adiabatic wall temperature and the external environment. External temperature was given as a boundary condition for the plastic cover (figure 7). Results show the behavior of the greenhouse environment when it is driven by natural convection and assisted by a heating system of this kind.

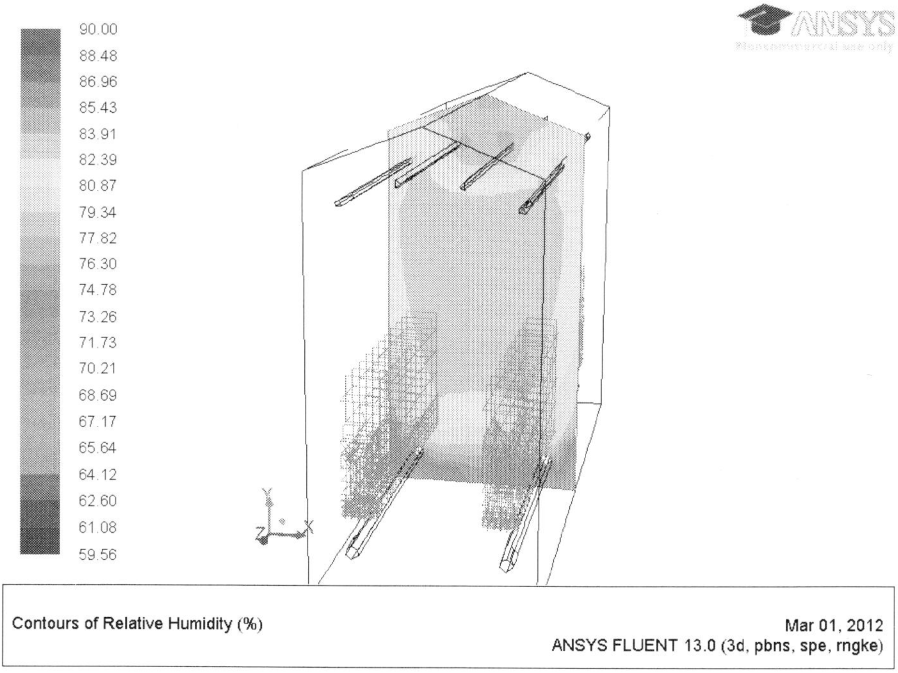

Figure 7. Contour of the Relative Humidity (%) in the center of cabinet

3.1.4. Mass diffusion within turbulent flows

Air is assumed as a mixture of liquid, vapor, and non-consumable gases. In this study, the species transport model available in FLUENT is possibly used to simulate the mass transport of different gases, such as CO_2, nitrogen, ammonia, etc., beginning from the diffusion flux of species *i*, which arises due to gradients of concentration and temperature. Such species model uses the dilute approximation (Ficks's law) to model mass diffusion. For turbulent flows, mass diffusion can be written as in Eq. 12 [18]

$$\overline{J}_i = \rho D_{i,m} \nabla Y_i - D_{T,i} \frac{\nabla T}{T} \tag{12}$$

where J_i (m^2/s) is the diffusion flux of species i (ammonia), ρ is the density of the mixture (kg/m^3), $D_{i,m}$ is the mass diffusion coefficient for specie i in the mixture m (m^2 /s), $D_{T,i}$ is the turbulent diffusion coefficient (m^2 s^{-1}), Yi is the mass fraction of specie i, and T is the temperature of the flow (K).

To achieve the management of environment in the greenhouse, is fundamental the simulation, to know which scenario would be the best to obtain a better production for the specific case. That is why, the gasses simulation are necessary. One specific case is the animal farm. Rabbits and hens are the most representative cases of toxic gasses production. On the other hand, in a greenhouse, the CO_2 production or extinction is a critical point in the crop production process.

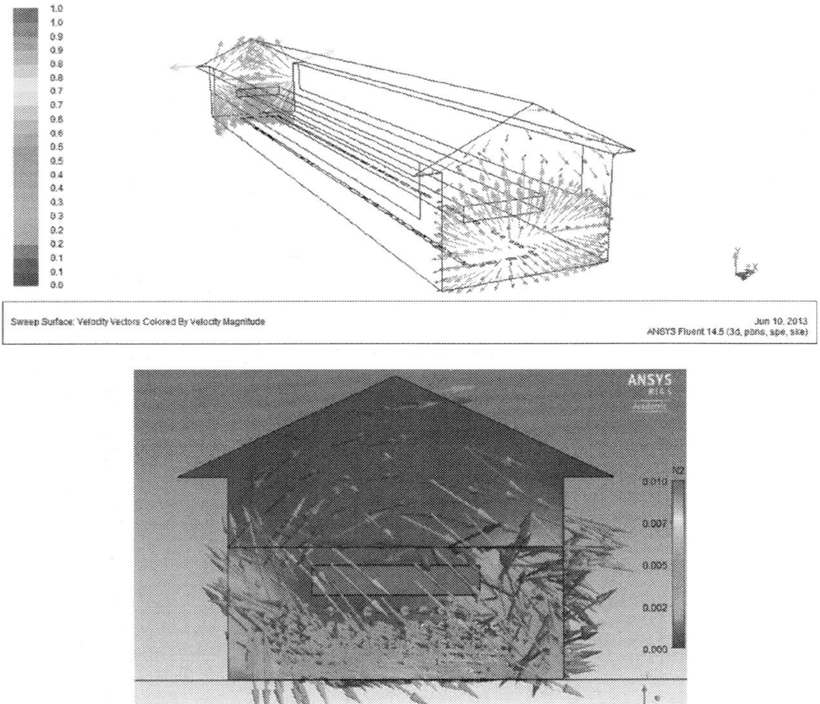

Figure 8. Wind velocity (up) and nitrogen distribution (dawn) in a domestic barn

As a result of the application of the diffusion equations and species model to simulate some gasses, specific simulations are shown in Figures 8 and 9.

Figure 9. Ammonia representation as a result of natural ventilation system

3.1.5. Distribution of ammonia

Wind direction is the most important factor to exchange air in a semi-closed farm. The concentration of ammonia is greater near the ground because it is where the ammonia is produced (source) and because air renovation is limited due to the vents being located at an elevated position (Figure 9).

Figures 9, 10, 11, and 12 represent ammonia distributions in a structure with rabbit production. If the wind is simulated parallel to the vents, the incoming air produced a lower concentration gradient at the zone beneath the cages (Figure 10). Lower concentration of ammonia results around the rabbit cages, even with the small ventilation rate produced. Thus, ammonia was more uniformly distributed when the wind direction was parallel.

Even though the air ventilation rate was much higher for the case when the wind direction was perpendicular to the vents, a notorious accumulation of ammonia in the area underneath the cages was produced (Figure 9). Ventilation rates increased at the vents level, but air circulation near the ground was limited. Therefore, the concentration of ammonia in the lower zone of the barn was higher than the concentration in the cages area and above. This was probably attributed to a limited air circulation within that zone.

If ammonia gets highly concentrated underneath, as it can be observed in Figure 10 for the scenario II, it represents a hazardous condition. The ammonia accumulated will eventually arise due to a flotation effect when high temperatures are present and that condition may be dangerous for the rabbits.

The ventilation system design is the restricted air circulation observed underneath the cages on the simulation. Thus, having both vent openings at an elevated position from the ground level may not be effective to promote enough ventilation rates and lower the ammonia concentrations sufficiently, especially beneath the rabbit cages. That could be alleviated by lowering the height of the vents in the rabbit barn. Therefore, a modification on the vents position was proposed and simulated.

The modification on the ventilation system of the rabbit barn consisted of having a difference in height between the vent inlet and the vent outlet (Figure 10). The vent inlet was at 0.2 m above the ground level and ran continuously 22 m along the barn. The position of the vent outlet remained as in the original design, at a height of 1.2 m from the ground level.

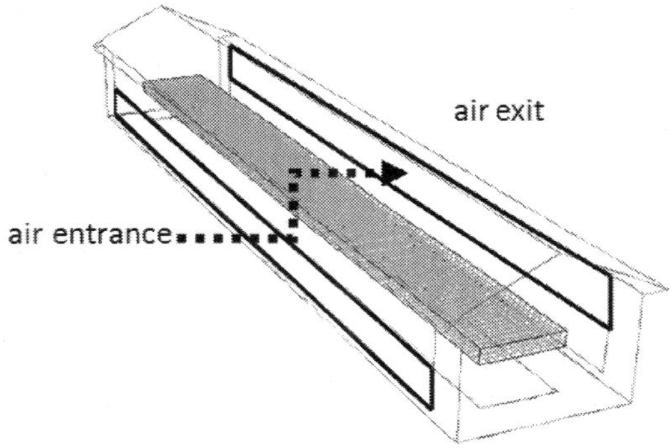

Figure 10. Design modification on the vent configuration of a barn. Vent inlet and vent outlet at a height of 0.2m and 1.2m above the ground, respectively

Air temperature and ammonia mass fraction on the vertical profile at the center of the barn are shown in Figures 11 and 12. Figure 11 shows the case when the wind is parallel and Figure 12, when the wind is perpendicular to the barn. For presentation purposes, the height of the barn is shown in the Y-axis (dependent variable). Results show an improvement on the vertical distribution of air temperature and the ammonia mass fraction with the inlet vent located at 0.2 m of height (Figure 11). Air circulation was particularly improved in the area underneath the rabbit cages, resulting in a reduction of ammonia concentration. A similar result can be seen in Figure 12 (parallel wind direction) where the thermal vertical gradient improved significantly and a reduction of ammonia concentration was achieved in the zone below the rabbit cages.

3.1.6. Theoretical approach of crop

An approach in the analysis of greenhouse ventilation consists in establishing a reliable methodology that describes the interactions that are taking place in the interface porous matrix-fluid.

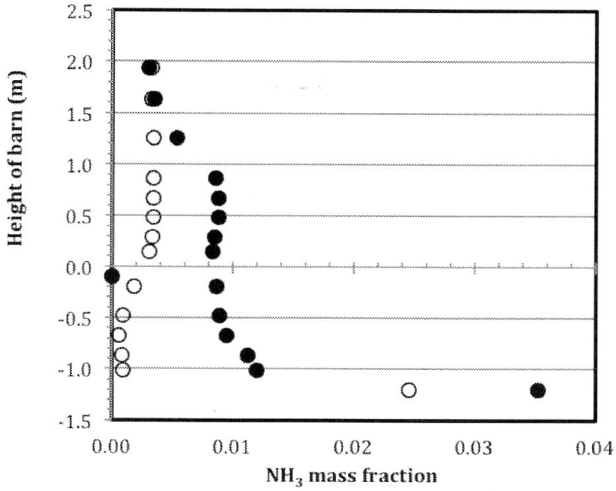

Figure 11. Ammonia distribution (mass fraction) wind direction parallel

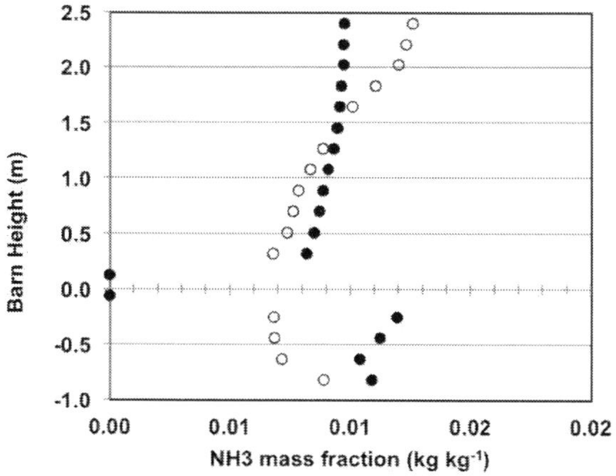

Figure 12. Ammonia distribution (mass fraction) wind direction orthogonal

The flow characteristics in a permeable material can be described in terms of permeability and porosity [38]. If a specific volume is considered, the porosity is represented by the space for which the fluid flows in relation with the total space contained in this volume. The porosity is 0 for zero flow and when it does not exist flow restrictions, the porosity will be 1. The permeability is defined as the ability of the material to flow across itself, analogous when a conduit is totally opened the permeability tends to infinite.

One of the first works to model the flow was published by [39] and later was used by Miguel et al. [38], who indicated that Eq. 13 can be used for analysis of one-dimensional mass transfer across a permeable material

$$\frac{\partial}{\varepsilon}\frac{\partial u}{\partial t}+\frac{\rho}{\varepsilon^2}u\frac{\partial u}{\partial x}=-\frac{\partial p}{\partial x}-\frac{u}{K}u-\rho\frac{Y}{\sqrt{K}}u|u|+\frac{\mu}{\varepsilon}\frac{\partial^2 u}{\partial x^2} \tag{13}$$

where u (u_i ε) is the superficial fluid speed (m/s), u_i is the speed across the material (m/s), ε is the porosity (m^2/m^2), ϱ is the density of the fluid (kg/m^3), P is the pressure (Pa), μ is the dynamic viscosity (kg/m·s), Y is the factor of inertia, x is the flow direction (m), and K is the media permeability (1/m).

Equation 14 is a general equation in explicit form that indicates how the flow speed is related to the pressure gradient by the force to the viscous resistance

$$\frac{\rho}{\varepsilon^2}u\frac{\partial u}{\partial x} \tag{14}$$

The effect of inertia due to the pore is included in the following relation

$$\rho\frac{Y}{\sqrt{K}}u|u| \tag{15}$$

Equation 15 represents the flow of high pressure inertial loss. The force of viscous resistance due to the transfer at the moment in the interface matrix-fluid is included as Eq. 16

$$\frac{\mu}{K}u \tag{16}$$

And the resistance for viscosity of the flow of the fluid as Eq. 17

$$\frac{\mu}{\varepsilon}\frac{\partial^2 u}{\partial x^2} \tag{17}$$

The equation of fluid movement across a porous media can be derived from this equation.

3.1.7. Crops (porous zone)

To study the effects of crop canopy in the airflow inside a greenhouse, Haxaire [40] made an experiment in a wind tunnel to determine the speed loss (drag effect) due to the presence of plants. He also analyzed the effect on the pressure drop (Dp) and the density of foliar area (L) due to changes in the air speed. The determination of the dragging coefficient has allowed new incursions in the study of the fluid dynamics; such studies have been conducted introducing this coefficient [41–44, 26].

The analysis and calculation of the effect of crops can be realized adding the term source into the Navier–Stokes Equation (Eq. 18):

$$\frac{\partial(\rho\varnothing)}{\partial t} + \nabla\left(\partial(\bar{u}\varnothing)\right) = \nabla(\Gamma\nabla\varnothing) + S_{\varnothing} \tag{18}$$

The source term $S_{\varnothing}$ contains the variable of interest, which represents the moment consumption due to the effect of dragging (drag effect) of the crop. This friction force can be expressed as the unit of volume of coverage using the Wilson formula [45] (Eq. 19).

$$S_{\varnothing} = -LC_D v^2 \tag{19}$$

where v is the speed of the air (m/s), L refers to the canopy density (m^2/m^3), and C_D is the friction coefficient (drag coefficient), which [40] experimentally established as 0.32. In order to include the effect of dragging or friction (drag effect) proportionally to the density of foliate area, the crop (cultures) was considered as porous media and it was described by Darcy-Forchheimer in [47], as shown in Eq. 20.

$$S_{\varnothing} = -\left[\frac{\mu}{K}v\right] + \rho\frac{C_f}{K}v^2 \tag{20}$$

where C_f = is the loss coefficient.

In greenhouse ventilation studies, the wind speed is such that the quadratic term of Eq. 20 dominates the linear term. This term can be neglected due to the small lost coefficient in the linear moment equation. The intrinsic permeability K can be deduced combining Eq. 19 and Eq. 20 by the following relation (Eq. 21)

$$\frac{C_f}{K} = LC_D \tag{21}$$

Dragging coefficients were estimated using experimental data for each crop as shown in Table 1, by applying Eq. 19 and Eq. 20. The flow reduction due to crop was simulated in the source term ($S_{\varnothing}$) in Eq. 19.

Crop	LAD (m^2m^{-3})	Cd	Sϕ
Tomato	5.6	0.26	2.9
Pepper	5.8	0.23	2.7
Egg Plant	3.7	0.23	1.7
Beans	3.0	0.22	1.6

Table 1. Characterization of four crops with different leave shape (Adapted from [46]).

The shape of the leaves determines air flow. Different crop shapes are shown in Figure 13.

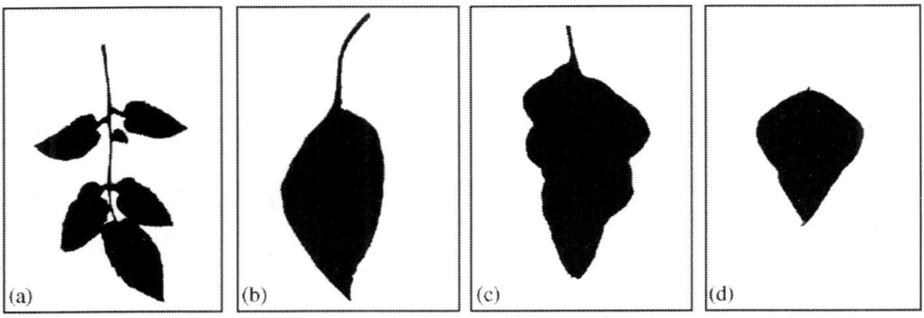

Figure 13. Morphology of the leaves of a) Tomato, b) Pepper, c) Egg plant, and c) Beans. Adapted from [46]

The unload coefficient is the factor that determines momentum loss according the transport equation; when insect mesh windows are used, the pressure drop is 80 %, with a variation of 8 % depending on the mesh porosity [48]. This implies that air speeds inside the greenhouse usually are less than 1 m/s. The reduction is accentuated along the cropped area from the air inlet, depending on cropped length and in a staked crop.

In this case, the wind speed drop for two crops studies is shown in Figure 14. The results show that even if the leaves configuration is different, the pressure drop depends principally on canopy density. This can be attributed that differences in wind speed are not significant. In addition, we have the hypothesis that at the first contact with the crop (Figure 14), airflow modifies the angle inducing a leaves "alignment" that favor the flow and becomes constant and less restrictive. The presence of crop canopy represents a reduction in airflow, which increases as moved away from the wind-inlet window. In this study, an immediate drop was of 0.1 m/s, but as becomes stable, the speed variations are of ± 0.05 m/s in 34 m of greenhouse length.

Due to presence of canopy, there is a speed reduction of 33 % between the inlet and outlet windows (above) through crop canopy (Figure 15A). Inside the canopy, the wind speed is

http://dx.doi.org/10.5772/61006

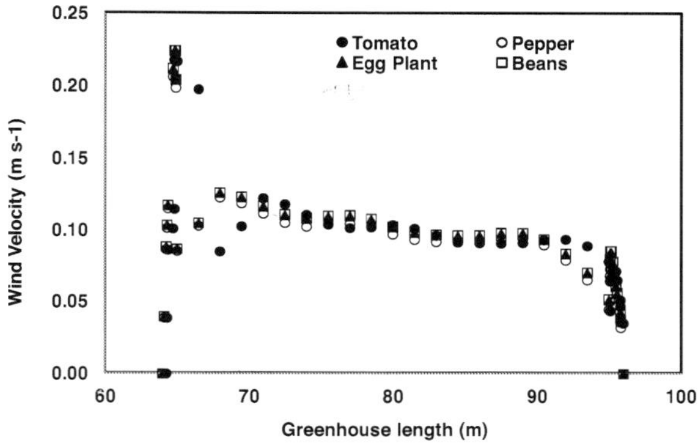

Figure 14. Longitudinal variation of the wind speed for four horticultural crops with different Cd

about 0.1 ms^{-1}, but above this, the increase is very small, up to 0.25 %; this can be very important for best placement of sensors to measure transpiration. This process is directly related to wind speed [49].

As a consequence of flow variation, the temperatures showed an increase toward the windows outlet. A 6 K increase was estimated (Figure 15B) with the particularity that this variable is dependent on the canopy height. This indicates the convective transport of the heat inside the greenhouse can be more homogeneous and independent from the air flow rate as it becomes stable.

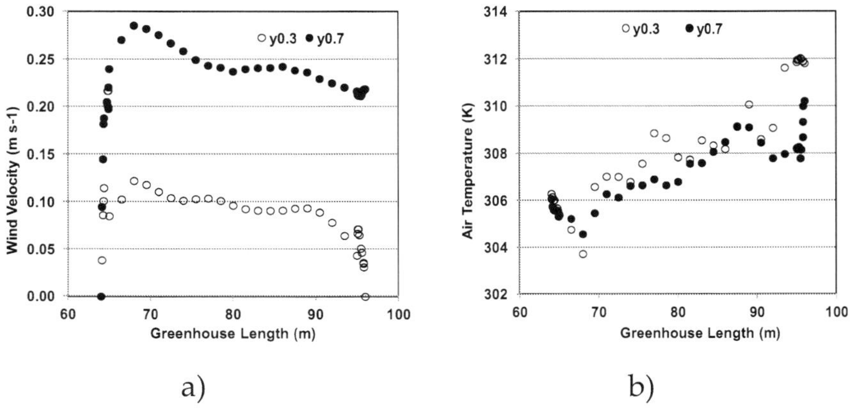

Figure 15. Variation of the wind velocity (m s-1) (A) and temperature (K) (B) in the cropped area (tomato) along the greenhouse (m)

Schematically, Figure 16 shows the distribution of the speed vectors in the dominant wind direction, and spatial distribution of the temperature in the greenhouse ambient.

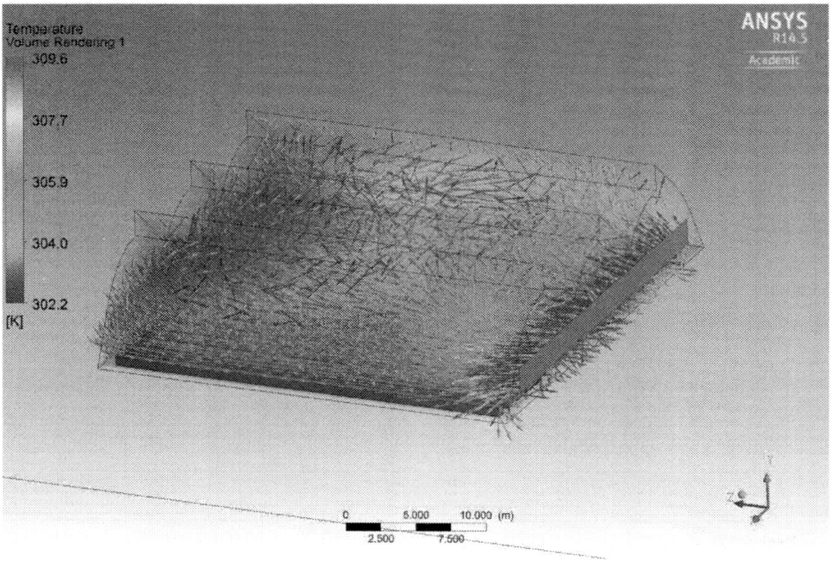

Figure 16. Spatial distribution of wind speed vectors (m/s) and temperature (K)

4. Conclusions

Most of the processes that occur in a greenhouse are related to mass transfer events (relative humidity, transpiration), momentum (velocity, species), and energy or exchange heat (convection, radiation) and consequently, the computational fluid dynamics is a robust tool for greenhouses and animal farm behavior simulation. The inclusion of culture and auxiliary climatic control systems are essential to the production process from an agricultural point of view. The results of this study indicate that once experimentally validated models, the use of these allows to know the spatial characteristics of the variables involved. From the mechanical point of view, the use of anti-insect screens and crop presence causes a drop of 80% in speed. A heating system for hot pipes under conditions of central Mexico is only necessary for specific periods before sunrise. A change in the position of the greenhouse windows in zigzag increases air movement and homogenized conditions inside, even if you keep the same ventilation ratio, but a variation in the porosity of the mesh has side effects in the air renewal ratio. The prospect of computational fluid dynamics applied to the protected agriculture, will focus on building a dynamic model with many variables to make efficient use of resources and reduce the use of fossil fuels in crop production.

Author details

Jorge Flores-Velázquez[1*], Abraham Rojano[2], Adriana Rojas-Rishor[3] and Waldo Ojeda Bustamante[1]

*Address all correspondence to: jorge_flores@tlaloc.imta.mx

1 Mexican Institute of Water Technology, Morelos, México

2 Autonomous University of Chapingo, Texcoco, México, México

3 University of Costa Rica, San José, Costa Rica

References

[1] Ljung L. System identification: theory for the user, 2nd ed. Prentice Hall, Linköping University, 1999, 375–380 p.

[2] Seginer I, Boulard T, Bailey B.J. Neural network models of the greenhouse climate. Journal of Agricultural Engineering Research. 1994; 53: 203–216.

[3] Boaventura J, Ruano A, Couto C. Identification of greenhouse climate dynamic models. In: Sixth international conference on computers in agriculture, Cancún, 1996, 161–171.

[4] Dayan J, Strassberg Y, Dayan E. Simulation and control of ventilation rates in greenhouses. Acta Horticulturae. 2001; 566: 67–74.

[5] Dayan J, Strassberg Y, Dayan E. The prediction of ventilation rates in greenhouses containing rose crops. Acta Horticulturae. 2002; 593: 55–62.

[6] Al-helal I. A computational fluid dynamics study of natural ventilation in arid region greenhouses. PhD thesis, Ohio State University, 1998.

[7] Romero-Gómez P, López-Cruz IL, Choi CY. Analysis of greenhouse natural ventilation under the environmental conditions of central Mexico. Transactions of the ASABE. 2008; 51: 1753–1761.

[8] Rico-García E, López-Cruz IL, Herrera-Ruiz G, Soto-Zarazúa GM, Castañeda-Miranda R. Effect of temperature on greenhouse natural ventilation under hot conditions: CFD simulations. Journal of Applied Sciences. 2008; 8: 4543–4551.

[9] Flores-Velázquez J, De la Torre-Gea G, Rico-García E, López-Cruz IL, Rojano-Aguilar A. Applied to the greenhouse environment. In: Hyoung Woo Oh (ed) Applied computational fluid dynamics. InTech, Croatia, 2012, 37–62.

[10] Anderson JD. Computational fluid dynamics. University of Maryland: McGraw Hill, 1995, 37–80 p.

[11] Li W. Airflow and contaminants in a swine barn with recirculation assisted slot inlets. PhD thesis, University of Saskatchewan, 1997.

[12] Mistriotis A, Bot G.P., Picuno P, Scarascia-Mugnozza G. Analysis of the efficiency of greenhouse ventilation using computational fluid dynamics. Agriculture and Forest Meteorology. 1997; 85: 217–228.

[13] Norton T, Sun D, Grant J, Fallon R, Dodd V. Application of computational fluid dynamics (CFD) in the modeling and design of ventilation systems in the agricultural industry: A review. Bioresource Technology. 2007; 98: 2386–2414.

[14] Norton T, Grant J, Fallon R, Sun DW. Assessing the ventilation effectiveness of naturally ventilated livestock buildings under wind dominant conditions using computational fluid dynamics. Biosystems Engineering. 2009; 103: 78–99.

[15] Bjerg B, Cascone G, Lee IB, Bartzanas T, Norton T, Hong SW, Seo IH, Banhazi T, Liberati P, Marucci A, Zhang G. Modelling of ammonia emissions from naturally ventilated livestock buildings. Part 3: CFD modelling. Biosystems Engineering. 2013; 116: 259–275.

[16] Romero-Gómez P, Choi CY, Lopez-Cruz IL. Enhancement of the greenhouse air ventilation rate under climate conditions of central Mexico. Agrociencia. 2010; 44: 1–15.

[17] Boulard T, Kittas C, Roy JC, Wang S. Convective and ventilation transfers in greenhouses, part 2: Determination of the distributed greenhouse climate. Biosystems Engineering. 2002; 83: 129–147.

[18] ANSYS-FLUENT. Fluent user guide. Lebanon: Ansys Inc., 1998.

[19] ASHRAE. Applications handbook: American Society of Heating, Refrigeration and Air Conditioning Engineers. Atlanta: ASHRAE, 1999.

[20] Willits DH, Li S, Yunker CA. The cooling performance of natural ventilated greenhouse in the southeastern US. Proceedings of the International Symposium on Greenhouse Cooling. Acta Horticulturae 2006; 719: 73–80.

[21] Montero JI, Hunt GR, Kamarudddin R, Anton A, Bailey BJ. Effect of ventilator configuration on wind driven ventilation in a crop protection structure for the tropics. Agricultural Engineering Research. 2001; 80: 99–107.

[22] Kittas C, Karamanci M, Katsoulas N. Air temperature in a forced ventilation greenhouse with rose crop. Energy and Buildings. 2005; 37: 807–812.

[23] Kittas C, Katsoulas N, Baille A. Influence of greenhouse ventilation regime on microclimate and energy partitioning of a rose canopy during summer conditions. Agricultural Engineering Research. 2001; 79: 349–360.

[24] Baeza EJ, Pérez-Parra JJ, Montero JI. Effect of ventilator size in natural ventilation in parral greenhouse by means of CFD simulations. Acta Horticulturae. 2005; 691: 465–472.

[25] Kacira M, Sase S, Okushima L. Effects of side vents and span numbers on wind-induced natural ventilation of a gothic multi-span greenhouse. Japan Agricultural Research Quarterly. 2004; 38: 227–233.

[26] Bartzanas T, Kittas C, Boulard T. Effect of vent arrangement on windward ventilation of a tunnel greenhouse. Biosystems Engineering. 2004; 88: 479–490.

[27] Ould K, Bournet PE, Migeon C, Boulard T, Chasseriaux G. Analysis of greenhouse ventilation efficiency based on computational fluid dynamics. Biosystems Engineering. 2006; 95: 83–98.

[28] Baeza EJ. Optimization of ventilation systems design for greenhouse parral-type. PhD thesis, University of Almeria, 2007.

[29] Teitel M, Tanny J. Radiative heat transfer from heating tubes in a greenhouse. Agricultural Engineering Research. 1998; 69: 185–188.

[30] Popovski K. Location of heating installations in greenhouses for low temperature heating fluids. In: Industrial thermal effluents for greenhouse heating, European cooperative networks on rural energy, 1st ed., Rome, 1986, 51–53.

[31] Roy JC, Boulard T, Bailley Y. Characterization of the heat transfer from heating tubes in a greenhouse. In: E-Proceedings Symposium AGENG, Warwick, 2000.

[32] Kempkes F, Van de Braak NJ. Heating system position and vertical microclimate distribution in a chrysanthemum greenhouse. Agricultural and Forest Meteorology. 2000; 104: 133–142.

[33] Piscia D, Montero JI, Baeza E, Bailey BJ. A CFD greenhouse night-time condensation model. Biosystems Engineering. 2012; 111: 41–154.

[34] Kim K, Yoon JY, Kwon HJ, Han JE, Son JE, Nam SW, Giacomelli GA, Lee IB. 3-D CFD analysis of relative humidity distribution in greenhouse with a fog cooling system and refrigerative dehumidifiers. Biosystems Engineering. 2008; 100: 245–255.

[35] Boulard T, Haxaire R, Lamrani MA, Roy JC, Jaffrin A. Characterization and modelling or de air fluxes induced by natural ventilation in a greenhouse. Agricultural Engineering Research. 1999; 74: 135–144.

[36] Tadj N, Draoui B, Theodoridis G, Bartzanas T, Kittas C. Convective heat transfer in a heated greenhouse tunnel. Acta Horticulturae. 2007; 747: 113–120.

[37] Alves I, Perrier A, Pereira LS. Aerodynamic cover and surface resistances of complete cover crops: how good is the "big leaf"? Transactions of the ASAE. 1998; 41: 345–351.

[38] Miguel AF, Vande-Braak NJ, Bot GP. Analysis of the airflow characteristic of greenhouse screening materials. Research in Agricultural Engineering. 1997; 67: 105–112.

[39] Hsu C, Cheng P. Thermal dispersion in a porous medium. International Journal of Heat and Mass Transference. 1990; 33: 1587–1597.

[40] Haxaire R. Characterization and modelling of the air flows within a greenhouse. PhD thesis, University of Nice Sophia Antipolis, 1999.

[41] Haxaire R, Boulard T, Mermier M. Greenhouse natural ventilation by wind forces. Acta Horticulturae. 2000; 534: 31–40.

[42] Lee IB, Short TH. Two-dimensional numerical simulation of natural ventilation in a multi-span greenhouse. Transaction of the ASAE. 2000; 43: 745–753.

[43] Fatnassi H, Boulard T, Bouirden L. Simulation of climatic conditions in full-scale greenhouse fitted with insect-proof screens. Agricultural and Forest Meteorology. 2003; 118: 97–111.

[44] Molina-Aiz FD, Valera DL, Alvarez AJ. Measurement and simulation of climate inside Almeria-type greenhouse using computational fluid dynamics. Agricultural and Forest Meteorology. 2004; 125: 33–51.

[45] Wilson JD. Numerical Studies of flow through a windbreak. Journal of Wind Engineering & Industrial Aerodynamics. 1985; 21: 119–154.

[46] Molina-Aiz FD, Valera DL, Alvarez AJ, Madueño A. A wind tunnel study of airflow through horticultural crops: determination of the drag coefficient. Biosystems Engineering. 2006; 93: 447–457.

[47] ANSYS-FLUENT. Fluent user guide. Lebanon: Ansys Inc., 1997.

[48] Flores-Velazquez J, Montero JI. Computational fluid dynamics CFD study of large-scale screenhouse. Acta Hortculturae. 2008; 797: 117–122.

[49] Stanghellini C. Transpiration of greenhouse crops: an aid to climate management. PhD thesis, Agricultural University of Wageningen, 1987.

Chapter 5

Technologic Appliance and Performance Concerns in Wheelchair Racing – Helping Paralympic Athletes to Excel

Pedro Forte, Tiago M. Barbosa and Daniel A. Marinho

Additional information is available at the end of the chapter

http://dx.doi.org/10.5772/61806

Abstract

Numerical simulations have provided useful evidence in helping several sportsmen to excel in their field. This methodology aims to have a deeper understanding on the influence of equipment and sports techniques on sports performance. In wheelchair racing, technology was used without considering specific sport (some of the Paralympic sports used the same technology of their Olympic counterparts). It has induced unique changes in prosthetic and wheelchair devices. Eventually, technology has become an essential part of Paralympic sports, wheelchair-racing being one of the most popular events. Numerical simulations can help us gather evidence on the effects of drag force acting upon the athlete-chair system.

Different types of wheelchairs are designed for racing (track and road races), net, and invasion sports. One of the various strategies to enhance performance is to minimize the aerodynamic drag of the frame, tires, helmet, sports outfit, and body posture.

Numerical simulations can be used to predict the fluid dynamics.

The goal of this chapter is to review the state-of-the-art numerical simulations and suggest further studies in wheelchair racing. The chapter will include sections covering: (i) main determinants in wheelchair racing; (ii) the effect of aerodynamic force in in wheelchair racing performance; (iii) analytical models, experimental testing, and numerical simulations in wheelchair racing; and (iv) numerical simulations on equipment and techniques.

Keywords: CFD, Paralympic, Performance

1. Introduction

Wheelchair racing is a major event in Paralympics. This sport holds short (100m) to long distance (42km) races. The athletes are classified according to their condition into classes (T5i, i.e., i=1, 2, 3 and 4 with injuries at C5-6, C7-8, T1-7 and T8-S4 respectively) [1, 2]. In wheelchair racing, sports science has been expressed as a need in the modeling and computer simulations of racing wheelchairs and the racing itself [3, 4].

The first game for disabled persons, in pair with the Olympic Games, was held in Rome in 1960. Pioneer research on wheelchair racing happened in the 1980s which was based on high-speed films at laboratory or running track settings (5). In the 1980s, manufacturers estimated that more than 10,000 racing wheelchairs commercialized worldwide [6]. Being increasingly popular, the competitions between the contenders in these sports became tighter and research is eagerly needed to help Paralympic athletes to excel.

Nowadays, most world-ranked Paralympic athletes, including wheelchair racers, develop an evidence-based practice with the help of coaches and sports analysts. Every detail of a race is deeply examined to have a deeper insight of the determinant factors that might help them excel. The resistance acting on the wheelchair racer is one of the major concerns for practitioners. The resistance force having an opposite direction to the displacement must be minimized so that for a given amount of thrust, the subject can reach a higher acceleration and speed. Lately, a few research projects were reported in the literature under this topic. So, the aim of this chapter is to review the state-of-the-art numerical simulations in wheelchair racing and suggest further studies in this sport.

2. The wheelchair race

The stroke cycle in wheelchair racing is divided into two phases: the propulsion and the recovery phases (Figure 1). The propulsion phase is characterized by the tangential force applied to the handrim. In the beginning of the race, this phase should be its highest and longest contact time so that the mechanical impulse will also be high [7]. It represents 33% to 35% of the full stroke cycle. The remaining 65% to 67% of the stroke time is the recovery phase in elite wheelchair racers [8, 9, 7, 10]. In this phase, we have the time period that the hand does not contact the rims. The hand tends to follow different paths until being positioned in the rim again to a new propulsion phase [11-13]. However the recovering phase may change, the free chosen push frequency (the preferred stroke frequency adopted by each subject) ranged from 32 to 86 pushes/min at a 6.58 m/s in wheelchair racing.

The stroke cycle can also be divided into 5 moments: catch, drive, release, lift and stretch, and finally, acceleration. Catch is the hand contact moment on the handrim, it usually occurs nearly the 1 and 2 o'clock positions. Drive phase is the moment of the hand and wrist acceleration on the handrim and it usually occurs nearly the 2 and 5 o'clock. Release phase is the moment of the contact breaking of the hand with the handrim, closer to the 6 o'clock. Lift and stretch phase is characterized by the elbow flexion and elevation in the sagittal plan. Acceleration moment is when the elbow extension is done before the handrim contact.

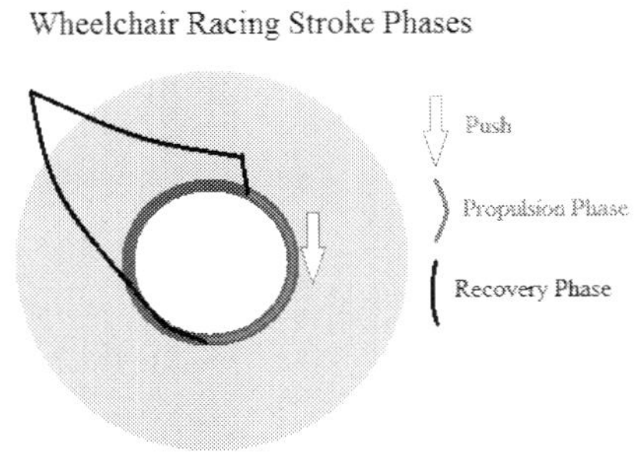

Figure 1. Stroke Cycle in Wheelchair Racing

During a race, the racers aim to reach the maximal speed as soon as possible and keep it for the remaining time (Figure 2). Figure 2 also depicts the intra-cycle speed (i.e. due to hand contact and recovery phase). This only proves that propulsive forces are higher than resistance forces (i.e. positive acceleration) at a given moment, and there are also moments that it will have higher intensity (i.e. negative acceleration).

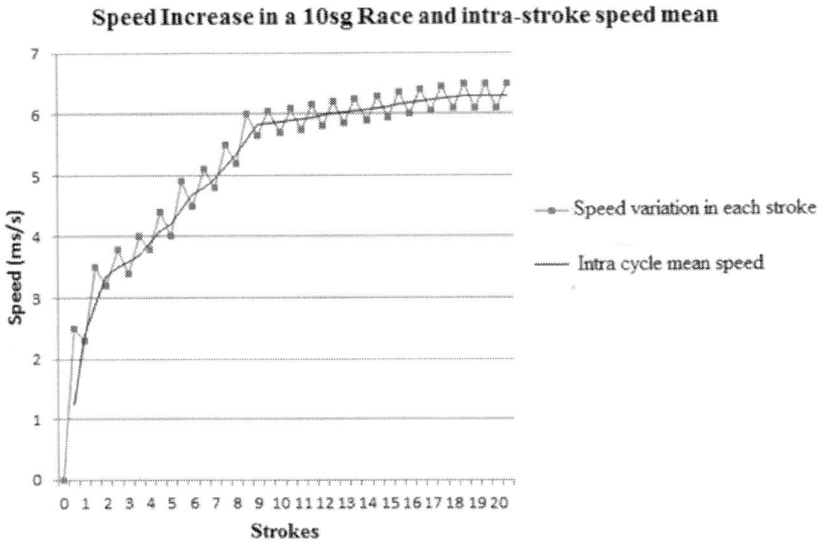

Figure 2. Theoretical Representation of the Speed over Time in 20 Stroke Cycles

Stroke kinematics is selected on a regular basis to report the stroke cycle. This includes the assessment of the average speed, stroke frequency, and stroke length.

The stroke frequency can be evaluated by measuring the number of cycles per minute and stroke speed given by the product of stroke length and stroke frequency. Stroke length is the difference between the contact distance of a stroke and the contact covered distance of the next stroke. Thus, stroke length can be explained as the covered distance with handrim contact in one stroke [14]. The average speed is calculated dividing the stroke length by the stroke frequency [14].

A higher stroke frequency combined with the hand linear velocity will lead to a greater energy cost, and probably will enhance cardiorespiratory stress affecting blood lactate and heart rate [15]. The increase of speed is related with blood lactate concentration and rise of heart rate. Despite this, the increase of the handrim diameter seems to reduce the blood lactate concentration at 20km/h, however it did not happen at 22km/h. [15].

An increased time of hand contact on the handrim without wrist acceleration will generate deceleration by friction resistance applied by the hand on the handrim. Therefore, to achieve a lower contact, it is recommended to produce a high speed soon after the start [16], which is possible to improve as demonstrated by sprinters in strength development [17]. The elbow motion seems to range from 60.9º to 5.2º in flexion to extension movement; and the maximum flexion velocity ranges from 515.4º/s and 572.8º/s, which is independent of the lightweight of the wheelchair. However, the propulsion arc reduces by 12º to 14º in the pumping-stroke technique that results from the shortest handrim contact possible, usually at the start of the race [18]. In the 1986 National Wheelchair Track and Field Championship in Illinois, in elite finalists both in paraplegic and quadraplegic wheelchair racers, the mean velocity-time ranged in each stroke from 5.1m/s to 5.5m/s and 3.7m/s to 4.27m/s respectively. The mean velocity in each stroke ranged 0.4m/s and 0.6m/s between the propulsion and recovery phase for the same subjects respectively. As the stroke frequency increases, velocity peak also enhances, despite the negative correlation between the velocity peak and the hand contact time with the handrim. The contact phase of the hand in handrim ranged from 13.58º to 15.6º for quadriplegics and paraplegics. The start acceleration phase at the rims starts at 48.54º and 71.1º. The break contact happens at 187.5º and 238.9º. The peak velocity for each stroke, occurs at 181.1º and 223.5º for quadriplegics and paraplegics respectively. These results were founded in a 10-second sprint with 20 to 25 propulsive cycles (strokes) for each subject [19].

The velocity peak seems to occur near the 10-second sprint in wheelchair track racers [7]. Wheelchair basketball players achieve 80% of the peak velocity in the first three strokes. On the other hand, wheelchair racers at the third stroke only achieve 55% of the peak velocity. The stroke frequency in short distances is greater than in long distances. In 10 seconds, sprints were obtained with 2 strokes per second [20]. However, in an 800m race, the stroke frequency ranged between 1.77 and 1.72 Hz [7].

http://dx.doi.org/10.5772/61806

3. Main determinants in wheelchair racing

The wheelchair racing athlete intends to reach the maximum acceleration as soon as possible. According to Newton's Second Law, the acceleration (a) is the ratio between force (f) and mass (m) (Equation 1).

$$a = \frac{F}{m} \tag{1}$$

In the wheelchair racing case, it is also possible to say that:

$$a = \frac{\left(F_{Prop} - F_{Resist}\right)}{m} \tag{2}$$

F_{Prop} and F_{Resist}, are the propulsive forces and the resistive ones respectively.

Positive acceleration is obtained by the applied propulsive forces on the wheels, overcoming the resistive ones (Figure 3). In wheelchair racing, the propulsive force is the push on the handrim, generating motion (i.e. the applied force by the ground in the wheel). The resistive forces are the rolling friction (Fr) and aerodynamic drag (Fd).

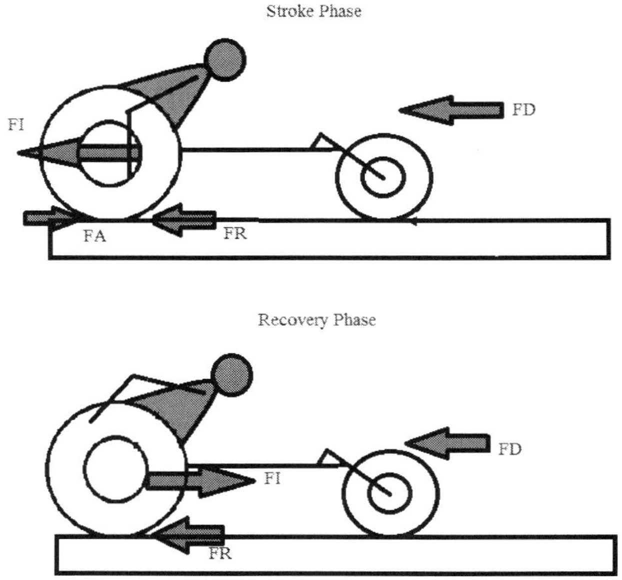

Figure 3. Free body diagram in wheelchair racing [21]; FI – Inertial force; FD – Drag force; FR – Rolling friction forces; FA – The applied force by the ground on the rear wheels.

Fuss [21] reported an analytical model to describe the performance based on these major determinants. The inertial force is the required force to change the body state and start the motion. The drag force is the air resistance to the wheelchair-athlete system. Rolling friction is the resistive force of the ground on the wheels' tires. Considering Newton's Third Law, the applied force by the ground on the rear wheel is the opposite one of the wheel in the ground derived by propulsion.

According to Fuss [21], the total energy expended by the athlete is taken by rolling friction and air drag. It is possible to say that the velocity is dependent on the kinetic energy of the system and its mass.

$$v = \sqrt{\frac{2\left(E_{in} - E_{loss}\right)}{m}} \tag{3}$$

In the equation, v is the velocity, E_{in} is the energy produced by the athlete and E_{loss} is the energy lost.

$$\mathrm{E}_{\mathrm{kin}} = \mathrm{E}_{\mathrm{input}} - \left(\mathrm{E}_{\mathrm{drag}} + \mathrm{E}_{\mathrm{friction}}\right) \tag{4}$$

The total kinetic energy is obtained from the sum of all mobile parts with a speed higher than zero. It is also considered that other movement oscillations do not contribute to the kinetic energy. Then:

$$E_{kin} = \frac{mv^2 + \sum_{i=1}^{3} I_i \omega_i}{2} \tag{5}$$

In Equation 5, Ii is the moment of inertia and ωi is the angular velocity of wheel i.

When the inertia force (FI) gets combined with distance, the kinetic energy (Ekin) can be formulated as:

$$E_{kin} = \int_{x1}^{x2} F_I d_x \tag{6}$$

Equations 7 and 8 consider that velocity comes from the work of the rear wheel (ωri) and acceleration is the angular velocity of the rear wheel (a = αri), then solving it for FI, we obtain:

$$F_I = a\left(m + \sum_{i=1}^{3} \frac{I_i}{r_i^2}\right) \tag{7}$$

In this formula, m is the mass of wheelchair-athlete system and I is obtained from the relation between the one wheel mass and the wheel radius gyration squared. The mass of one wheel multiplied by its radius of gyration squared when accelerated, an equivalent mass results in FI:

$$M = m + \sum_{i=1}^{3} \frac{I_i}{r_i^2} \tag{8}$$

The propulsion cycle is given by the stroke and recovery phases, the equilibrium forces in the stroke one are given by:

$$F_A + F_I + F_D + F_R = 0 \tag{9}$$

$$F_A + M_a + C_{Dv^2} + mg\mu_R + mgk_{fv^2} = 0 \tag{10}$$

C_D is the drag coefficient, v is the instantaneous velocity (considering distance and time at a given moment) and the acceleration ($d^2x/\ dt^2$). The C_D cluster is obtained from 0.5ϱcDA, being the product of air density (ϱ) with the drag coefficient of the air (C_D) and the frontal surface area (A). Considering that wheelchairs are in straight line and according to Equation 14, 12 and 13, FR is given by:

$$F_R = \mu_R mg + k_f mgv^2 \tag{11}$$

Where μ_R is the rolling friction coefficient, m is the mass, g the gravitational acceleration and k_f the coefficient of speed.

$$F_A + M_a + c_1 v^2 + c_2 = 0 \tag{12}$$

C_1 is given by,

$$c_1 = C_D + mgk_f \tag{13}$$

And C2,

$$c_2 = mg\mu_R \tag{14}$$

Thereby, in a wheelchair-athlete system, the responsible variables for the energy losses are FD and FR and both forces are equal to zero in a loss-free environment.

4. Propulsion

In the propulsion phase, Fuss [21] described four forces as the main ones. The ground applied one at the rear wheel, aerodynamic drag force and the rolling friction one. When the mass gets accelerated, it produces an inertial force (FI). The kinetic energy derived from FI, is stored on the wheels, it depends of the moment of inertia and the angular speed of the wheels creating the system motion. The four forces in equilibrium are described as,

$$F_A + F_I + F_D + F_R = 0 \tag{15}$$

F_A is the force applied to the ground on the wheels derived from the tangential force applied on the handrim (F_I). For movement occurring, F_A must be higher than F_I, F_D and F_R.

Several investigations have been made to better understand the physiological factors and demands that may affect the wheelchair propulsion. Others, for instance, [10] presented the motion of the racing wheelchair derived by the sum of all external forces by:

$$F\left(R/r_{pr}\right) = My + Iv/R + I_f v/r + Fa + M_B/r_{ax} + M_b/r_{axf} + F_R + F_r + W\sin 0\left(x\right) \tag{16}$$

Where, "R/r_{pr}" represents the gear ratio of the handring to the wheel. F - The tangential force in the handrings; R and r_{pr} - The radius of the rear wheels and handrings; M - The mass of the system athlete-wheelchair; v - The velocity of the athlete-wheelchair system, given by: v = x = dx/dt; I and I_f - The inertia of the rear wheels and the wheels inertia; r - The radius of the front wheels; Fa - The air resistance of the athlete-wheelchair system; M_B - The rear hubs bearing resistance; M_b - The front hubs bearing resistance; r_{ax} and r_{axf} - The radius of the rear and front axles respectively; F_R and Fr- The rolling resistance of the rear and front wheels respectively; W - The weight of the athlete-wheelchair system; Θ (x) - The inclination angle (i.e., changes in elevation); x - The covered distance;

5. Inertia

The sum of the kinetic energy to all mobile parts is given by Equation 5. It includes the energy of the three wheels and the translational energy. The kinetic energy is equal to the inertial force combined with distance (Equation 6). Thus, the inertial force (F_I) is given by:

$$F_I = aM \tag{17}$$

M represents the sum of the inertial moments.

The Inertial Force is the required force for a body state change. At the beginning of the wheelchair race, the inertial force is the required force for starting the wheelchair. Therefore, it is clear that the mass influences the inertial moment in the system and that the mass reduction will lead to a sooner and/or higher speed. Fuss [21] reported that a reduction of 1kg in a wheelchair-athlete system will improve the winning time by 0.132sg.

6. Air drag and rolling resistance

In wheelchair racing, the air drag results from the air resistance in the surface area of the athlete-wheelchair system. In contrast, the rolling resistance results from the friction of the tyres on the ground. Aerodynamic resistance can be minimized by reducing the frontal area of the subject, intending to improve the winning time of a racing wheelchair. At speeds greater than 5m/s aerodynamic drag represents 90% of the resistive forces [22]. Between 5.32m/s and 6.83m/s, rolling resistance is greater than aerodynamic drag with a partial contribution of 65% to 75% and 25% to 35% respectively. At greater speeds, aerodynamic drag starts to represent a greater contribution of the resistive forces [23].

Air drag and rolling resistance can be acceded from Equation 18 for rolling resistance and Equation 19 for air drag:

$$F_R = \mu_g mg + k_f mgv^2 \tag{18}$$

$$F_D = 0.5 \rho A_d v^2 C_D \tag{19}$$

In these two equations, F_R is the rolling friction, μ_g is the rolling friction coefficient, m is the mass of the athlete-wheelchair system, g is the gravity acceleration and k_f is the coefficient of the speed influence in rolling resistance force, ϱ is the air density, A_d is the frontal area, v^2 is the velocity, and C_D is the drag coefficient.

Lightweight clothes should be worn by wheelchair racers to reduce the aerodynamic drag [23]. A flexed upper trunk position also reduces the aerodynamic drag, as does reducing the rear wheel spokes to 24 [22]. In cycling, a flexed upper trunk position reduces the frontal area in 20% to 29% [24]. In wheelchair racing, the same position reduces the frontal area in 3% to 4% [25].

Mass would also influence the rolling friction; reducing the wheelchair-athlete system would also improve the speed, mainly in short distances. A 1kg reduction would improve 1-2.3% of the winning time in wheelchair racing [21].

7. Analytical models, experimental testing, and numerical simulations in wheelchair racing

7.1. Experimental testing: Coast-down technique

Aerodynamic and rolling resistance can be tested providing realistic opportunities that cannot be achieved in laboratory. The coast-down methods founded in literature are: (i) roll distance applying the ramp methods; (ii) the timing gate method that measures the velocity decrease in different points between two marks (gates); and (iii) the velocity method that measures directly the speed in each time.

The roll distance applying the ramp method consists rolling, a wheelchair from a ramp to the field, measuring the covered distance by the wheelchair [26]. The timing gate method was developed as an alternative to the roll distance, produced in less time and space compared with the whole roll until the wheelchair stops. Two marks should be performed counting the time when the wheelchair passes between them [27]. In the velocity method, the racer reaches the maximum speed, stops the propulsion and the velocity is recorded in each instance [27, 21]. A dynamometer-based coast-down test was used for the wheel deceleration calculus [28]. For wheel deceleration in each trial, hub (Hubxyz) and handrim (Rimxyz) markers were placed. The wheel rotation on each time was assessed by the law of cosines:

$$\theta = \cos^{-1}\left(\frac{A \times B}{\|A\| \ \|B\|}\right) \tag{20}$$

A = Rimxyz(i) – Hubxyz(i); B = Rimxyz (i + 1) – Hubxyz (i + 1).

To avoid wheel size differences, the angular speed was converted to linear one. It has fitted a line to the linear velocity (v) data intending to determine the wheel deceleration (ad), the time to coast down from 2 m/s to 1 m/s is represented by "t" and the initial wheel velocity is represented by "$v_{0''}$.

$$v = a_d t + v_0 \tag{21}$$

The estimation for rolling resistance can be done based on Cooper's method [10]. Intending to simplify the method, aerodynamic drag, wheel and roller bearing resistances, as other external resistances were neglected. Thus, the rear wheel motion was described by:

$$T_w = \left(I_r \frac{R}{r} I_w\right) a_w + F_{RR} R \tag{22}$$

Where, "T_w" was the torque applied by the hand to the rear wheel; "I_r" and "I_w" were the moment of inertia of the roller and the rear wheel, respectively; "r" and "R", the roller and rear wheel radius, respectively; "a_w" and "F_{RR}" as the angular acceleration of the rear wheel and the rolling resistance force respectively.

Considering that the propulsive force was discontinued, "T_w" is equal to zero, and "a_w" represents the wheel angular deceleration. The equation is being used to describe the rear wheel rotation. The roller inertial moments (0.87 ± 0.15 kg-m^2) and wheels (0.12 ± 0.02 kg-m^2) were experimentally acceded resourcing the acceleration method by DiGiovine et al. [29].

Applying the "I_r" and "r" values into Equation 22, and rewriting "a_w" as the ratio between linear deceleration and radius (a_d/R), rolling resistance force is obtained from the wheel radius, inertia, and deceleration, calculated by:

$$F_{RR} = \frac{-a_d}{R^2}\left(I_w + 5.47R\right) \tag{23}$$

Coast-down distance (CDD) calculation was computed using the equation of motion, where acceleration is assumed to be constant and t is the time for coast-down technique ends, when the wheelchair stops:

$$CDD = \frac{1}{2}a_d t^2 + v_0 t \tag{24}$$

Another method exists in accelerating the wheelchair (e.g., from 2.5 to 12.8m/s). The rider stops the propulsive phases reaching the target speed counting the length and/or time until the wheelchair stops [30]. In this method both aerodynamic drag and rolling resistance are obtained. Figure 4 depicts a speed decay of wheelchair considering the velocity method, measuring the speed in each time.

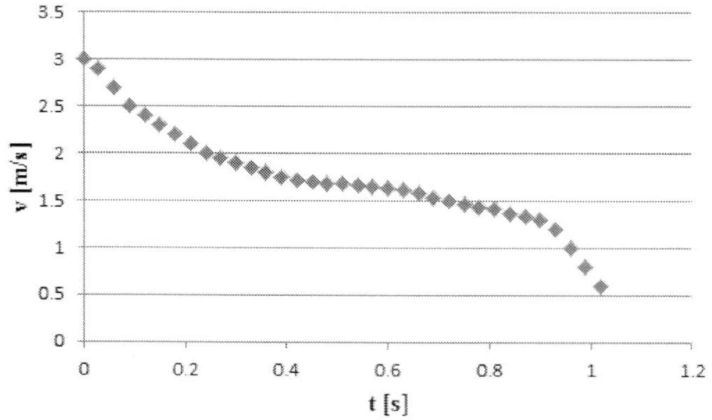

Figure 4. The Velocity Decay of a Manual Wheelchair over a Typical Trial

7.2. Wind tunnel testing

At least one paper can be found in the literature reporting the assessment of the aerodynamics in wind tunnel between two wheelchair models [31]. The drag forces were measured for speeds between 54.86km/h and 164.60km/h with intervals of 18.28km/h. However, air viscosity and ground motion effect are not taken into account with this methodology. Aside from this, a better wind tunnel test should be performed intending to determine the best helmet and clothes to use in races [31]. The Drag was proportional with the speed in the two models A and B. Model A presented a strict nose, with fenders at the wheels and an exceeded front angle attack. The drag force was acceded in grams (gr) and ranged from 26(gr) to 360(gr) in both models, with and without pilot. With and without pilot, Model B presented a lower drag force for velocities lower than 91.44km/h. At higher speeds, Model A presented lower drag force with and without pilot.

From this experimental test, it is possible to confirm that specific wheelchairs should be made for specific racers, and a sprint race wheelchair should be different to a long race wheelchair.

7.3. Analytical method for drag assessment

For rolling friction and aerodynamic drag, Burton, Fuss & Subic [32] presented an analytical procedure to estimate rolling friction and aerodynamic drag. FR is non-linear when calculated from visco-elastic models, considering the deformation of the tyres, reduces in the ground higher speed in parabolic function. The equations for these two forces are expressed in Equations 18 and 19.

According to Fuss [21], based on vehicles' data [33], consider that the μ_g of a racing wheelchair is 0.01, K_f is $5x10^{-6}$ s^2m^{-2}, a mean speed of 10 m/s and a mass athlete-wheelchair system of 80kg the first and second term of the equation 22 is 7.85 and 0.35N.

In a partial contribution assessment of air drag, Barbosa et al. [23] assumed the air density of 1.2041 kg/m3 in the sea level at 20º C. The surface area was measured with the photogrammetric technique in the frontal plane and the drag coefficient was assumed to be 0.7. All estimations were completed in each speed moment between 0 m/s and 13m/s increasing in every 0.1m/s. considering the world speed record, the air drag represents 34.89%.

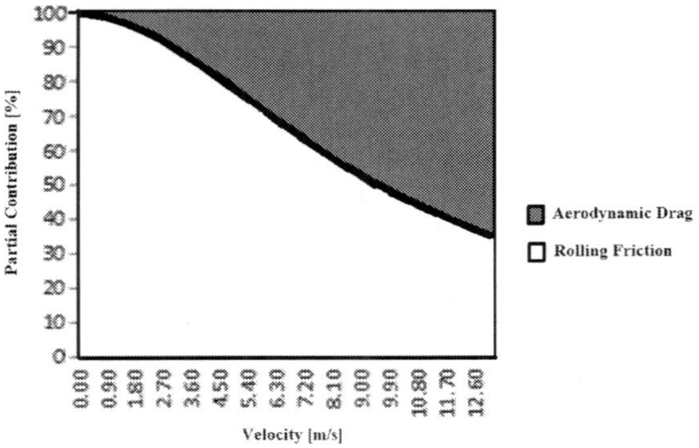

Figure 5. Partial Contribution of Aerodynamic Drag and Rolling Friction by Barbosa et al. [23].

7.4. Numerical simulations: Computer fluid dynamics

Computational Fluid Dynamics (CFD) has been used over the last 20 years. There are a lot of benefits on CFD and tools, such as FLUENT, CFX, STAR-CD and FiDAP, are all commercially available and used in industrial settings in the engineering community since the 1990s. Engineers and scientists started to use these simulations in competitive sports to reach a performance advantage, to improve sports equipment design and elite athletes' aerodynamics or hydrodynamic enhancements [34].

In sport sciences, the CFD presented concordance between numerical simulations and in vivo tests. For the simulation, a 3D body model scan is required and the images processing can be made with recourse to Anatomics Pro (Anatomics, Kannapolis, NC, USA) and FreeForm (Sensable Technologies, Woburn, MA, USA). The scan files are saved as IGES (*.igs) format, intending to be executable in Gambit/FLUENT (FLUENT Inc, Hanover, NH, USA). In Gambit/ FLUENT, it is possible to generate the greed and to define the finite elements in 3D areas.

The numerical simulation consists discretization of Navier-Stokes equations by the finites volumes methods. These equations come from Newton's Second Law in fluid mechanics, assuming that the fluid stress is the sum of diffusion of its viscosity, resulting from an applied pressure term. The equation resolution determines the fluid speed in a determined point at space and time. CFD is based on an approximated finite volume. In this approximation, space is divided into small cells to form a mesh or greed, applying a solver algorithm for the equations of fluid volume motion resolution [35, 36].

The Reynolds-Averaged Navier-Stokes (RANS) comes from decomposing the instantaneous values into means and/or fluctuating components. Fluid flow behavior (Equation 25), Reynolds stresses (Equation 26), temperature (Equation 27) and mass transfer (Equation 28) can be solved in this methodology.

$$\frac{\partial_{\mathrm{Ui}}}{\partial_{\mathrm{xi}}} = 0 \tag{25}$$

$$\frac{\partial_{Ui}}{\partial_t} \pm U_j \frac{\partial_{Ui}}{\partial_{x_j}} = -\frac{1}{\rho}\frac{\partial P}{\partial x_j} + \frac{\partial}{\partial x_j}\left(2vS_{ij} - \overline{\mu_j'\mu_i'}\right) \tag{26}$$

$$\frac{\partial_{\theta i}}{\partial_t} \pm U_j \frac{\partial_{\theta}}{\partial_{x_j}} = \frac{1}{\rho_{cp}}\frac{\partial}{\partial x_j}\left(k\frac{\partial \theta}{\partial x_j} - \overline{\mu_j'\theta'}\right) \tag{27}$$

$$\frac{\partial_{C}}{\partial_t} \pm U_j \frac{\partial_{C}}{\partial_{x_j}} = \frac{\partial}{\partial x_j}\left(D\frac{\partial C}{\partial x_j} - \overline{\mu_j'c'}\right) \tag{28}$$

The μ_i and x_i are the instantaneous velocity and the position, p the instantaneous pressure, t is the time, ρ the fluid density, v is the molecular kinematic viscosity, c_p heat capacity, k is the thermal conductivity and S_{ij} the strain-rate tensor, c is the instantaneous concentration, and D is the molecular diffusion coefficient.

The Reynolds stresses component ($\overline{\mu_j'\mu_i'}$), describes the turbulence of the mean flow being the exchange of momentum by the change of the fluid parcels. In a laminar flow, the molecules are the responsible for momentum exchange (molecular viscosity). However, in a turbulent flow (turbulent viscosity) the parcels of flow are the ones that exchange the momentum. To finish this calculus, it is also necessary to use a turbulence model to represent flow scales. The first order Boussinesq eddy-viscosity hypothesis to model the Reynolds stress in function of velocity and time is used. However, it is also a necessary model for the linear or non-linear eddy-viscosity distinction. The Reynolds stress is given by:

$$\overline{\mu_j'\mu_i'} = 2v_t S_{ij} - \frac{2}{3}k\delta_{ij} \tag{29}$$

v_t is the turbulent viscosity and the mean strain rate S_{ij} is given by,

$$S_{ij} = \frac{1}{2}\left(\frac{\partial U_i}{\partial_{xj}} + \frac{\partial j}{\partial_{xi}}\right) \tag{30}$$

The turbulent kinetic energy (k) is given by,

$$k = \frac{1}{2}\overline{\mu_i'\mu_i'} \tag{31}$$

And the Kronecker delta (δ_{ij}),

$$\delta_{ij} = \begin{cases} 1; if\ i = j \\ 0; if\ i \neq j \end{cases} \tag{32}$$

Assuming the gradient diffusion (Gradient-diffusion Assumption) for heat and mass fluxes, as function of the temperature gradients in the mean flow, the turbulence hot flow is given by,

$$-\overline{\mu_j'\theta'} = D_{\theta t}\frac{\partial \vartheta}{\partial_{xi}} \tag{33}$$

$D_{\theta t}$ is the turbulent heat diffusivity (turbulent Prandtl number), and $\frac{\partial \vartheta}{\partial_{xi}}$ is the temperature gradient in the mean flow. The turbulent Prandtl number is obtained by,

$$Pr_t = \frac{v_t}{D_{\theta,t}} \tag{34}$$

And the turbulent mass flow $-\overline{\mu_j' c'}$ is given by,

$$-\overline{\mu_j' c'} = D_{c,t} \frac{\partial C}{\partial x_i} \tag{35}$$

$D_{c,t}$ is the turbulent mass diffusivity (turbulent Schmidt number) and $\frac{\partial C}{\partial x_i}$ the concentrated gradient in the mean flow. The turbulent Schmidt number is calculated from Equation 36,

$$Sc_t = \frac{v_t}{D_{\theta,t}} \tag{36}$$

Once RANS needed a turbulence model, less expensive equations are created with additional variables, transforming in meanings of the instantaneous equations calculations. This results from removing several small equations and adding other unknown variables, determined by the turbulence models. Standard K – ε turbulence model used by [37] in a computational simulation on a counter-clock cyclist helmet, however, it is only valid in a completely turbulent fluid. The same mode is assumed for wheelchair racing. In FLUENT, the turbulence models available are: (i) Standard K – psilon; (ii) Standard K – ε; (iii) Spalart – Allmoras; (iv) Reynolds Stress (RSM) [38].

In this model (Standard K – ε) the Bussinesq hypothesis is given by,

$$-\rho\overline{\mu_i' \mu_j'} = 2\mu_t S_{ij} - \frac{2}{3}\rho k \delta_{ij} \tag{37}$$

Where the turbulent viscosity,

$$\mu_t = \rho C_\mu \frac{k^2}{\varepsilon} \tag{38}$$

The mean tension rate is given by Equation 30. The kinetic energy of turbulent fluctuation and the dissipation of the kinetic energy (m^2/s^2) are given by Equations 31 and 40 respectively.

$$k = \frac{1}{2}\overline{\mu_i' \mu_i'} = \frac{1}{2}\overline{\mu'^2} + \overline{v'^2} + \overline{w'^2} \tag{39}$$

k, is the energy measuring associated with the turbulent fluctuations in the flow.

$$\varepsilon = v\overline{\frac{\partial \mu'_i \partial \mu'_i}{\partial x_j \partial x_j}} \tag{40}$$

ε, is caused by the work of the smallest eddies against the viscous stresses in the flow.

Then the determination of k (Equation 41) and ε (Equation 42) by their transport equations are,

$$\rho\frac{\partial k}{\partial t} + \rho\frac{\partial}{\partial x_i}(k\mu_i) = \frac{\partial}{\partial x_j}\left(\left(\mu + \frac{\mu_t}{\sigma_k}\right)\frac{\partial k}{\partial x_j}\right) + G_k + G_b - \rho\varepsilon \tag{41}$$

$$\rho\frac{\partial \varepsilon}{\partial t} + \rho\frac{\partial}{\partial x_i}(\epsilon\mu_i) = \frac{\partial}{\partial x_j}\left(\left(\mu + \frac{\mu_t}{\sigma_\varepsilon}\right)\frac{\partial \varepsilon}{\partial x_j}\right) + C_{1\varepsilon}\frac{\varepsilon}{k}(G_k + C_{3\varepsilon}G_b) - \rho C_{2\varepsilon}\frac{\varepsilon^2}{k} \tag{42}$$

Where, $\rho\frac{\partial \varepsilon}{\partial t}$ and $\rho\frac{\partial k}{\partial t}$ are the variation of the local in time, $\rho\frac{\partial}{\partial x_i}(k\mu_i)$ and $\rho\frac{\partial}{\partial x_i}(\mu_i)$ the adjective term, $\frac{\partial}{\partial x_j}\left(\left(\mu + \frac{\mu_t}{\sigma_k}\right)\frac{\partial k}{\partial x_j}\right)$ and $\frac{\partial}{\partial x_j}\left(\left(\mu + \frac{\mu_t}{\sigma_\varepsilon}\right)\frac{\partial \varepsilon}{\partial x_j}\right)$ are the diffusion, G_k is the generation of k by the gradients mean velocity, G_b the generation of k due the fluctuation and ϱε the dissipation of k. σk and σε are the turbulent Prandtl numbers for k and ε respectively. And the constants $C_{1\varepsilon}$, $C_{2\varepsilon}$, C_μ, σ_k e σ_ε were experimentally determined, $C_{1\varepsilon} = 1.44$; $C_{2\varepsilon} = 1.92$; $C_\mu = 0.09$; $\sigma_k = 1.0$; $\sigma_\varepsilon = 1.3$.

The Gambit software allows the building of a representative graphic model of the volume subdivided in sub-volumes, trying to make the process as realistic as possible. This software also allows defining the frontiers. In solid frontiers and close to them, the FLUENT software computes the Reynolds tension and ε. It applies solid specific frontiers conditions for Reynolds tension using balance hypotheses, without considering the convection and diffusion of tension transport (Equation 25). In a local coordinate system, T is tangential coordinate, η the normal and λ the binomial one. The Reynolds tension in the adjacent cells to the frontier, are calculated by the equation:

$$\frac{\overline{\mu'^2_\lambda}}{k} = 1.098, \frac{\overline{\mu'^2_\eta}}{k} = 0.247, \frac{\overline{\mu'^2}}{k} = 0.655, \frac{\overline{\mu'_t\mu'_\eta}}{k} = 0.255 \tag{43}$$

FLUENT solves the transport equation (41) for k obtaining. For calculus convenience, the equation is globally solved, albeit the calculus of the k values is only necessary near the frontier. In the rest of the domain, the k is calculated by the equation (26).

The mesh can be constructed by quadrangular elements, with a space size of 0.1mm. The resulting data from the computational simulation of the determined flow regime and the visualization of the pressure profiles and speed are obtained by FLUENT. The data processing allows calculating the drag coefficient in the diverse forms [39].

7.5. Numerical simulations on wheelchair racing

CFD methodology starts being used for equipment tests. In wheeler riders' helmets, CFD has shown that airflow velocities could be improved with grooves in polymer foam liner and also improving the sweat evaporation. A top helmet hole would improve the velocity in that point, however velocities at the back became lower [40].

The ideal posture and some changes in the wheelchair could be possible to define the three first places. The tests revealed that the most drag negative influence in performance came from the athlete and not from the wheelchair. With a subtle modification in the sitting position, it could save 10% of the aerodynamic drag. No results were presented by the authors relatively to the wind tunnel test [41].

CFD methodology must be applied in different fabrics, helmets, and wheelchairs. The frame design, as the tube sizes must influence the fluid flow behavior. Also, different positions should be considered to be analyzed the flexed head in the start of the race or the look forward position. In each stroke phase, the different fluid flows should also be analyzed intending to reach the ideal motion in the stroke phase and recovery one, as the effect of the wheelchair designs and materials in the fluid flow [23].

Thus, fluid dynamics analysis should be performed intending to evaluate possible different designs of the wheelchairs and fulfill the lack of literature in this area. There is a need of results presentation such as drag and drag coefficient in different positions, helmets, and cloths at different speeds.

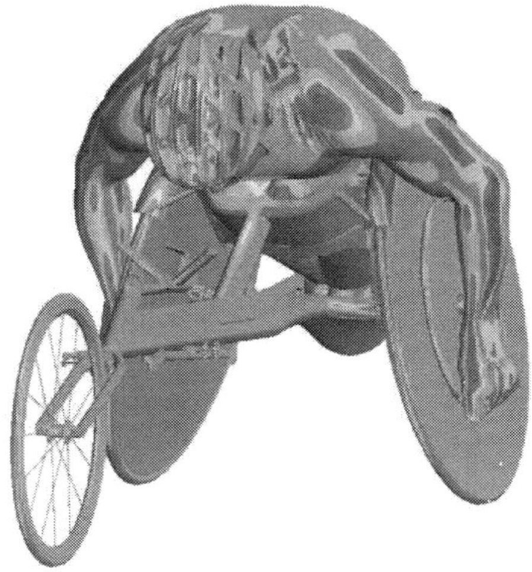

Figure 6. CFD Methodology of a Scanned Model in TotalSim (http://www.totalsimulation.co.uk/wp/cfd-image-gallery/)

8. Summary and conclusions

The sports science of wheelchair racing lacks research. However, there is some information that could positively contribute to increase performance in athletes. Managing the mathematical models based on Physics laws, coaches could identify the mechanic performance obstacles and try to minimize them.

The drag reduction by the rolling and air resistance access also contributed to increase the performance in athletes. Better propulsion could be obtained by reducing the total drag combined with a mass reduction and other possible aerodynamic positions. The purpose of computational simulations and/or computer fluid dynamics is the time improvement achieved by reducing the aerodynamic drag. No data was found about CFD tests in wheelchair racing.

The stroke technique should also be focused, despite the lack of indications about the number of hours off training after an injury occurrence. It is defined that the high angular velocities near the shoulder and the elbow generated by the strokes induce an overuse stress increasing the risk for joint injuries. There are also some indications about the contact zones, force applying, and contact break at the rear wheels.

Physiological variables should also be studied in wheelchair racing athletes, an area that lacks information in this sport. Thus, a precise control of the total drag and the efficiency of the stroke technique, which is related to high levels of strength derived by the strength and physical condition training, will positively contribute for a better performance.

Author details

Pedro Forte[1,2], Tiago M. Barbosa[2,3] and Daniel A. Marinho[1,2*]

*Address all correspondence to: dmarinho@ubi.pt

1 University of Beira Interior, Department of Sport Sciences, Covilhã, Portugal

2 Research Centre in Sports, Health and Human Development, Covilhã, Portugal

3 National Institute of Education. Nanyang Technological University, Singapore

References

[1] IWAS (2008) IWAS Athletics, worldwide ranking. http://athletics.iwasf.com/. Accessed 14 Aug 2009

[2] IPC International Paralympic Committee (2007) IPC athletics classification manual for physical impairments 2008–2010. http://www.paralympic.org/release/Summ-

er_Sports/Athletics/News/2007_12_07_ClassificationManual.pdf. Accessed 14 Aug 2009.

[3] Depauw, K. P. (1986). Research on sport for athletes with disabilities. Adap Phys Activ Q, 3(4): 292-299.

[4] Depauw, K. P. (1988). Sports for individuals with disabilities: Research opportunities. Adap Phys Activ Q, 5: 80-89.

[5] Cooper, R. A. (1990c). Wheelchair racing sports science: a review. Journal of rehabilitation research and development, 27(3), 295-312.

[6] Lakomy, H. K. A., Cambell, I. & Williams, C. (1987). Treadmill performance and selected physiological characteristics of wheelchair athletes. Br J Sports Med, 21(3), 130-133.

[7] Ridgway, M., Pope, C., Wilkerson, J. (1988). A kinematic analysis of 800-meter wheelchair racing techniques. Adapt Phys Act Q 5(2): 96-107.

[8] Bymes, D.P. (1983). APlm and EMG drive phase analysis of the competitive wheelchair stroke. Unpublished master's thesis, University of Alberta.

[9] Cooper, R. A. (1990b) An exploratory study of racing wheelchair propulsion dynamics. Adapt Phys Act Q, 7(1): 74-85.

[10] Sanderson, D. J., Sommer, H. J. (1985). Kinematic features of wheelchair propulsion. J Biomech 18(6): 423-429.

[11] Cooper, R. A. (1990). A systems approach to the modeling of racing wheelchair propulsion. J Rehabil Res Dev, 27(2): 151-62.

[12] Davis, R., Ferrara, M., & Byrnes, D. (1988). The competitive wheelchair stroke. NSCA Journal, 10(3), 4-10.

[13] Higgs, C. (1985). Propulsion of racing wheelchairs. In C. Sherrill (Ed.), Sport and disabled athletes (pp. 165-172). Champaign, IL: Human Kinetics

[14] Chow, J. W., Chae, W. (2007). Kinematic analysis of the 100 m wheelchair race. J Biomech, 40: 2564–2568.

[15] Costa, G. B., Rubio, M. P., Belloch, S. L., & Soriano, P. P. (2009). Case study: effect of handrim diameter on performance in a paralympic wheelchair athlete.Adapted physical activity quarterly, 26(4), 352-363.

[16] O'Connor, T., Robertson, R. N., Cooper, R. A. (1998). Three-dimensional kinematic analysis and physiologic assessment of racing wheelchair propulsion. Adapt Phys Activ Q, 15: 1–14.

[17] Keogh, J. W. L. (2011). Paralympic sport: an emerging area for research and consultancy in sports biomechanics. Sports Biomech, 10: 249–268.

[18] Rudins, A., Laskowski, E. R., Growney, E. S., Cahalan, T. D., & An, K. N. (1997). Kinematics of the elbow during wheelchair propulsion: a comparison of two wheelchairs and two stroking techniques. Archives of physical medicine and rehabilitation, 78(11), 1204-1210.

[19] Gehlsen, G. M., Davis, R. W., & Bahamonde, R. (1990). Intermittent velocity and wheelchair performance characteristics. Adapted Physical Activity Quarterly, 7(3), 219-230.

[20] Coutts, K. D. (1990). Kinematics of sport wheelchair propulsion. Journal of rehabilitation research and development, 27(1), 21-26.

[21] Fuss, F. K. (2009). Influence of mass on the speed of wheelchair racing. Sports Engineering, 12(1): 41-53.

[22] LaMere, T.J., & Labanowich, S. (1984). The history of sports wheelchairs: Part III, The racing wheelchair 1976-1983. Sports 'n Spokes, 10 (1), 12-16.

[23] Barbosa, T. M., Forte, P., Morais, J. E., Coelho, E. (2014). Partial contribution of rolling friction and drag force to total resistance of an elite wheelchair athlete. 1st International Conference in Sports Sciences & Technology. ICSST, Advanced Materials for Sport Technology, 749-753.

[24] Burke, E. (Ed.). (1986). Science of cycling. Champaign, IL: Human Kinetics Publishers.

[25] Hedrick, B., Wang, Y. T., Moeinzadeh, M., & Adrian, M. (1990). Aerodynamic positioning and performance in wheelchair racing. Adapted Physical Activity Quarterly, 7(1).

[26] FIFA. (2001). Quality Concept for Football Turf. Available: http://www.fifa.com/mm/document/afdeveloping/pitchequip/fqc_football_turf_folder_342.pdf.

[27] Kolitzus, H. J. (2003). Ball Roll Behavior: The Functional Relationship of the Ball Roll Distance and the Timing Gate Method; How to Calculate the Ball Roll Distance from Timing Gate Measurements. ISSS - International Association for Sports Surface Sciences, Eschenz / Switzerland, Available: http://www.isss-sportsurfacescience.org/downloads/documents/ZPKPAJJUWY_Ball_Roll_BehaviorKS.pdf.

[28] Kwarciak, A. M., Yarossi, M., Ramanujam, A., Dyson-Hudson, T. A., & Sisto, S. A. (2009). Evaluation of wheelchair tire rolling resistance using dynamometer-based coast-down tests. J Rehabil Res Dev, 46(7), 931-38.

[29] DiGiovine CP, Cooper RA, Boninger ML. Dynamic calibration of a wheelchair dynamometer. J Rehabil Res Dev. 2001;38(1):41–55.

[30] Candau, R. B., Grappe, F., Menard, M., Barbier, B., Millet, G.Hoffman, M. D. 1999. Simplified deceleration method for assessment of resistive forces in cycling. Medicine and Science in Sports and Exercise, 31: 1441–1447.

[31] MacLeish, M. S., Cooper, R. A., Harralson, J., & Ster, J. F. (1993). Design of a composite monocoque frame racing wheelchair. Journal of rehabilitation research and development, 30, 233-233.

[32] Burton, M., Fuss, F. K., & Subic, A. (2010). Sports wheelchair technologies. Sports Technology, 3(3), 154-167.

[33] Petrushov, V. A. (1998). Improvement in vehicle aerodynamic drag and rolling resistance determination from coast-down tests. Proceedings of the Institution of Mechanical Engineers, Part D: Journal of Automobile Engineering, 212(5), 369-380.

[34] Hanna, R. K. (2012). CFD in Sport-a Retrospective; 1992-2012. Procedia Engineering, 34, 622-627.

[35] Marinho, D. A., Barbosa, T. M., Mantha, V., Rouboa, A. I., & Silva, A. J. (2012). Modelling propelling force in swimming using numerical simulations. In: Juarez LH (Ed). Fluid Dynamics, Computational Modeling and Applications. Pp. 439-448. InTech. Rijeka.

[36] Marinho, D.A., Silva, A.J., Reis, V.M., Barbosa, T. M., Vilas-Boas, J.P., Alves, F.B., Machado, L., Rouboa, A. (2011) Three-dimensional CFD analysis of the hand and forearm in swimming. Journal of Applied Biomechanics. ISSN 1065-8483. 27:1, p. 74-80.

[37] Caboz, E. B. (2010). Simulação computacional do escoamento em torno de um capacete de ciclista usado nas provas de contra-relógio. Dissertação de mestrado. Faculdade de Engenharia da Universidade do Porto.

[38] White, F. (1999). Mecanica dos Fluidos, Ed. McGraw -Hill, 4ª edicao. ISBN : 85-86804-24-X.

[39] Carvalho, F. (2008). A aplicação da dinâmica computacional de fluidos no estudo do arrasto hidrodinâmico na natação. Dissertação de Mestrado em Educação Fisica e Desporto. Universidade de Trás os Montes. Vila Real.

[40] Pinnoji, P. K., & Mahajan, P. (2006, September). Impact analysis of helmets for improved ventilation with deformable head model. In Proceeding of IRCOBI conference, Madrid (pp. 159-70).

[41] Rushby-Smith, T., & Douglas, L. (2012). Paralympic Technology. Ingenia, 51, p.33.